PLANT FOOD BY-PRODUCTS

Industrial Relevance for
Food Additives and Nutraceuticals

Postharvest Biology and Technology

PLANT FOOD BY-PRODUCTS

Industrial Relevance for
Food Additives and Nutraceuticals

Edited by

J. Fernando Ayala-Zavala, PhD

Gustavo González-Aguilar, PhD

Mohammed Wasim Siddiqui, PhD

APPLE
ACADEMIC
PRESS

Apple Academic Press Inc.
3333 Mistwell Crescent
Oakville, ON L6L 0A2 Canada

Apple Academic Press Inc.
9 Spinnaker Way
Waretown, NJ 08758 USA

Library and Archives Canada Cataloguing in Publication

Plant food by-products : industrial relevance for food additives and nutraceuticals / edited by J. Fernando Ayala-Zavala, PhD, Gustavo González-Aguilar, PhD, Mohammed Wasim Siddiqui, PhD.

(Postharvest biology and technology book series)
Includes bibliographical references and index.
Issued in print and electronic formats.
ISBN 978-1-77188-640-6 (hardcover).--ISBN 978-1-315-09933-0 (PDF)
1. Food industry and trade--By-products. 2. Functional foods. 3. Food additives.
I. Ayala-Zavala, Jesús Fernando, editor II. González-Aguilar, Gustavo A., editor
III. Siddiqui, Mohammed Wasim, editor IV. Series: Postharvest biology and technology book series
TD899.F585P63 2017   664'.08   C2017-907031-2 C2017-907032-0

Library of Congress Cataloging-in-Publication Data

Names: Ayala-Zavala, Jesāus Fernando, editor. | González-Aguilar, Gustavo A., editor. | Siddiqui, Mohammed Wasim., editor.
Title: Plant food by-products : industrial relevance for food additives and nutraceuticals / editors, J. Fernando Ayala-Zavala, PhD, Gustavo González-Aguilar, PhD, Mohammed Wasim Siddiqui, PhD.
Description: Toronto ; [Waretown] New Jersey : Apple Academic Press, 2018. | Series: Postharvest biology and technology | Includes bibliographical references and index.
Identifiers: LCCN 2017050507 (print) | LCCN 2017051538 (ebook) | ISBN 9781315099330 (ebook) | ISBN 9781771886406 (hardcover : alk. paper)
Subjects: LCSH: Vegetable trade--By-products--Recycling. | Plant products industry--By-products--Recycling. | Agricultural processing--By-products--Recycling. | Food additives. | Food--Composition. | Functional foods.
Classification: LCC TP373.8 (ebook) | LCC TP373.8 .P54 2018 (print) | DDC 338.1/75--dc23
LC record available at https://lccn.loc.gov/2017050507

Apple Academic Press also publishes its books in a variety of electronic formats. Some content that appears in print may not be available in electronic format. For information about Apple Academic Press products, visit our website at **www.appleacademicpress.com** and the CRC Press website at **www.crcpress.com**

ABOUT THE EDITORS

J. Fernando Ayala-Zavala, PhD

J. Fernando Ayala-Zavala, PhD, is a professor and research leader of the Laboratory of Emerging Technologies at CIAD in Hermosillo, Sonora, Mexico. He is the author of more than 59 peer-reviewed journal articles and 20 book chapters and is the editor of several books and scientific journals. Dr. Ayala-Zavala has also participated in more than 35 international studies. He is a professor of master and doctoral studies, tutoring research theses and teaching courses on topics related to postharvest of fruits and vegetables, minimal processing, bioactive compounds in plant tissues, functional foods, and science communication. Since the beginning of his career, Dr. Ayala-Zavala has been interested in the study and creation of basic and applied knowledge in the postharvest preservation of plant foods. Specifically, he has participated in projects related to the use of emerging technologies to ensure the safety and quality of horticultural products. His research interests include the use of natural compounds, irradiation, active packing, and edible films and conservation techniques. In addition, Dr. Ayala-Zavala has been working with countries such as Italy, Spain, France, USA, Brazil, Argentina, Uruguay, Chile, and Panama to share experiences on the aforementioned topics.

Gustavo González-Aguilar, PhD

Gustavo González-Aguilar, PhD, is a research scientist at the Center of Research in Food Development (Mexico), specializing in fresh-cut tropical and subtropical fruits, postharvest technology, antioxidants and functional foods, food preservation, and development of new technologies. His primary research interest is to study the effects of abiotic and biotic stress on antioxidant status of fruits and vegetables. Another area of interest is the study of different postharvest treatments to prolong the shelf-life of fruits and vegetables and how these factors affects the bioaccessibility and bioavailability of bioactive molecules after simulated digestion *in vitro* and *in vivo* and their possible health benefits. He has authored over 150 scientific publications (JCR), including over 42 book chapters and eight books. He has participated in different networks on food preservation, fresh-cut, antioxidants, and functional foods. His extension work includes participation in training and design of fresh-cut facilities in Mexico and Latin America. He is leader of the network antioxidants and functional foods in Mexico supported by the Mexican council (CONACYT).

Mohammed Wasim Siddiqui, PhD

Mohammed Wasim Siddiqui, PhD, is an Assistant Professor and Scientist in the Department of Food Science and Post-Harvest Technology, Bihar Agricultural University, Sabour, India. His contribution as an author and editor in the field of postharvest biotechnology has been well recognized. He is an author or co-author of 34 peer reviewed research articles, 32 book chapters, 2 manuals, and 18 conference papers. He has 11 edited books and an authored book to his credit. Dr. Siddiqui has established an international peer-reviewed journal, the *Journal of Postharvest Technology*. Dr. Siddiqui is a Senior Acquisitions Editor in Apple Academic Press for Horticultural Science and is editor of the book series Postharvest Biology and Technology. He has been serving as an editorial board member and active reviewer of several international journals, such as *PLoS ONE*, (PLOS), *LWT- Food Science and Technology* (Elsevier), *Food Science and Nutrition* (Wiley), *Acta Physiologiae Plantarum* (Springer), *Journal of Food Science and Technology* (Springer), *Indian Journal of Agricultural Science* (ICAR), etc.

Recently, Dr. Siddiqui was conferred with several awards, including the Best Citizen of India Award (2016); Bharat Jyoti Award (2016); Best Young Researcher Award (2015) by GRABS Educational Trust, Chennai, India; and the Young Scientist Award (2015) by the Venus International Foundation, Chennai, India. He was also a recipient of the Young Achiever Award (2014) for the outstanding research work by the Society for Advancement of Human and Nature (SADHNA), Nauni, Himachal Pradesh, India, where he is an Honorary Board Member and Life Time Author. He has been an active member of the organizing committee of several national and international seminars, conferences, and summits. He is a key member in establishing the World Food Preservation Center (WFPC), LLC, USA. Presently, he is an active associate and supporter of WFPC, LLC, USA. Considering his outstanding contribution in science and technology, his biography has been published in *Asia Pacific Who's Who* and *The Honored Best Citizens of India*.

CONTENTS

LIST OF CONTRIBUTORS

Emilio Alvarez-Parrilla
Departamento de Ciencias Químico-Biológicas, Instituto de Ciencias Biomédicas, Universidad Autónoma de Ciudad Juárez, Anillo Envolvente del PRONAF y Estocolmo s/n, Ciudad Juárez, Chihuahua, CP 322310, México

Jesús Fernando Ayala-Zavala
Centro de Investigación en Alimentación y Desarrollo, A. C. (CIAD, AC), Carretera a la Victoria Km. 0.6, La Victoria, Hermosillo, Sonora, CP 83000, México. E-mail: jayala@ciad.mx

Vasudha Bansal
Department of Civil & Environmental Engineering, Hanyang University, 222 Wangsimni-Ro, Seoul 133791, South Korea

Ariadna Thalía Bernal-Mercado
Centro de Investigacion en Alimentacion y Desarrollo, A.C. (CIAD, AC), Carretera a la Victoria Km 0.6, La Victoria, Hermosillo, Sonora 83000, Mexico

Laura A. Contreras-Angulo
Centro de Investigación en Alimentación y Desarrollo A.C. Carretera a El Dorado km 5.5 Campo El Diez, 80110, Culiacán, Sinaloa, México

M. R. Cruz-Valenzuela
Centro de Investigacion en Alimentacion y Desarrollo, A.C. (CIAD, AC), Carretera a la Victoria Km 0.6, La Victoria, Hermosillo, Sonora 83000, Mexico. E-mail: reynaldo@ciad.mx

Jorge Esteban Davila-Avina
Univ Autonoma Nuevo Leon, Fac Ciencias Biol, Apdo Postal 124-F, Ciudad Univ, San Nicolas De Los Garza 66451, Nuevo Leon, Mexico

Laura Alejandra de la Rosa
Departamento de Ciencias Químico-Biológicas, Instituto de Ciencias Biomédicas, Universidad Autónoma de Ciudad Juárez, Anillo Envolvente del PRONAF y Estocolmo s/n, Ciudad Juárez, Chihuahua, CP 322310, México. E-mail: ldelaros@uacj.mx

J. Abraham Domínguez-Avila
Centro de Investigación en Alimentación y Desarrollo, AC (CIAD, AC), Carretera a la Victoria Km 0.6, La Victoria CP 83000, Hermosillo, Sonora, Mexico

Alexis Emus-Medina
Centro de Investigación en Alimentación y Desarrollo A.C. Carretera a El Dorado km 5.5 Campo El Diez, 80110, Culiacán, Sinaloa, México

Laura Grecia Flores-Acosta
Functional and Nutraceutical Foods Laboratory, Centro de Investigación en Alimentación y Desarrollo A. C., Unidad Culiacán, AP 32-A, Sinaloa 80129, México

L. E. García-Amezquita
Centro de Investigación en Alimentación y Desarrollo, Av. Rio Conchos S/N, Parque Industrial, Cd. Cuauhtémoc, Chihuahua 31570, México

Gustavo Adolfo Gonzalez-Aguilar
Centro de Investigación en Alimentación y Desarrollo, A. C. (CIAD, AC), Carretera a la Victoria Km. 0.6, La Victoria, Hermosillo, Sonora, CP 83000, México. E-mail: gustavo@ciad.mx

Mirian González-Ayón
Centro de Investigación en Alimentación y Desarrollo A.C., Culiacán, Sinaloa, CP 80110, México

M. G. Goñi
Grupo de Investigación en Ingeniería en Alimentos, Facultad de Ingeniería, Universidad Nacional de Mar del Plata, Juan B, Justo 4302, 7600 Mar del Plata, Buenos Aires, Argentina; Consejo Nacional de Investigaciones Científicas y Técnicas (CONICET), CABA, Buenos Aires, Argentina

Erick Paul Gutiérrez-Grijalva
Functional and Nutraceutical Foods Laboratory, Centro de Investigación en Alimentación y Desarrollo A. C., Unidad Culiacán, AP 32-A, Sinaloa 80129, México

M. M. Gutierrez-Pacheco
Centro de Investigación en Alimentación y Desarrollo, A.C. (CIAD, AC), Carretera a la Victoria Km 0.6, La Victoria, Hermosillo, Sonora 83000, México

José Basilio Heredia
Functional and Nutraceutical Foods Laboratory, Centro de Investigación en Alimentación y Desarrollo A. C., Unidad Culiacán, AP 32-A, Sinaloa 80129, México. E-mail: jbheredia@ciad.mx

Daniel Lira-Morales
Centro de Investigación en Alimentación y Desarrollo A.C., Culiacán, Sinaloa, CP 80110, México

C. A. Mazzucotelli
Grupo de Investigación en Ingeniería en Alimentos, Universidad Nacional de Mar del Plata, Buenos Aires, Argentina

Magaly B. Montoya-Rojo
Centro de Investigación en Alimentación y Desarrollo A.C., Culiacán, Sinaloa, CP 80110, México

José Alberto Núñez-Gastélum
Departamento de Ciencias Químico-Biológicas, Instituto de Ciencias Biomédicas, Universidad Autónoma de Ciudad Juárez, Anillo Envolvente del PRONAF y Estocolmo s/n, Ciudad Juárez, Chihuahua, CP 322310, México

L. A. Ortega-Ramirez
Centro de Investigacion en Alimentacion y Desarrollo, A.C. (CIAD, AC), Carretera a la Victoria Km 0.6, La Victoria, Hermosillo, Sonora 83000, Mexico

R. Pacheco-Ordaz
Centro de Investigacion en Alimentacion y Desarrollo, A.C. (CIAD, AC), Carretera a la Victoria Km 0.6, La Victoria, Hermosillo, Sonora 83000, Mexico

H. Palafox-Carlos
Herbalife, Camino al Iteso No. 8900 Int. 1 A, Col. El Mante, CP 45609, Tlaquepaque, Jalisco, Mexico

A. E. Quirós-Sauceda
Centro de Investigación en Alimentación y Desarrollo, AC (CIAD, AC), Carretera a la Victoria Km 0.6, La Victoria CP 83000, Hermosillo, Sonora, Mexico

Joaquín Rodrigo-García
Departamento de Ciencias de la Salud, Instituto de Ciencias Biomédicas, Universidad Autónoma de Ciudad Juárez, Anillo Envolvente del PRONAF y Estocolmo s/n, Ciudad Juárez, Chihuahua, CP 322310, México

G. Rojas-Verde
Univ Autonoma Nuevo Leon, Fac Ciencias Biol, Apdo Postal 124-F, Ciudad Univ, San Nicolas De Los Garza 66451, Nuevo Leon, Mexico

Jacqueline Ruiz-Canizales
Centro de Investigación en Alimentación y Desarrollo A.C. Carretera a El Dorado km 5.5 Campo El Diez, 80110, Culiacán, Sinaloa, México

J. Adriana Sañudo-Barajas
Centro de Investigación en Alimentación y Desarrollo A.C., Culiacán, Sinaloa, CP 80110, México. E-mail: adriana@ciad.mx

Mohammed Wasim Siddiqui
Department of Food Science and Postharvest Technology, Bihar Agricultural University, Sabour, Bhagalpur, Bihar, India

Brenda Adriana Silva-Espinoza
Centro de Investigacion en Alimentacion y Desarrollo, A.C. (CIAD, AC), Carretera a la Victoria Km 0.6, La Victoria, Hermosillo, Sonora, 83000, Mexico. E-mail: bsilva@ciad.mx

L. Siqueira-Oliveira
Department of Biochemistry and Molecular Biology, Federal University of Ceará, Av. Mr. Hull 2297 Bl. 907, Campus do Pici, 60455-760 Fortaleza, CE, Brazil

L. Solís-Soto
Univ Autonoma Nuevo Leon, Fac Ciencias Biol, Apdo Postal 124-F, Ciudad Univ., San Nicolas De Los Garza 66451, Nuevo Leon, Mexico

M. R. Tapia-Rodriguez
Centro de Investigacion en Alimentacion y Desarrollo, A.C. (CIAD, AC), Carretera a la Victoria Km 0.6, La Victoria, Hermosillo, Sonora 83000, Mexico

Nancy Varela-Bojórquez
Centro de Investigación en Alimentación y Desarrollo A.C., Culiacán, Sinaloa, CP 80110, México

Francisco Javier Vázquez-Armenta
Centro de Investigacion en Alimentacion y Desarrollo, A.C. (CIAD, AC), Carretera a la Victoria Km 0.6, La Victoria, Hermosillo, Sonora 83000, Mexico

Alma Angelica Vazquez-Flores
Departamento de Ciencias Químico-Biológicas, Instituto de Ciencias Biomédicas, Universidad Autónoma de Ciudad Juárez, Anillo Envolvente del PRONAF y Estocolmo s/n, Ciudad Juárez, Chihuahua, CP 322310, México

Gabriela Vázquez-Olivos
Functional and Nutraceutical Foods Laboratory, Centro de Investigación en Alimentación y Desarrollo A. C., Unidad Culiacán, AP 32-A, Sinaloa 80129, México

G. R. Velderrain-Rodriguez
Centro de Investigación en Alimentación y Desarrollo, AC (CIAD, AC), Carretera a la Victoria Km 0.6, La Victoria CP 83000, Hermosillo, Sonora, Mexico

Rosabel Vélez-de la Rocha
Centro de Investigación en Alimentación y Desarrollo A.C., Culiacán, Sinaloa, CP 80110, México

Mercedes Verdugo-Perales
Centro de Investigación en Alimentación y Desarrollo A.C., Culiacán, Sinaloa, CP 80110, México

Abraham Wall-Medrano
Departamento de Ciencias de la Salud, Instituto de Ciencias Biomédicas, Universidad Autónoma de Ciudad Juárez, Anillo Envolvente del PRONAF y Estocolmo s/n, Ciudad Juárez, Chihuahua, CP 322310, México

C. Zoellner
Department of Population Medicine and Diagnostic Sciences, Cornell University, S2-072 Schurman Hall, Ithaca, NY 14853, USA

LIST OF ABBREVIATIONS

ALA	α-linolenic acid
CVD	cardiovascular diseases
DF	dietary fiber
DHA	docosahexanoic acid
EPA	eicosapentanoic acid
EU	European Union
FA	ferulic acid
FAO	Food and Agriculture Organization of the United Nations
FOSHU	Food for Specified Health Uses
FWRA	Food Waste Reduction Alliance
GAE	gallic acid equivalent
GSF	grape seed flour
ICO	International Coffee Organization
IDF	insoluble dietary fiber
LA	linoleic acid
LCA	life cycle assessment
MAE	microwave-assisted extraction
MDG	Millennium Development Goals
OSI	oxidative stability index
PCA	*p*-coumaric acid
PEFs	pulsed electric fields
PUFAs	polyunsaturated fatty acids
ROS	reactive oxygen species
RSC	radical scavenging capacity
SA	sinapic acid
SCG	spent coffee ground
SDF	soluble dietary fiber
SFE	supercritical fluid extraction
SWE	subcritical water extraction
TAE	tannic acid equivalent
TEAC	Trolox equivalent antioxidant capacity
TPC	total phenolic content

ABOUT THE BOOK SERIES: POSTHARVEST BIOLOGY AND TECHNOLOGY

As we know, preserving the quality of fresh produce has long been a challenging task. In the past, several approaches were in use for the postharvest management of fresh produce, but due to continuous advancement in technology, the increased health consciousness of consumers, and environmental concerns, these approaches have been modified and enhanced to address these issues and concerns.

The Postharvest Biology and Technology series presents edited books that address many important aspects related to postharvest technology of fresh produce. The series presents existing and novel management systems that are in use today or that have great potential to maintain the postharvest quality of fresh produce in terms of microbiological safety, nutrition, and sensory quality.

The books are aimed at professionals, postharvest scientists, academicians researching postharvest problems, and graduate-level students. This series is a comprehensive venture that provides up-to-date scientific and technical information focusing on postharvest management for fresh produce.

Books in the series address the following themes:

- Nutritional composition and antioxidant properties of fresh produce
- Postharvest physiology and biochemistry
- Biotic and abiotic factors affecting maturity and quality
- Preharvest treatments affecting postharvest quality
- Maturity and harvesting issues
- Nondestructive quality assessment
- Physiological and biochemical changes during ripening
- Postharvest treatments and their effects on shelf life and quality
- Postharvest operations such as sorting, grading, ripening, de-greening, curing, etc.
- Storage and shelf-life studies
- Packaging, transportation, and marketing
- Vase life improvement of flowers and foliage

- Postharvest management of spice, medicinal, and plantation crops
- Fruit and vegetable processing waste/byproducts: management and utilization
- Postharvest diseases and physiological disorders
- Minimal processing of fruits and vegetables
- Quarantine and phytosanitory treatments for fresh produce
- Conventional and modern breeding approaches to improve the post-harvest quality
- Biotechnological approaches to improve postharvest quality of horticultural crops

We are seeking editors to edit volumes in different postharvest areas for the series. Interested editors may also propose other relevant subjects within their field of expertise, which may not be mentioned in the list above. We can only publish a limited number of volumes each year, so if you are interested, please email your proposal wasim@appleacademicpress.com at your earliest convenience.

We look forward to hearing from you soon.

Editor-in-Chief:
Mohammed Wasim Siddiqui, PhD
Scientist-cum-Assistant Professor | Bihar Agricultural University
Department of Food Science and Technology | Sabour | Bhagalpur | Bihar | INDIA
AAP Sr. Acquisitions Editor, Horticultural Science
Founding/Managing Editor, *Journal of Postharvest Technology*
Email: wasim@appleacademicpress.com
wasim_serene@yahoo.com

BOOKS IN THE POSTHARVEST BIOLOGY AND TECHNOLOGY SERIES

Postharvest Biology and Technology of Horticultural Crops: Principles and Practices for Quality Maintenance
Editor: Mohammed Wasim Siddiqui, PhD

Postharvest Management of Horticultural Crops: Practices for Quality Preservation
Editor: Mohammed Wasim Siddiqui, PhD, Asgar Ali, PhD

Insect Pests of Stored Grain: Biology, Behavior, and Management Strategies
Editor: Ranjeet Kumar, PhD

Innovative Packaging of Fruits and Vegetables: Strategies for Safety and Quality Maintenance
Editors: Mohammed Wasim Siddiqui, PhD, Mohammad Shafiur Rahman, PhD, and Ali Abas Wani, PhD

Advances in Postharvest Technologies of Vegetable Crops
Editors: Bijendra Singh, PhD, Sudhir Singh, PhD, and Tanmay K. Koley, PhD

Plant Food By-Products: Industrial Relevance for Food Additives and Nutraceuticals
Editors: J. Fernando Ayala-Zavala, PhD, Gustavo González-Aguilar, PhD, and Mohammed Wasim Siddiqui, PhD

Emerging Postharvest Treatment of Fruits and Vegetables
Editors: Kalyan Barman, PhD, Swati Sharma, PhD, and Mohammed Wasim Siddiqui, PhD

PREFACE

This book describes and analyzes the main advantages of the integral exploitation of the use of plant food byproducts as food additives and nutraceuticals. The mass of byproducts obtained as a result of processing plant food may approach or even exceed that of the corresponding valuable product affecting the economics of crops. In the past, this costly problem has been mitigated to some extent by processing the byproducts further to yield a product that presents less of a disposal problem or that has some marginal economic value. The economics of processing crops could be improved by developing higher-value use for their byproducts. For instance, several patents have been published relating the use of crops as a source of nutraceutical compounds. It has now been reported that the byproducts of plant food contains high levels of various health-enhancing substances that can be extracted from the byproducts to provide nutraceuticals.

This book analyzes the potential uses of plant byproducts; where one of the majors can be as food additives (antioxidants, antimicrobials, colorants, flavorings, and thickener agents) or nutraceuticals (antioxidant, anticarcinogenic, antiinflamatory, prebiotics, antiobesogenic, inmunomodulators, among others). The antimicrobial power of plant and herb extracts has been recognized for centuries, and mainly used as natural medicine, however, the trends in using these compounds as food preservatives is increasing nowadays. In addition, plants produce a wide range of volatile compounds, some of which are important for flavor quality factors in fruits, vegetables, spices, and herbs. It is well known that agro industrial byproducts are rich in dietary fibers (DF). The DF additive provides economic benefits to the food and pharmaceutical industries. Apart from the well-known health effects, DF shows some functional properties as food additives, such as water-holding capacity, swelling capacity, increasing viscosity or gel formation which are essential in formulating certain food products. The main goal of this book is to ascribe the relevance of an integral exploitation of the food chain, since the perspective of food security and sustainability of the population, as well as the main plant food processing industries as generators of byproducts, state of the art of the advances in every industry on the claimed uses.

CHAPTER 1

INTEGRAL EXPLOITATION OF THE PLANT FOOD INDUSTRY: FOOD SECURITY AND SUSTAINABLE DEVELOPMENT

ARIADNA THALÍA BERNAL-MERCADO,
FRANCISCO JAVIER VÁZQUEZ-ARMENTA,
GUSTAVO A. GONZÁLEZ-AGUILAR, and
BRENDA ADRIANA SILVA-ESPINOZA*

Centro de Investigacion en Alimentacion y Desarrollo, A.C. (CIAD, AC), Carretera a la Victoria Km 0.6, La Victoria, Hermosillo, Sonora 83000, Mexico

Corresponding author. E-mail: bsilva@ciad.mx

CONTENTS

ABSTRACT

Nowadays, the population is growing more, and, with it, the challenge of feeding all habitants in an adequate way at the right time. The lack of food security is one of the main problems worldwide, especially in undeveloped countries, who do not have the opportunity to access, dispose, and use the nutrients of food that are necessary for life. One of the most important reasons for the lack of food security is the great waste of food that occurs in production, industry, and/or consumption. Food waste refers to any food or inedible part of a food that has been removed from the food supply chain to be recovered or disposed in compost, incineration, energy production, or at sea. It is estimated that about 1.3 trillion tons per year, equivalent to one-third of all food production per year, is wasted. In addition, the waste of food generates a great problem of environmental contamination. Among the foods that are most wasted in the industry, fruits and vegetables occupy 44% because of their perishable nature and their susceptibility to mechanical damage and attack of microorganisms. To reduce food loss and waste, it is necessary to find an alternative use of wastes and prevent the overproduction of food. The solution proposed during this chapter is the exploiting of bioactive compounds such as polyphenols, fiber, vitamins, and minerals with multiple beneficial health activities that can be obtained from fruit and vegetable wastes. This would promote less waste, high economic potential for the production of ingredients and functional products with added value, greater sustainable development, and environmental protection.

1.1   INTRODUCTION

Over the next decades, the world has to face an important challenge and a huge opportunity concerning food security, economic development, and environment (Hanson, 2013). In addition, world population is growing considerably that increased the risk of compromising food security, increment in food wastage, and environmental changes. The link between all these factors is to find solutions to assure food security in a sustainable manner for the entire world population. Global organizations have worked in relation to food security but have not yet managed to eradicate hunger in the world's poorest countries. There are many causes of food insecurity, some of them are involved with economic and social factors such as a lack of access, availability, utilization, and stability of food.

One of the factors that affect food security is food waste, which varies according to regions, being more vulnerable in developing countries. Food waste is generated throughout all the food chain production, processing, and consumption. Moreover, people behavior and trends in consumerism contributes to the generation of large amounts of food waste, which could compromise food security. All this leads to major environmental problems such as greenhouse gas emissions, contamination of soil and water, reducing landfills, bacterial contamination, increment in oxygen demand, among others. In addition to environmental factors, food waste also impacts the economy, as millions of dollars are spent on food that is not consumed by humans.

Food industries are largely responsible for generating large quantities of waste, particularly agro-industries, with negative impacts on the environment and the economy. Many processing fruit and vegetable industries generate a lot of by-products. These wasted products are important sources of useful bioactive compounds (phenolic compounds, antioxidants, fiber) as nutraceuticals and pharmaceuticals. In this context, the use of by-products of plants industries is proposed as a tool of sustainability to fight against hunger, climate change, and food waste that can be used as coproducts with added value contributing to the social–economic and environmental benefits.

1.2 FOOD SECURITY CHALLENGES

In a world that produces enough food for everyone, people should not be hungry. During the past decades, a remarkable progress in food production that allowed a reduction in the world's population suffering from hunger has been observed (FAO, 2015). However, efforts have not been enough since there are a considerable amount of people that still do not have access to sufficient safe and nutritive food (Hanson, 2013). In this sense, the world needs food security and this is the challenge that governments, politicians, scientists, farmers, industries, environmentalists, and economists have to face (Poppy et al., 2014).

The Food and Agriculture Organization of the United Nations (FAO) defined food insecurity as a "situation that exists when people lack secure access to sufficient amounts of safe and nutritious food for normal growth and development and an active and healthy life." Food security is achieved when everyone is capable to physically, socially, and economically access, at all time, to enough, safe, and nutritive food to satisfy their dietary needs for a good and healthy life (Charles et al., 2010). Everyone in the world must have the right to food security without discrimination of sex, nationality, economic status, religious beliefs, among others kind of discrimination (FAO, 2015).

Food insecurity can be transitory, seasonal, and chronic depending of the exposure time. Chronic food insecurity is persistent through time and is produced when people are not able to satisfy their need for food for an extended period of time due to situation of poverty and to the lack of resources and access to productive or financial capitals (FAO, 2008). In contrast, transitory food insecurity occurs when there is an unexpected drop in the ability to produce or access enough food to maintain a good nutritional-innocuous status and takes place in short terms (FAO, 2008). Seasonal food security falls between chronic and transitory food insecurity because it occurs for limited period of time but it can be recurrent. It arises when there is a cyclical pattern of inadequate availability and access to food. This is associated with seasonal fluctuations in climate, cropping patterns, work opportunities, and diseases (FAO, 2008).

Furthermore, it is important to know how intense or severe is the impact of the identified problem on the overall food security and nutrition status of the population. For this purpose, different "scales" or "phases" to "grade" or "classify" food security have been developed by food-security analysts using different indicators and cut-off points or "benchmarks," such as: generally food secure, chronically food insecure, acute food and livelihood crisis, humanitarian emergency, and at least famine (FAO, 2008).

According to the FAO, food security has four dimensions defined as food availability, safety, economic, and physical access to consumption and stability over time (Hanson, 2013). Food availability means adequate amounts of food that are produced and ready for people's use, this includes safety and quality. Economic and physical access refers to have enough resources to obtain foods. To achieve food security access, policy has to focus on incomes, expenditure, markets, and prices. Food utilization is when human is able to ingest and metabolize food to avoid diseases this is the result of good care and feeding practices, food preparation, diversity of the diet, and intrahousehold distribution of food (FAO, 2008). Combined with good biological utilization of food consumed, this determines the nutritional status of individuals. Finally, stability refers to maintain the others dimensions over time. Actually, even if the food intake is adequate, it is still considered to be food insecure if there is inadequate access to food on a periodic basis, risking a deterioration of the safe-nutritional status (FAO, 2008). Some authors have proposed a fifth dimension, environmental sustainability, where food is produced and consumed without compromise natural resources and ensure sufficient food for future generations (Hanson, 2013).

One of the most important concerns of humanity is to fight global poverty and all its dimensions; in this sense, world leaders gathered at the beginning of the new millennium in the United Nations. At this meeting, eight Millennium Development Goals (MDG) were established, which were the predominant development framework for the world over the last 15 years (ONU, 2015). Also, the world has another objective related to hunger, 182 representatives of governments who participated in the World Food Summit (WFS) celebrated in Rome in 1996, pledged "…To eradicate hunger in all countries, with the aim immediately reduce the number of undernourished people to half their present level no later than 2015."

According to data from FAO (2015) derived from the MDG targets, the current state of world food security is about 795 million people that are undernourished globally (18.6%), down 167 million over the last decade (14.9%), and 216 million less than in 1990–1992 (10.9%) (Fig. 1.1). This decrement has been more obvious in developing regions, regardless growth population (Fig. 1.2). Regarding to developing regions, undernourished people within total population has decreased to 12.9% compared with 23.3% in 1990–1992. More than half of the countries monitored (129 countries) have reached the MDG first target related to food security. Regions such as Latin America, the east and south eastern regions of Asia, the Caucasus and Central Asia, and the northern and western regions of Africa have made fast progress; while others could not reach the MDG first target, such as southern Asia, Oceania, the Caribbean, and southern and eastern Africa.

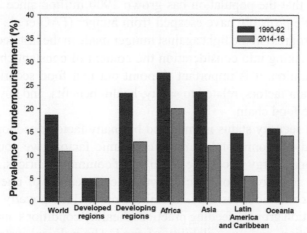

FIGURE 1.1   Percentage of prevalence of undernourishment around the world.

Note: Data for 2014–2016 refer to provisional estimates.

Source: FAO. *The State of Food Insecurity in the World 2015.* FAO: Rome, 2015.

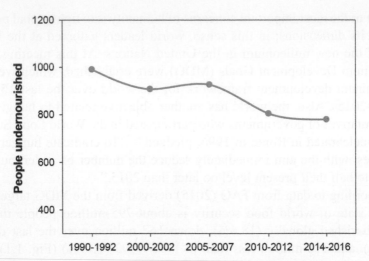

FIGURE 1.2   Number of people undernourished in the developing regions.
Note: Data for 2014–2016 refer to provisional estimates.
Source: FAO. *The State of Food Insecurity in the World 2015.* FAO: Rome, 2015.

However, the objective of WFS has not been reached yet, although the number of people suffering from hunger is not halved since 1990–2015. To achieve this objective, the number of undernourished people should have been reduced from 1000 million in 1990 to 515 million in 2015 and actually there are 795 million people suffering from hunger. Nevertheless, if we consider that the population has grown 1900 million since 1990, 2000 million people probably have escaped from hunger (FAO et al., 2015). The significant progress in the fight against hunger made in the last decade should be analyzed taking into consideration the context of constant change in the world. In addition, it is important to point out that food security includes other important factors, related to safety, health benefits, and environmental impact of the food chain.

The food security status is impacted by many factors: economic, social, political, and environmental. Among economic factors, we could find the rise in food and energy prices, the volatility of commodity prices, an unstable market, less inclusive economic growth, the increment of unemployment and global economic recessions. Social factors include poverty conditions, rapid population growth, feeding practices, healthy conditions, inappropriate food distribution, and unavailability of food. The political instability, civil conflicts, food distribution, and unavailability are political factors. Also, the environmental conditions such as weather extreme events, climate change, and natural disasters have been slowing the progress against food insecurity (FAO et al., 2015).

The consequences of food insecurity are associated with a wide range of health outcomes in children and adults (Cook and Frank, 2008). Among their significant finds, food insecurity cause several damages in children such as higher risks of some birth defects when pregnancy mother is in a food insecure state (Carmichael et al., 2007), anemia when children are in a households suffering from food insecurity (Skalicky et al., 2006), lower nutrient intakes (Cook et al., 2004), higher levels of aggression and anxiety (Whitaker et al., 2006), higher probabilities of being hospitalized (Cook et al., 2006), higher probabilities of dysthymia and other mental health issues (Alaimo et al., 2002), higher probabilities of asthma (Kirkpatrick et al., 2010), and higher probabilities of behavioral problems (Huang et al., 2010).

Regarding adults, food security can also have negative health effects. Households suffering from food insecurity are more likely to have adults who have lower nutrient intakes (McIntyre et al., 2003), long-term physical health problems (Tarasuk, 2001), higher levels of depression (Whitaker et al., 2006), diabetes (Seligman et al., 2007), higher levels of chronic disease (Seligman et al., 2010), and more instances of oral health problems (Muirhead et al., 2009). Food insecure seniors have lower nutrient intakes (Lee and Frongillo, 2001; Eicher-Miller et al., 2009) and are more likely to have limitations in activities of daily living (Ziliak et al., 2008).

To improve the nutrition aspect of food security, much effort needs to be made in the world. First of all, it has to be an economic change. Since population of the rural sector represents a high percentage of the people suffering from hunger and malnutrition in developing countries, the economic growth has to be inclusive, offer more opportunities of employment to rural people and improve productivity in small-holder farmers for reducing undernourishment (Hanson, 2013). In addition, the promotion of social protection programs around the world that help to diminish the malnutrition, hunger, and poverty by encouraging good health, better nutrition, and education are key to overcoming hunger and food insecurity (Gundersen et al., 2011; FAO et al., 2015). Also, social programs with people education and capacitation, mainly aimed to the poorest people, will help them to improve their life and participate in global economy (Food and Agriculture Organization, 2013). Nevertheless, more important than food production and social protection, it is necessary to reach the unequal distribution of food around different regions in the world and improve food access in the poorest population.

According to United Nations projections, global population will continue to grow to 9.3 billion in 2050, 2.3 more billion than in 2012 (Charles et al.,

2010; Hanson, 2013; Service, 2014). However, coupled with this growth also will increase wealth and consequently the middle class population that will demand more processed food. In this sense, food available will need to increase, it is suggested that in 2050, the food needs to increase 70–100% more than today (Hanson, 2013). Growing that amount of needed food and the industry impacts will put a significant strain on the planet, linking agricultural intensification with biodiversity conservation and hunger reduction is a great challenge for the future (Tscharntke et al., 2012).

Another challenge is the climate change. By the end of this century, the global mean temperature could be warmer than the one at the end of the previous century. In their review, Wheeler and von Braun (2013) suggested that climate change will compromise the food security in all its dimensions. Food accessibility will be affected through the negative effects of climate change in crops (Chakraborty and Newton, 2011), most tropical areas being more affected than those at higher latitudes. Furthermore, large parts of the world where crop productivity is expected to decline under climate change coincide with countries that currently have a high problem of hunger (Wheeler and von Braun, 2013). Food access connects to climate change through indirect pathways. Access to food is largely a matter of economic and physical capacity to obtain food. Prices of the basic resources, such as land and water, include expectations of climate change, such as revaluation of land with access to water. Concerning food utilization, attaining nutritional well-being depends upon water and sanitation, and it is obvious that it will be affected by any impact of climate change. Stability of whole food systems may be at risk under climate change, as climate can be an important determinant for future price trends, as well as the short-term variability of prices, and cost of decontamination (Wheeler and von Braun, 2013). At the same time, food producers are experiencing greater competition for land, water, and energy, and the need to curb the many negative effects of food production on the environment is becoming increasingly clear (Charles et al., 2010).

With this in mind, our food system needs to adapt to new demands of society and the world has the challenge to supply food for all population (Tscharntke et al., 2012), with no exception and being more empathetic with the poorest regions to end the hunger, but at the same time, the world needs to be more environmental friendly and sustainable. The Science and Technology Options Assessment, of the European Parliament, proposed some options to feed 10 billion people: (1) interactions between climate change and agriculture and between biodiversity and agriculture; (2) plant breeding and innovative agriculture; (3) options for sustainable food processing; (4)

options for cutting food waste; (5) recycling agricultural, forestry, and food wastes and residues for sustainable bioenergy and biomaterials (Service, 2014). Most of these points are related to the main goal of the food technology area and the topic of this book.

1.3 FOOD AND AGRO-INDUSTRIAL WASTES AND SUSTAINABILITY

Modernization of industries, trends in consumerism and the high growth population, and their food demand have resulted into generation of large amounts of diverse wastes (Ajila et al., 2012). According to the FAO, there exists a difference between food loss and food waste, although the definitions are not universal worldwide. Food loss occurs usually when food is not used, destroyed, or degraded during production, postharvest, and processing stages of the supply chain, whereas food waste is generated principally at the end of the supply chain once the food has been processed. By contrast, the project funded by the European Commission Framework Programme 7 named Food Use for Social Innovation by Optimising Waste Prevention Strategies (FUSIONS) and the UK Waste and Resources Action Programme (WRAP) refer to both of these concepts as food waste (Garcia-Garcia et al., 2015).

According to these organizations, food waste is defined as "any food and inedible part of food removed from (lost to or diverted form) the food supply chain to be recovered or disposed (including composted, crops ploughed in/not harvested, anaerobic digestion, bio-energy production, co-generation, incineration, disposal to sewer, landfill or discarded to sea)" (Bos-Brouwers et al., 2014). To consider food loss or waste, FAO only take into consideration the parts of the food that could have been eaten by people. By contrast, FUSIONS and WRAP also include inedible parts of food such as bones or egg shells in the definition of food waste; however, the food wasted in redistribution and packaging and inedible parts of food sent to animal feed are not included in this definition (Garcia-Garcia et al., 2015).

FAO (2013) estimates that each year, about a third of all food produced for human consumption is lost or wasted, which amounts to about 1.3 billion t/year. The World Resources Institute, a US-based global research organization, found that about 24% of all the calories produced for human consumption do not actually end up reaching human mouths (Lipinski, 2014) and there are 198 million hectares (the size of Mexico) used to produce food that is not eaten. The Food Waste Reduction Alliance (FWRA) (2013) estimated that 25–40% of food that is grown, processed, and transported in the United States will never be consumed. Overall, the economic costs of food loss are generally thought to be massive.

The FAO estimates that food worth over US$750 billion (based on 2009 producer price) is lost or wasted annually. In the United States, nearly 61 million tons of food waste are generated every year (GMA, 2012), while Europe generate an estimate of 90 million tons annually (EC, 2013). Footprint analyses suggest that North America and Europe consume resources (energy, land, materials, etc.) as though they inhabit multiple planets, to be more precise; the United States consumes as though it inhabits five planets and Europe three (Lang and Barling, 2013). This situation is contradictory, since a little bit more than 1 of every 10 humans still suffering from hunger. This represents an opportunity to ensure food security and mitigate environmental impact. These solutions should be capable of exploiting the precious resources represented by food waste to achieve social, economic, and environmental benefits.

There are several reasons why food loss and waste are important, according to Buzby and Hyman (2012), there are three principals reason. The first reason is that population will continue to grow; then, the world will need more food to feed people. In addition, most of this population growth will occur in developing countries, where they also face issues of food security. The second reason why food loss is important is that food waste represents significant amounts of money, and the last reason is the adversely impact on society and on the environment.

The production of food waste covers all the food life cycle: from agriculture, up to industrial manufacturing and processing, retail, and household consumption (Parfitt et al., 2010; Mirabella et al., 2014). Food losses in industrialized countries are as high as in developing countries, but in developing countries, more than 40% of the food losses occur at postharvest and processing levels, while in industrialized countries, more than 40% of the food losses occur at retail and consumer levels. Food waste at consumer level in industrialized countries (222 million ton) is almost as high as the total net food production in sub-Saharan Africa (230 million ton) (Gustavsson et al., 2011).

It is estimated that the per capita food waste by consumers in Europe and North-America is 95–115 kg/year, while this figure in sub-Saharan Africa and South/Southeast Asia is only 6–11 kg/year (Gustavsson et al., 2011). In United Kingdom about 7.0 million tons of food and drink were wasted thrown away from homes in 2012; 4.2 million tons were avoidable, worth £12.5 billion, 1.2 million tons was considered "possibly avoidable" and 1.6 million tons was unavoidable waste (Quested et al., 2012). A study in the United States estimated that in 2008, total food loss at the retail and consumer levels in that country was $165.6 billion. On a per capita basis, food lost from the food supply at the consumer level is equivalent to 124 kg of food per year at an estimated retail price of $390/year at retail prices. The per capita estimates may be helpful to consumers to make them more mindful of their daily and yearly of food loss (Buzby and Hyman, 2012).

Causes of food wastes in developing countries are associated to financial and technical limitations in harvesting techniques, storage, cooling facilities, infrastructure, packaging, and marketing systems (Gustavsson et al., 2011). Food waste in low-income countries is a reflection of weaknesses in the food supply chain and it is directly related to food insecurity since it reduces availability of nutritious food (Unit, 2014). By the other side, in developed countries, the causes of food wastes are associated to the final stages of the food supply chain such as consumer and industries actions.

Consumer behavior is the main cause in medium/high income countries due to the large amounts of food purchased that are not consumed and to the expiring dates of products, in combination with the inconsiderate approach of those consumers who can afford to waste food (Gustavsson et al., 2011). Despite high amount of food waste is generated in developed countries, it does not contribute to food insecurity, in contrast is the result of higher food availability. Also, the lack of management in the supply chain contributes to food waste in developed countries. Food can be wasted by farmers and industries due to quality standards, which reject food items that are not perfect in shape or appearance (Gustavsson et al., 2011). Food waste in industrialized countries can be reduced only if food industries, retailers, and consumers are conscious of this world-impacting problem (Tamer and Çopur, 2014).

In developed countries, 42% of food waste is produced by households, while 39% losses occur in the food-manufacturing industry, 14% in food service sector, and remaining 5% in retail and distribution (Mirabella et al., 2014). It can be observed that a high percentage is discarded from food industries. Losses are largest for fruit and vegetable industries from production to the end of the food supply chain estimated in 44%, followed by roots and tubers (20%), cereals (19%), milk (8%), and meat, fish, and sea food with less than 4%. (Gustavsson et al., 2011; Lipinski et al., 2013). In fruits and vegetables, food loss occur due to their highly perishable characteristics and susceptibility to mechanical damage and spoilage and if they do not have good appealing, they are discarded (Ayala-Zavala et al., 2010). In addition, they can be unsuitable for consumption because of the presence of pests, bacteria, or rots. Moreover, during industrial processing, loss may also occur if food or parts of food is not suitable for processing and is thus discarded during preparation or is improperly processed. For example, during juice production, once the juice of a fruit or vegetable is squeezed, pulp, peels, and seeds are typically discarded, generating a huge amount of by-products.

The agro-food supply chain covers a broad variety of manufacturing processes that generate accumulative quantities of different waste, especially organic residues that causes environmental pollution. Management of agro-industrial residues is one of the complicated problems in agriculture and in agro-industry (Ajila et al., 2012). This waste, in addition to being a great loss of valuable materials, such as food, feed, fuel, and various chemicals and bioactive compounds, also raises serious management problems, both from the economic and environmental point of view. Protection of environment toward the aim of sustainability is one of the mostly discussed topics in the area of waste management. The sustainable solutions for waste management range from waste prevention, waste minimization, cleaner production, and also zero-emission systems (Ajila et al., 2012; Lin et al., 2013).

Some environmental problems associated with organic wastes include severe pollution problems due to high associated chemical and biological oxygen demand, uncontrolled greenhouse emissions (CO_2), varying pH and chemical composition due to seasonal variations and changes in food processing, materials prone to bacterial contamination, and high accumulation rate that decrease land options to dispose (Lin et al., 2013). Other environmental impacts associated with food wastes are natural resource depletion in terms of soil, nutrients, water, and energy; disruption of biogenic cycles due to intensive agricultural activities; and all other characteristic impacts at any step of the food chain (Girotto et al., 2015).

In general, during all food's life cycle, food production and food waste can cause negative environmental effects, such as greenhouse gas emissions from cattle production (Lundqvist et al., 2008); air pollution caused by farm machinery and trucks that transport food, water pollution and damage to marine and freshwater fisheries from agricultural chemical run-off during crop production; and soil erosion, salinization, and nutrient depletion that arise from unsustainable production and irrigation practices (Nellemann, 2009). Disposing of uneaten food at the end of food's life cycle also can create emissions that can negatively impact human health and the environment. Landfilling food waste also negatively impacts the environment through the methane gas generated when food-waste decomposes anaerobically (Buzby and Hyman, 2012).

Depending on the classification of waste, edible–nonedible, eatable–noneatable, and source of waste generation, the management will be different

(Garcia-Garcia et al., 2015). During this process, relevant regulations must be consulted, and environmental, social, and economic impacts must be taken into account. Prevention of food becoming waste is the preferred option, followed first by redistribution to people and second by redistribution to animals. If surplus food cannot be reused, the valuable compounds contained in the waste, as well as the technologies available, should be assessed. Thermal treatments with energy recovery can be used with all types of waste; however, it presents important environmental ramifications, such as gas emissions and ash. Landspreading is typically used with plant-based products at the agricultural stage. Thermal treatments without energy recovery and landfilling should be avoided where at all possible because of their environmental impact and lack of positive outcomes (Garcia-Garcia et al., 2015).

In addition to environmental effects, food waste also has important economic and social impact. The total cost of food waste can reach USD 1 trillion each year; to this number should be added an additional USD 700 billion related to the environmental impact and USD 900 billion associated with social costs (Garcia-Garcia et al., 2015). Socially, there is a conflict between food waste and food security.

The definitions of sustainable development indicated that there must be concern about the intergenerational transfer of natural capital (El-Ramady, 2014). There are four basic objectives included in most definitions of sustainability: concern to keep the environment intact in the long run, concern with regard to the well-being of coming generations, rejection of rapid population growth, and concern whether it will be possible to maintain economic growth at the sight of diminishing resources (El-Ramady, 2014).

Garnett (2014) has proposed three perspectives on achieving food system sustainability: efficiency oriented, demand restraint perspective, and food system transformation. Efficiency is orientated that the food security is associated to a food supply problem. The main responsibility of efficiency is for governments and food industry actors such as agricultural businesses, farming unions, manufacturers, and retailers. With this in mind, efficiency is achieved when more food is produced with technological innovations and administrative improvements to satisfy increasing and changing demand by growing populations and in the same way enhance nutrition and produce food without compromising the environment.

According to Garnett (2014) for the demand restraint, the problem lies on consumers and with companies who promote unsustainable consumption patterns. Consumers are the main actors in this problem because of consumerism behavior; in addition, industries allow the excessive consumption that is leading an environmental crisis. For the other side, the food system transformation contemplates both production and consumption linked to the relation between them in the food system, being the problem, creating an imbalance among them. The imbalance in this relationship is associated to the extreme problems: excess and insufficiency in the environment and in relation to health (obesity and hunger) (Garnett, 2014).

Environmental sustainability through the reduction of food waste can only be achieved through structural change. First, people need to be more empathetic with the world problems such as climate change, hunger, and poverty. To reduce food waste at consumer level the idea of "if I consume more, I am happier" must be eradicated. Consumers must have the social compromise to avoid the consumerism buying only the food they will eat. Moreover, people should be more conscious about quantities of food they need, and reaching this, also family economy will be improved. While efforts to change consumer behavior may result in reduction in food waste in developed countries, modifications in legislation and business behavior on the way to more sustainable food production and consumption will be necessary to reduce waste from its current high levels (Parfitt et al., 2010).

Industry-led initiatives or government-led policies have the greatest potential to reduce food loss in the next decade (Buzby and Hyman, 2012). Food industries should be more aware about the waste and environmental impact. They are responsible for the increase in the consumerism through their publicity, to produce huge amounts of food that will not be consumed, and for the trend to pack the food in huge containers. All this contributes to increase food waste. Industries should be more environmentally committed to the correct waste disposal and search for the recovery, and reuse of food waste and by-products. With this in mind, the food industries have raised an economic objective denominated "zero-waste economy" in which waste are used as raw material for new products and applications. In the other side, governments should make and apply laws and regulations to encourage economically food industries to reduce or reuse food by-products and wastes.

Countries around the world are concerning about how to feed the growing population without compromising the environment, and at the same, they are promoting the prevention of food loss and waste. The European Union (EU) commitment to halve the disposal of edible food in the EU by 2020, and for them, 2014 was the "year against food waste." Also, the 7th Environment Action Programme guides European Environment Policy until 2020, and by 2050, they will have a prosperity and healthy environment stem from an innovative, circular economy where nothing is wasted and where natural resources are managed sustainably, and biodiversity is protected, valued, and restored in ways that enhance the society's resilience (Underwood, 2013; European Commission, 2015).

In 2015, Argentina received the Bioeconomy Symposium (Anonymous, 2015). The bioeconomy concept emphasizes the development of production systems in which biomass obtained sustainably plays a key role in meeting the demands for food, energy, raw materials, and industrial inputs. This view emerges in the early years of this century as a possible response to the growing demand for food (both quantity and quality), restrictions on the availability of fossil fuels and the impacts of climate change on natural resources.

In Mexico, there exist the general law for prevention and integral management of wastes (the waste law) (Mexicano, 2015). This law indicate the right of everyone to live in a suitable environment for their development and welfare. In addition, this law is the responsible to hold activities related to the generation and integrated management of waste modalities issued the order and public interest for achieving sustainable national development. This law applies the principles of recovery, shared responsibility, and integral waste management, under the criteria of environmental, technological, economic, and social efficiency, which should be considered in the design of tools, programs, and plans of environmental policy for waste management. In addition, the waste law encourages the recovery of waste and the development of markets for products, under the criteria of environmental, technological, and economic efficiency; prevents contamination of sites for materials handling and waste and define the criteria to be subject to remediation; strengthen research and scientific development and technological innovation to reduce waste generation and design alternatives for treatment aimed at cleaner production processes; and promotes utilization of waste through actions aimed to restore economic value of waste through reuse, remanufacturing, redesign, recycling, and recovery of materials or energy seconded.

In the United States, the Food Waste Reduction Alliance (FWRA) was created, an industry-wide effort launched in 2010 and focused on addressing food waste, in particular within the section of the supply chain from the food manufacturer to retail grocery and restaurant. It is led by the Grocery Manufacturers Association, Food Marketing Institute, and National Restaurant Association, with active leadership and participation by its members. The alliance seeks to reduce food waste throughout the supply chain. The goals of the FWRA are threefold: (1) To avoid and reduce food waste wherever possible within members operations and supply chains. (2) To increase the donation of safe and healthy foods that would have gone to waste and to send food to food banks to help address hunger issues. (3) To divert unavoidable food waste away from landfills toward higher value uses, such as animal feed, composting, and waste-to-energy.

Moreover in the United States, the Environmental Protection Agency (EPA) is responsible to protect, through law, from significant risks to human health and the environment where they live, learn, and work; protect the environment concerning natural resources, human health, economic growth, energy, transportation, agriculture, industry, and international trade, and these factors are similarly considered in establishing environmental policy; contribute to make the communities and ecosystems diverse, sustainable and economically productive; and work with other nations to protect the global environment (EPA, 2014).

1.4 LOOKING FOR INTEGRAL EXPLOITATION IN THE PLANT FOOD INDUSTRY

To reduce food loss and waste, it is necessary to find an alternative use of the undesirable food excess and prevent the overproduction of food. The utilization of food waste can include animal feed, energy production, and production of specific chemical compounds as precursors for plastic material production, chemical, or pharmaceutical applications (Ajila et al., 2012). The increasing demand for chemicals and fuels are encouraging the reuse and efficient valorization of organic waste from the food supply chain for the production of novel value-added materials, chemicals, and fuels, as a complementary approach to the conventional practices (Lin et al., 2013). In this sense, it is no longer practical to discard by-products and wastes, especially when a significant amount of valuable raw materials have a strong economic potential like the production of new products and functional ingredients with a significant added value (Tamer and Çopur, 2014). Moreover, the recovery process of by-products is part of the current existing sustainable development and environmental protection (Ajila et al., 2012).

Due to legislation and environmental reasons, the industry is more and more forced to find an alternative use for the residual matter. Recovery of bioactive compounds is an elegant way to reuse waste streams, while being economically interesting on the other hand (Babbar and Oberoi, 2014). More specifically, agro-processing industry generates large amount of waste therefore, recovery of bioactive compounds from agro-industrial residues is a good alternative to obtain high-value components (Babbar and Oberoi, 2014). Wasted investment and political should provide an incentive to push the food industry to reduce or reuse food waste generation to gain benefits on both the financial and environmental fronts. The measures taken to reduce or reutilize food waste are linked to improve food security to satisfy the nutritional needs of the population. In this context, integral exploitation of fresh produce can reduce the food insecurity always contemplating the sustainability (economy, social, and environmental).

The full utilization of agro-industrial produce is a requirement and a demand that needs to be encountered by industries wishing to implement low-waste technology (Ayala-Zavala et al., 2011). Fruit and vegetables industries, for example, fresh-cut and juice production, separate the desired value product from other constituents of fresh produce, such as peels, seeds, and unused pulps, known as by-products (Ayala-Zavala et al., 2010). The nonedible parts of vegetables and fruits yield between 25% and 30% (Ajila et al., 2010), which generates a huge amount of by-products that normally are discarded. In addition, there is a lack of information in the type and quantities of by-products that are discarded from the global fruit and vegetables industries; however, it is estimated that is a large amount.

In addition, some by-products have been found to be a better source of bioactive compounds than the own edible pulp (Vega-Vega et al., 2013a). Peels from apples, peaches, and pears were found to contain twice the amount of total phenolic compounds as that contained in fruit pulp (Gorinstein et al., 2001). In this context, the integral exploitation of plants represents a successful opportunity to obtain nutrients, fibers, and phenolic compounds and, at the same time, to have economic benefits and beneficial impact on the environment. Some examples of horticultural by-products that can show the profitability of the extraction of bioactive compounds with great potential for biotechnology industry are tropical fruits, citrus, apples, and grapes.

For example, sliced apples produce 10.91% of pulp and seed (core) as by-products and 89.09% of the final products, whereas in the apple juice industry, the main by-product is apple pomace that represents 25% of the total processed fruit (Ayala-Zavala et al., 2010). The apple pomace consists of a heterogeneous mixture of peels, core, and seed, with high content of water and insoluble carbohydrates, such as hemicellulose, cellulose, and lignin. Also, it includes vitamins, minerals, and phytochemicals. Apple pomace can be converted into various food and industrial ingredients such as dietary fiber, phenolic compounds, and antioxidants. A study showed that incorporating 22% of apple pomace to extruded products, the total dietary fiber content increased from 0.8 to 14 g/100 g product, showing good amounts of soluble dietary fiber (8–20 g/100 g pomace) and good amount of phenolic compounds (93 mg/100 g product) compared to control fruits (Paraman et al., 2015).

Citrus fruits are processed to obtain juice, jam, segments, and canned products, discarding large amounts of peels and seeds (Marín et al., 2007). Peeled mandarins, for example, produce 16.05% of peels and 83.95% of final products (Ayala-Zavala et al., 2010). Citrus by-products are a source of several compounds, principally water, soluble sugars, fiber, organic acids, proteins, minerals, oils, and lipids and also contains flavonoids and vitamins (Ignat et al., 2011). Citrus by-products from industrial processing are used by chemical industry to extract phytochemicals and essential oils (Marín et al., 2007). In a study, the orange seed extract added to orange wedges increased the total phenolic content (6 times), total flavonoids (5.3 times), and antioxidant capacity (2.2 times and 6.8 for 2,2-diphenyl-1-picrylhydrazyl (DPPH) and trolox equivalent antioxidant capacity (TEAC), respectively) in the treated fruits (Cruz-Valenzuela et al., 2013).

Mango is one of the most cultivated fruits in world and its processing involves juices, fresh-cut, jams, canned products, frozen products, among others. Their processing yields large amounts of by-products including peels (13.5%), seeds (11%), and unusable pulp (17.94%) (Ayala-Zavala et al., 2010). The major phenolic compounds found in mango-seed extract are tannins, gallic, cumaric, ferulic and caffeic acids, vanillin, and mangiferin (Abdalla et al., 2007), which are able to neutralize free radicals and inhibit cellular oxidation, in addition to other benefic properties. Mango by-products have shown a higher antioxidant capacity and phenolic content than the own pulp. Mango "Haden" seed extract showed a flavonoids content of 164.6 mg quercetin equivalent (EQ)/g and a total phenolic content of 875.1 mg gallic acid equivalent/g) (Vega-Vega et al., 2013b). These properties were profited and the mango seed extract was added to fresh-cut mango to increase antioxidant capacity tested by DPPH, TEAC, and oxygen radical absorbance capacity (2.9, 2.3, and 2.8 times, respectively), phenolic (7.4 times), and flavonoid (3.1 times) content than the control fruits (Vega-Vega et al., 2013a).

Grapes are one of the most produced and consumed fruits in the world and about 80% is used by the winemaking industry, leading about 30% of solid wastes. By-products of wine industry are constituted by pomace, skins, seeds, and stems and are a good source of phenolic compounds, such as anthocyanins, flavan-3-ols, flavonols, stilbenes, and phenolic acids (Drosou et al., 2015), which have been associated with the maintenance of endothelial function, increase in antioxidant capacity, protection against LDL oxidation, and neuroprotective effects (Gollucke and Ribeiro, 2012).

In the other hand, vegetable processing also generates a huge amount of by-products. Potatoes are processed to formulate, principally, mashed potatoes and chip potatoes. These industries discard some by-products such as the peel, being the major waste from the potato processing and a potential source of functional and bioactive compounds, including antioxidants, dietary fiber, vitamins, and minerals (Amado et al., 2014). A study have shown that chlorogenic and ferulic acid were the major phenolic compounds in potato peel and were able to stabilize soybean oil presenting antioxidant properties. In another study, four peel and flesh potato varieties have shown the presence of chlorogenic, caffeic, p-coumaric, and ferulic acid and the antioxidant capacity were more potent in peels than in the flesh (Albishi et al., 2013). Both studies suggested that the potential of peel waste contributes to the revalorization of these kinds of by-products to be employed as an effective source of antioxidants in food systems.

Tomato-processing industry generates some residues that are a potential source of dietary fiber and bioactive compounds. These by-products have shown to be a good alternative to recovery lycopene for the use in food and cosmetics industries, also tomato peel fiber can be used as a food supplement, improving the different physical, chemical, and nutritional properties of foods (Machmudah et al., 2012; Papaioannou and Karabelas, 2012). A study, using three methods (enzyme hydrolysis, maceration, and ultrasonic assistance) to extract phenolic compounds and dietary fiber of tomato peel, demonstrated that the main phenolic compounds were rutin, naringenin, rutin derivatives, and chlorogenic acid derivatives. In addition, this extract shows a content of total dietary fiber of 84.16% and a content of lycopene between 3 and 4 mg/100 g (Navarro-González et al., 2011). All these studies demonstrate the efficacy of compounds, mainly phenolics and flavonoids, from by-products agro-industry in antioxidant capacity and improvement of health. More detailed information of specific cases will be given in the next chapters of this book.

1.5 CONCLUSION

The world had to feed a continuous growing population through a rise in food production increasing productivity; however, it seems that this will be not enough due to consumer and industrial behavior. The reduction of food waste and the use of the best alternatives in treatment of wastes generated are a way to contribute to food security and to world sustainability. It is necessary to make some changes; the world needs an inclusive agriculture, improve the food chain supply to minimize wastes, and compromise of consumers, industries, and governments. Taking into account all these facts, this represents an opportunity to exploit integrally all the plants used in industry achieving the goal to reduce food waste to ensure food security and sustainability.

KEYWORDS

- **food security**
- **food losses**
- **fruit and vegetable industry**
- **polyphenols**
- **antioxidant capacity**
- **waste management**

REFERENCES

Abdalla, A. E.; Darwish, S. M.; Ayad, E. H.; El-Hamahmy, R. M. Egyptian Mango By-product 1. Compositional Quality of Mango Seed Kernel. *Food Chem.* **2007,** *103* (4), 1134–1140.

Ajila, C.; Brar, S. K.; Verma, M.; Rao, U. P. Sustainable Solutions for Agro-Processing Waste Management: An Overview. *Environmental Protection Strategies for Sustainable Development*; Springer: Berlin-Heidelberg, 2012; pp 65–109.

Ajila, C.; Rao, L. J.; Rao, U. P. Characterization of Bioactive Compounds from Raw and Ripe *Mangifera indica* L. Peel Extracts. *Food Chem. Toxicol.* **2010,** *48* (12), 3406–3411.

Alaimo, K.; Olson, C. M.; Frongillo, E. A. Family Food Insufficiency, but Not Low Family Income, Is Positively Associated with Dysthymia and Suicide Symptoms in Adolescents. *J. Nutr.* **2002,** *132* (4), 719–725.

Albishi, T.; John, J. A.; Al-Khalifa, A. S.; Shahidi, F. Phenolic Content and Antioxidant Activities of Selected Potato Varieties and their Processing By-products. *J. Funct. Foods* **2013,** *5* (2), 590–600.

Amado, I. R.; Franco, D.; Sánchez, M.; Zapata, C.; Vázquez, J. A. Optimisation of Antioxidant Extraction from *Solanum tuberosum* Potato Peel Waste by Surface Response Methodology. *Food Chem.* **2014,** *165*, 290–299.

Anonymous. *Bioeconomía Argentina*, 2015. http://www.bioeconomia.mincyt.gob.ar/bioeconomia-argentina/ (retrieved Sept. 15, 2015).

Ayala-Zavala, J.; Vega-Vega, V.; Rosas-Domínguez, C.; Palafox-Carlos, H.; Villa-Rodriguez, J.; Siddiqui, M. W.; Dávila-Aviña, J.; González-Aguilar, G. Agro-industrial Potential of Exotic Fruit Byproducts as a Source of Food Additives. *Food Res. Int.* **2011,** *44* (7), 1866–1874.

Ayala-Zavala, J.; Rosas-Domínguez, C.; Vega-Vega, V.; González-Aguilar, G. Antioxidant Enrichment and Antimicrobial Protection of Fresh-Cut Fruits Using their Own Byproducts: Looking for Integral Exploitation. *J. Food Sci.* **2010,** *75* (8), R175–R181.

Babbar, N.; Oberoi, H. S. Potential of Agro-residues as Sources of Bioactive Compounds. *Biotransformation of Waste Biomass into High Value Biochemicals*; Springer: Berlin-Heidelberg, 2014; pp 261–295.

Bos-Brouwers, H.; Timmermans, A.; Soethoudt, J.; Ostergren, K.; Gustavsson, J.; Hansen, O.; Möller, H.; Anderson, G.; O'Connor, C.; Quested, T. *FUSIONS Definitional Framework for Food Waste*, 2014.

Buzby, J. C.; Hyman, J. Total and Per Capita Value of Food Loss in the United States. *Food Policy* **2012,** *37* (5), 561–570.

Carmichael, S. L.; Yang, W.; Herring, A.; Abrams, B.; Shaw, G. M. Maternal Food Insecurity Is Associated with Increased Risk of Certain Birth Defects. *J. Nutr.* **2007,** *137* (9), 2087–2092.

Cook, J. T.; Frank, D. A. Food Security, Poverty, and Human Development in the United States. *Ann. N.Y. Acad. Sci.* **2008,** *1136* (1), 193–209.

Cook, J. T.; Frank, D. A.; Berkowitz, C.; Black, M. M.; Casey, P. H.; Cutts, D. B.; Meyers, A. F.; Zaldivar, N.; Skalicky, A.; Levenson, S. Food Insecurity Is Associated with Adverse Health Outcomes among Human Infants and Toddlers. *J. Nutr.* **2004,** *134* (6), 1432–1438.

Cook, J. T.; Frank, D. A.; Levenson, S. M.; Neault, N. B.; Heeren, T. C.; Black, M. M.; Berkowitz, C.; Casey, P. H.; Meyers, A. F.; Cutts, D. B. Child Food Insecurity Increases Risks Posed by Household Food Insecurity to Young Children's Health. *J. Nutr.* **2006,** *136* (4), 1073–1076.

Cruz-Valenzuela, M. R.; Carrazco-Lugo, D. K.; Vega-Vega, V.; Gonzalez-Aguilar, G. A.; Ayala-Zavala, J. F. Fresh-Cut Orange Treated with its Own Seed By-products Presented Higher Antioxidant Capacity and Lower Microbial Growth. *Int. J. Postharv. Technol. Innov.* **2013,** *3* (1), 13–27.

Chakraborty, S.; Newton, A. C. Climate Change, Plant Diseases and Food Security: An Overview. *Plant Pathol.* **2011,** *60* (1), 2–14.

Charles, J.; Godfray, H.; Beddington, J. R.; Crute, I. R.; Haddad, L.; Lawrence, D.; Muir, J. F.; Pretty, J.; Robinson, S.; Thomas, S. M.; Toulmin, C. Food Security: The Challenge of Feeding 9 Billion People. *Science* **2010,** *327* (5967), 812–818.

Drosou, C.; Kyriakopoulou, K.; Bimpilas, A.; Tsimogiannis, D.; Krokida, M. A Comparative Study on Different Extraction Techniques to Recover Red Grape Pomace Polyphenols from Vinification Byproducts. *Ind. Crops Prod.* **2015,** *75,* 141–149.

EC. *Food Waste in Europe*, European Commission, 2013. http://ec.europa.eu/dgs/health_ food-safety/information_sources/docs/speeches/speech-food-wasteexpo-07022013_en.pdf (retrieved Sept. 15, 2015).

Eicher-Miller, H. A.; Mason, A. C.; Weaver, C. M.; McCabe, G. P.; Boushey, C. J. Food Insecurity Is Associated with Iron Deficiency Anemia in US Adolescents. *Am. J. Clin. Nutr.* **2009,** *90* (5), 1358–1371.

El-Ramady, H. R. Integrated Nutrient Management and Postharvest of Crops. *Sustainable Agriculture Reviews*; Springer: Berlin-Heidelberg, 2014; pp 163–274.

EPA. *Our Mission and What We Do*, October 2014. http://www2.epa.gov/aboutepa/our-mission-and-what-we-do (retrieved Sept. 15, 2015).

European Commission. *Environment Action Programme to 2020*, 2015. http://ec.europa.eu/ environment/action-programme/ (retrieved Sept. 15, 2015).

FAO. *An Introduction to the Basic Concepts of Food Security*, 2008. www.foodsec.org/docs/ concepts_guide.pd.

FAO. *The Post-2015 Development Agenda and the Millennium Development Goals*, 2015. http://www.fao.org/post-2015-mdg/14-themes/food-security-and-the-right-to-food/en/ (retrieved Sept. 15, 2015).

FAO; IFAD; WFP. *The State of Food Insecurity in the World 2015. Meeting the 2015 International Hunger Targets: Taking Stock of Uneven Progress.* Rome, FAO, 2015.

Food and Agriculture Organization. *Food Wastage Footprint: Impacts on Natural Resources— Summary Report.* FAO: Rome, Italy, 2013.

Food Waste Reduction Alliance. *Analysis of US Food Waste Among Food Manufacturers, Retailers, and Restaurants,* BSR, 2013.

FWRA. *Food Waste Reduction Alliance,* 2013. http://www.foodwastealliance.org/ (retrieved Sept. 15, 2015).

Garcia-Garcia, G.; Woolley, E.; Rahimifard, S. A Framework for a More Efficient Approach to Food Waste Management. *Int. J. Food Eng.* **2015,** *1* (1). DOI:10.18178/ijfe.1.1.65-72.

Garnett, T. Three Perspectives on Sustainable Food Security: Efficiency, Demand Restraint, Food System Transformation. What Role for Life Cycle Assessment? *J. Clean. Prod.* **2014,** *73,* 10–18.

Girotto, F.; Alibardi, L.; Cossu, R. Food Waste Generation and Industrial Uses: A Review. *Waste Manage.* **2015,** *45,* 32–41.

GMA. *Grocery Manufacturers Association. Food Waste: Tier 1 Assessment,* 2012. http://www.foodwastealliance.org/wp-content/uploads/2013/06/FWRA_BSR_Tier1_FINAL.pdf (retrieved Sept. 15, 2015).

Gollucke, A. P. B.; Ribeiro, D. A. Use of Grape Polyphenols for Promoting Human Health: A Review of Patents. *Rec. Pat. Food Nutr. Agric.* **2012,** *4* (1), 26–30.

Gorinstein, S.; Martín-Belloso, O.; Park, Y.-S.; Haruenkit, R.; Lojek, A.; Číž, M.; Caspi, A.; Libman, I.; Trakhtenberg, S. Comparison of Some Biochemical Characteristics of Different Citrus Fruits. *Food Chem.* **2001,** *74* (3), 309–315.

Gundersen, C.; Kreider, B.; Pepper, J. The Economics of Food Insecurity in the United States. *Appl. Econ. Perspect. Policy* **2011,** *33* (3), 281–303.

Gustavsson, J.; Cederberg, C.; Sonesson, U.; Van Otterdijk, R.; Meybeck, A. *Global Food Losses and Food Waste.* Food Agriculture Organization of the United Nations: Rome, 2011.

Hanson, C. Food Security, Inclusive Growth, Sustainability, and the Post-2015 Development Agenda. *Background Research Paper Submitted to the High Level Panel on the Post-2015 Development Agenda, World Resource Institute,* 2013.

Huang, J.; Oshima, K. M. M.; Kim, Y. Does Food Insecurity Affect Parental Characteristics and Child Behavior? Testing Mediation Effects. *Soc. Serv. Rev.* **2010,** *84* (3), 381.

Ignat, I.; Volf, I.; Popa, V. I. A Critical Review of Methods for Characterisation of Polyphenolic Compounds in Fruits and Vegetables. *Food Chem.* **2011,** *126* (4), 1821–1835.

Kirkpatrick, S. I.; McIntyre, L.; Potestio, M. L. Child Hunger and Long-Term Adverse Consequences for Health. *Arch. Pediatr. Adolesc. Med.* **2010,** *164* (8), 754–762.

Lang, T.; Barling, D. Nutrition and Sustainability: An Emerging Food Policy Discourse. *Proc. Nutr. Soc.* **2013,** *72* (01), 1–12.

Lee, J. S.; Frongillo, E. A. Nutritional and Health Consequences Are Associated with Food Insecurity among US Elderly Persons. *J. Nutr.* **2001,** *131* (5), 1503–1509.

Lin, C. S. K.; Pfaltzgraff, L. A.; Herrero-Davila, L.; Mubofu, E. B.; Abderrahim, S.; Clark, J. H.; Koutinas, A. A.; Kopsahelis, N.; Stamatelatou, K.; Dickson, F. Food Waste as a Valuable Resource for the Production of Chemicals, Materials and Fuels. Current Situation and Global Perspective. *Energy Environ. Sci.* **2013,** *6* (2), 426–464.

Lipinski, B. *Reducing Food Loss and Waste: An Overlooked Strategy for Creating a Sustainable Food System.* World Resources Institute, 2014. http://www.wri.org/blog/2014/10/reducing-food-loss-and-waste-overlooked-strategy-creating-sustainable-food-system (retrieved Sept. 15, 2015).

Lipinski, B.; Hanson, C.; Lomax, J.; Kitinoja, L.; Waite, R.; Searchinger, T. *Installment 2 of "Creating a Sustainable Food Future" Reducing Food Loss and Waste.* Working Paper, World Resource Institute, 2013.

Lundqvist, J.; de Fraiture, C.; Molden, D. Saving Water: from Field to Fork: Curbing Losses and Wastage in the Food Chain. *SIWI Policy Brief.* SIWI, 2008.

Machmudah, S.; Winardi, S.; Sasaki, M.; Goto, M.; Kusumoto, N.; Hayakawa, K. Lycopene Extraction from Tomato Peel By-product Containing Tomato Seed Using Supercritical Carbon Dioxide. *J. Food Eng.* **2012,** *108* (2), 290–296.

Marín, F. R.; Soler-Rivas, C.; Benavente-García, O.; Castillo, J.; Pérez-Alvarez, J. A. By-products from Different Citrus Processes as a Source of Customized Functional Fibres. *Food Chem.* **2007,** *100* (2), 736–741.

McIntyre, L.; Glanville, N. T.; Raine, K. D.; Dayle, J. B.; Anderson, B.; Battaglia, N. Do Low-Income Lone Mothers Compromise their Nutrition to Feed their Children? *Can. Med. Assoc. J.* **2003,** *168* (6), 686–691.

Mexicano, C. d. D. d. H. C. d. l. U. *Ley general para la prevención y gestión integral de los residuos.* S. General, 2015.

Mirabella, N.; Castellani, V.; Sala, S. Current Options for the Valorization of Food Manufacturing Waste: A Review. *J. Clean. Prod.* **2014,** *65*, 28–41.

Muirhead, V.; Quiñonez, C.; Figucircdo, R.; Locker, D. Oral Health Disparities and Food Insecurity in Working Poor Canadians. *Comm. Dentistry Oral Epidemiol.* **2009,** *37* (4), 294–304.

Navarro-González, I.; García-Valverde, V.; García-Alonso, J.; Periago, M. J. Chemical Profile, Functional and Antioxidant Properties of Tomato Peel Fiber. *Food Res. Int.* **2011,** *44* (5), 1528–1535.

Nellemann, C. *The Environmental Food Crisis: The Environment's Role in Averting Future Food Crises. A UNEP Rapid Response Assessment,* UNEP/Earthprint, 2009.

ONU. *Objetivos de Desarrollo del Milenio: Informe de 2015*, 2015.

Papaioannou, E. H.; Karabelas, A. J. Lycopene Recovery from Tomato Peel under Mild Conditions Assisted by Enzymatic Pre-treatment and Non-ionic surfactants. *Acta Biochim. Pol.* **2012,** *59* (1), 71.

Paraman, I.; Sharif, M. K.; Supriyadi, S.; Rizvi, S. S. Agro-food Industry Byproducts into Value-Added Extruded Foods. *Food Bioprod. Process.* **2015,** *96*, 78–85.

Parfitt, J.; Barthel, M.; Macnaughton, S. Food Waste within Food Supply Chains: Quantification and Potential for Change to 2050. *Philos. Trans. R. Soc. B: Biol. Sci.* **2010,** *365* (1554), 3065–3081.

Poppy, G.; Jepson, P.; Pickett, J.; Birkett, M. Achieving Food and Environmental Security: New Approaches to Close the Gap. *Philos. Trans. R. Soc. B: Biol. Sci.* **2014,** *369* (1639), 20120272.

Quested, T.; Ingle, R.; Parry, A. *Executive Summary: Household Food and Drink Waste in the United Kingdom.* WRAP, 2012. http://www.wrap.org.uk/sites/files/wrap/hhfdw-2012-summary.pdf (retrieved Sept.; 2015).

Seligman, H. K.; Bindman, A. B.; Vittinghoff, E.; Kanaya, A. M.; Kushel, M. B. Food Insecurity Is Associated with Diabetes Mellitus: Results from the National Health Examination and Nutrition Examination Survey (NHANES) 1999–2002. *J. Gen. Intern. Med.* **2007,** *22* (7), 1018–1023.

Seligman, H. K.; Laraia, B. A.; Kushel, M. B. Food Insecurity Is Associated with Chronic Disease among Low-Income NHANES Participants. *J. Nutr.* **2010,** *140* (2), 304–310.

Service, E. P. R. *How to Feed the World in 2050?*, 2014 http://epthinktank.eu/2014/01/08/ how-to-feed-the-world-in-2050/ (retrieved Sept. 15, 2015).

Skalicky, A.; Meyers, A. F.; Adams, W. G.; Yang, Z.; Cook, J. T.; Frank, D. A. Child Food Insecurity and Iron Deficiency Anemia in Low-Income Infants and Toddlers in the United States. *Matern. Child Health J.* **2006,** *10* (2), 177–185.

Tamer, C. E.; Çopur, Ö. U. Development of Value-Added Products from Food Wastes. *Food Processing: Strategies for Quality Assessment*; Springer: Berlin-Heidelberg, 2014; pp 453–475.

Tarasuk, V. S. Household Food Insecurity with Hunger Is Associated with Women's Food Intakes, Health and Household Circumstances. *J. Nutr.* **2001,** *131* (10), 2670–2676.

Tscharntke, T.; Clough, Y.; Wanger, T. C.; Jackson, L.; Motzke, I.; Perfecto, I.; Vandermeer, J.; Whitbread, A. Global Food Security, Biodiversity Conservation and the Future of Agricultural Intensification. *Biol. Conserv.* **2012,** *151* (1), 53–59.

Underwood, E.; Baldock, D.; Aiking, H.; Buckwell, A.; Dooley, E.; Frelih-Larsen, A.; Naumann, S.; O'Connor, C.; Poláková, J.; Tucker, G. *Options for Sustainable Food and Agriculture in the EU.* Synthesis Report of the STOA Project 'Technology Options for Feeding 10 Billion People', Institute for European Environmental Policy: London/Brussels, 2013.

Unit, T. E. I. *Global Food Security Index 2014.* Special Report: Food Loss and Its Intersection with Food Security, 2014.

Vega-Vega, V.; Silva-Espinoza, B. A.; Cruz-Valenzuela, M. R.; Bernal-Mercado, A. T.; González-Aguilar, G. A.; Ruíz-Cruz, S.; Moctezuma, E.; Siddiqui, M. W.; Ayala-Zavala, J. F. Antimicrobial and Antioxidant Properties of Byproduct Extracts of Mango Fruit. *J. Appl. Bot. Food Quality* **2013a,** *86* (1), 205–211.

Vega-Vega, V.; Silva-Espinoza, B. A.; Cruz-Valenzuela, M. R.; Bernal-Mercado, A. T.; González-Aguilar, G. A.; Vargas-Arispuro, I.; Corrales-Maldonado, C. G.; Ayala-Zavala, J. F. Antioxidant Enrichment and Antimicrobial Protection of Fresh-Cut Mango Applying Bioactive Extracts from their Seeds By-products. *Food Nutr. Sci.* **2013b,** *4*, 197–203.

Wheeler, T.; von Braun, J. Climate Change Impacts on Global Food Security. *Science* **2013,** *341* (6145), 508–513.

Whitaker, R. C.; Phillips, S. M.; Orzol, S. M. Food Insecurity and the Risks of Depression and Anxiety in Mothers and Behavior Problems in their Preschool-Aged Children. *Pediatrics* **2006,** *118* (3), e859–e868.

Ziliak, J. P.; Gundersen, C.; Haist, M. *The Causes, Consequences, and Future of Senior Hunger in America*; UK Center for Poverty Research, University of Kentucky: Lexington, KY, 2008; p 71.

CHAPTER 2

ECONOMIC AND ENVIRONMENTAL BENEFITS OF UTILIZING PLANT FOOD BY-PRODUCTS

J.E. DÁVILA-AVIÑA[1]*, C. ZOELLNER[2], L. SOLÍS-SOTO[1], G. ROJAS-VERDE[1], and L.E. GARCÍA-AMEZQUITA[3]

[1]Universidad Autónoma de Nuevo León, Facultad de Ciencias Biológicas. Apdo. Postal 124-F, Ciudad Universitaria, San Nicolás de los Garza, Nuevo León 66455, México

[2]Department of Population Medicine and Diagnostic Sciences, Cornell University, S2-072 Schurman Hall, Ithaca, NY 14853, United States

[3]Centro de Investigación en Alimentación y Desarrollo, Av. Rio Conchos S/N, Parque Industrial, Cd. Cuauhtémoc, Chihuahua, 31570, México

*Corresponding author. E-mail: jorge.davilavn@uanl.edu.mx

CONTENTS

ABSTRACT

Until recently, food waste was not considered to be beneficial to the world population, it was used mostly for animal feeding; however, nowadays it is considered as a potential source of energy in several areas. To increase the eco-sustainability of food-processing industry, it is necessary to exploit the use of its by-products. The objective of this chapter is to discuss the economic and environmental benefits of supporting value addition of plant food by-products. Besides from the dual economic and environmental benefits of reducing the volume of by-products generated during food manufacturing, this chapter will cover the economic opportunities for industrial stakeholders to enter into new markets by reusing existing surplus materials, the environmental benefits to deferring these waste streams from landfills or land application, and the challenges and research needed to accomplish these aforementioned benefits. The rise in consumption of processed fruit and vegetable products brings this sector of the food industry to the forefront of environmental sustainability; however, the interdisciplinary nature of this topic will require collaboration between industries, regulators, and consumers to maximize the utilization of these commodities.

2.1 INTRODUCTION

As waste management across industries is currently a globally important economic and environmental issue to solve, a review of the current considerations with respect to the plant food industry is especially appropriate due to its grand volume and utility. The fruit and vegetable industry relies on a variety of processing methods to produce desirable consumer packaged goods, such as beverages, snacks, and prepared foods, utilizing the components of the raw material. However, depending on the commodity, the percentage of nonutilized raw material can range from <25% for temperate climate crops up to 20–60% for tropical climate crops (Kammerer et al., 2014; Anal, 2013; Schieber et al., 2001). Volumes of this material, that is, typically discarded as solid and liquid wastes, elicits both economic and environmental sustainability concerns. Traditional treatments for solid waste of the agronomic sector include thermal processing, evaporation, membrane processing, anaerobic digestion, among others, due to their high organic content (proteins, oils, sugars, vitamins, colorants, antioxidants) of these wastes and residues, resulting in high treatment costs prior to disposal (Murugan et al., 2013; Kao and Chen, 2012). Actually, whether this material should be called "waste," or not, is the topic of discussion. Usually disposed, or used for animal feed after thermal processing, these plant materials can also be minimally processed for isolation, or enhancement of their bioactive compounds or functional properties, making the investment in alternative options an attractive economic and environmental opportunity.

To put some perspective of this topic around plant foods, "by-products" are referred to as nonutilized subsequent material flow from the primary manufacturing of raw edible plant products (i.e., pits, pomace, and peels from fruit juice operations), whereas "waste" accumulates due to postharvest losses (culled or spoiled products). Food waste is mainly generated at retail and consumer levels in industrialized countries on contrast to developing countries in which waste is generated at postharvest and processing levels (Arancon et al., 2013). Ideally the most economic and environmentally sustainable option for dealing with such materials is to minimize their creation by focusing resources in production, storage, and processing efficiency. However, there is a threshold to this efficiency, mainly due to cost and physiology of the raw materials, at which point, waste is still produced. Subsequently, these materials still contain useful components that can serve as raw materials in another context. Research on processing technology for separation and purification of such components has opened up the market for value-added products and additives.

The objective of this chapter is to discuss the economic and environmental benefits of supporting value addition of plant food by-products. Apart from the dual economic and environmental benefits of reducing the volume of by-products generated during food manufacturing.

2.2   ECONOMIC BENEFITS

Last year, the Food and Agriculture Organization of the United Nations (FAO) estimated that up to a third of the food aimed for human consumption globally is wasted every year (Gustavsson et al., 2011). Food wastes are produced throughout all the food chain. In 2006, food waste produced in the European Union (EU) was about 89 million tons per year or 179 kg per capita; however, estimates suggest that by 2020, this mass will increase to 126 million tons based on increases in population and affluence (Monier et al., 2011). The current world population of 7.2 billion is projected to increase by 1 billion over the next 12 years and reach 9.6 billion by 2050 (DeSA, 2013). As a consequence of the efforts to promote the consumption of fresh green products, since they provide many benefits such as the presence of antioxidants and other compounds that enhance human health, consumption of fruit and vegetables has increased so much so that the food waste produced by the global agroindustry is now estimated to be 800,000 t of fresh matter per year, without considering the wastage during food processing (Ayala-Zavala et al., 2010).

The annual consumption of fruits, vegetables, and grains has grown exponentially due to population growth. This means that every year tons of by-products are generated. The fruit and vegetable wastes include trimmings, peelings, stems, seeds, shells, bran, residues remaining after extraction of oil, starch, sugar, juice, off-specification, damaged, out-of-date, or returned products (European Commission). Their composition, based on cellulose, hemicellulose, pectin, and traces of lignin, could allow for a large amount of value-added products to be recovered or produced, such as enzymes, reducing sugar, furfural, ethanol, protein, amino acid, carbohydrates, lipids, organic acids, phenols, activated carbon, cosmetics, resins, medicines, and other miscellaneous products (Ubalua, 2007). Biomolecules and by-products of the fruit- and vegetable-processing industry have potential as functional foods or as adjuvants in food processing or in medicinal and pharmaceutical preparations (Spatafora and Tringali, 2012).

The solid waste (called "pomace") is obtained by pressing of fruits or vegetables and can be pulp, peels, and seeds (Schieber et al., 2003). In some cases, pomace represents more than 40% of total plant food (such as the artichoke, asparagus, mango, citrus fruits, papaya, pineapple) (Goñi and Hervert-Hernández, 2011). For example, apple pomace is a by-product derived from juice pressing and it is considered a source of dietary fiber (over 50% of dry weight) and phenolic compounds (1200–4000 mg/kg dry weight), including flavanols (catechin, epicatechin, procyanidins) and hydroxycinnamates (Schieber et al., 2003). Millions of tons of various pomace wastes originate from viticulture, olive oil production, tomato processing, and citrus processing, among others (Laufenberg et al., 2003).

Several industries produce different kinds of food by-products, rich in valuable compounds. Their valorization could convert them into high-value products with application in diverse biotechnological fields such as pharmaceuticals, food, or cosmetics. By obtaining high-value components such as proteins, polysaccharides, fibers, flavor compounds, and phytochemicals, these wastes can be repurposed as nutritionally and pharmacologically functional ingredients. Reuse of these compounds would also reduce the waste environmental impact and the related treatment costs (Barbulova et al., 2015). For example, depending on the by-product generated, the costs of drying, storage, and shipment of them are economically limiting factors and in some cases are processes required by legislation. Therefore, by-products of agroindustry and food process industry utilization have received attention as an alternative source of animal feedstuffs. The use of vegetable waste for animal feed without pretreatments is complicated by animal intolerance to some waste components; therefore, it is necessary to eliminate this interference by using several processes such as bioconversion or distillation, which are additional costs (Barbulova et al., 2015).

The use of by-products of the fruit and vegetable-processing industry, as a source of functional compounds is a promising economic opportunity. The discarded products are usually managed as refuse or used to produce animal feed. In the past, food wastes were considered neither a cost nor a benefit (they were used as animal feed or brought to landfills or for composting). This sentiment has recently changed due to the growing environmental concerns in the EU, the need to minimize the impact of waste on human health, which is bringing more stringent regulations; the high disposal costs that are changing the already low profits of the food industry, and, the growing awareness of the benefits deriving from potentially marketable components present in foods wastes and by-products (Laufenberg et al., 2003).

2.2.1 FOOD PLANT BY-PRODUCT USES IN BIOFUELS, COSMETIC, AND ANIMAL FEED

Every year, the processing industries of citrus and tobacco products and the industries of manufacture food products, such as vegetables and fruits, grains and starch, generate tons of waste. The low cost of these kinds of residues increase their multiple applications in several processes, such as biofuel and enzyme production, bioactive compounds, in cosmetic manufacture, and even in animal feed. In addition, nowadays, the consumers prefer natural products as an alternative to synthetic substances. The food industry produces large amounts of waste from which the sector of vegetables and fruits represent 14.8%, followed by the grain and starch products with 12.9% (Baiano, 2014).

The rate of vegetable waste has been recently addressed to discharge mainly in south Europe, an area rich in agro-industrial production, with special reference to some fruit crops such as citrus, grape, olive, and almond (Spatafora and Tringali, 2012).

2.2.2 FRUITS

Edible parts of fruit comprise 78% juice and 22% by-products, these latter constitute approximately 52% of the total weight of fruit (Kulkarni and Aradhya, 2005), finally these by-products conformed of inedible parts such as seed, husk, peel has many bioactive components such as polyphenols, ellagitannins, vitamins, minerals, and polyunsaturated fatty acids are sent to waste and environmental pollution. However, by-products of pomegranate have antioxidant, anticarcinogenic, and antimicrobial properties particularly (Gil et al., 2000; Seeram et al., 2005; Ferreira, 2007).

Fruits, such as apple, grape, citrus, peach, apricot, mango, banana, pineapple, guava, kiwifruit, papaya, passion fruit, tomato, carrot, and other, offer by-products as pectin, polyphenols, antioxidants, essences, D-limonene, organic acids, juice pulp, ethanol, flavonoid, lycopene, among others (Schieber et al., 2001). In this sense, grapes and apples are the most important fruits in the temperate climate and are characterized by a large edible portion and moderate amounts of waste material, such as peels, seeds, and stones. Insofar as oranges, pineapples, bananas, watermelons, and mangos are the common fruits of tropical and subtropical areas and had higher ratios of by-products arise from tropical and subtropical fruit processing (Askar, 1998).

Of particular interest is to know that grape is the world's largest fruit crop, with production of more than 65 million metric tons per year. About 80% of the total crop is used in wine making, and pomace represents approximately 20% of the weight of grapes processed (Mazza and Miniati, 1993). According to statistics from the FAO, different compounds such as ethanol, tartrates, citric acid, grape seed oil, hydrocolloids, and dietary fiber are recovered from grape pomace. Citrus plants are one of the major fruit crops around the world with global availability and popularity contributing human diet. Fresh citrus fruits are a source of dietary fibers, associated with gastrointestinal disease prevention and lowered circulating cholesterol; they are rich in vitamins C and B (thiamin, pyridoxine, niacin, riboflavin, pantothenic acid, and folate) and phytochemicals, such as carotenoids, flavonoids, and limonoids (Liu et al., 2012). Due to the large amounts being processed into juice, residues of citrus juice production are a source of dried pulp and molasses, fiber–pectin, cold-pressed oils, essences, D-limonene, juice pulps and pulp wash, ethanol, pectin, limonoids, and flavonoids (Ozaki et al., 2000).

Pineapple is the most important tropical fruit generating a high amount of by-products. These by-products represent around 35% of the pineapple fruit, being the rind and the core the predominant ones (Reinhardt and Rodriguez,

2007). Pineapple rind and core have a high content of compounds with anti-oxidant activity, which give them a huge potential concerning their valorization in the cosmetics industry (Wu et al., 2012).

2.2.2.1 FRUIT PEELS

Fruit peels are produced around the world in large quantities from the processing of agricultural products. These residues are primarily composed of pectin, cellulose, hemicellulose, and small amount of lignin. They represent an important source of sugars and for this reason makes them a choice to produce ethanol (Salazar-Ordóñez et al., 2013).

Orange juice is one of the most widely consumed beverages today. To prevent the environment of this waste, the orange peel has just been utilized in the manufacture of cattle feed. Citrus peel contains essential oils (0.8%), pectin (4%), cellulose (5%), hemicellulose (4%), soluble sugars, such as glucose, fructose, sucrose, and galactose (6%), and water (80%) (Sánchez Orozco et al., 2012; Zhou et al., 2008). Due to this, waste citrus peel can be used for ethanol production, enzymes, microbial biomass, flavoring, organic acids, and antioxidants.

To use the peel of citrus in the production of ethanol, the removal of limonene is crucial, because it is a compound released into the environment when the peel dry for later use, and their recovery being more feasible and economically viable to the ethanol production (Zhou et al., 2007). In addition, the limonene has an inhibitory effect on the growth of microorganisms used in late steps. The production of ethanol from citrus peel waste is carried out using commercial enzymes, this represents a high cost but can be reduced if in-house-produced enzymes are used for saccharification. This means the use of microorganisms producing hydrolytic enzymes (cellulases, pectinases, and xylanases) (Choi et al., 2015).

One of the determining factors in the cost of producing ethanol from citrus peel is the place where the plant is located. It is recommended that is near to the juice processing plant, which would employ machinery used for the production of juices. Limtong et al. (2007) suggests that an ideal microorganism used for ethanol production must have rapid fermentation potential, appreciable thermotolerance, ethanol tolerance, and high osmotolerance. It has been suggested that the use of crude extracts of enzyme and producing the combined use of native strains of the same, allowing to carry out simultaneous saccharification fermentation, improving yields in ethanol (Sandhu et al., 2012).

Banana peel is a fruit residue, it accounts for 30–40% of the total fruit weight. The main components are carbohydrates, proteins, and fibers in significant amounts. Due to their high carbohydrate content, it is suitable to use in fermented processes (potential growth medium for yeast strains). The annual production of banana is around 107 million metric tons that were produced in 2011, from around 130 countries. In 2012, a quantity of 19,550,339 t was exported worldwide. The United States is the largest consumer of this kind of fruit. Approximately, 1 t of waste is produced for every 10 t of bananas (Hossain et al., 2011). Using pectinases, cellulases, and amylases, the banana peel can be converted into fermentable sugars which can be used as feedstock to produce ethanol by fermentation and distillations. The ethanol produced by this route provides favorable conditions to help maintain energy security from a waste agricultural product (Bhatia and Paliwal, 2010).

2.2.3 BIOETHANOL PRODUCTION

Nowadays, bioethanol is mainly produced from crops used for human consumption, which affects the food balance, causing the price of food and feed to rise. In addition, there are other negative environmental impacts due to changes in land use, such as extending the use of energy crops and generating monoculture areas, especially in developing countries (Salazar-Ordóñez et al., 2013). It has been reported that the total of crop residues and wasted crops can produce 491 billion liters of bioethanol per year, about 16 times higher than the actual world bioethanol production. Although the use of lignocellulosic biomass material is still limited, it is a new alternative in bioethanol production. Additionally, this kind of biomass can be supplied on a large-scale basis from different low-cost raw materials such as food plant residues (citrus peel, sugarcane bagasse, corn and grains by-products, among others) (Mtui, 2009; Gupta and Verma, 2015). Agricultural residues are easier than wood to use as feed stocks for biofuels due to their lower lignin and higher hemicellulose contents (Tang et al., 2011). For bioethanol production, three processes are required: pretreatment, enzyme hydrolysis, and fermentation.

The conversion of lignocellulosic material for the production of bioethanol must be economically feasible; this means that the production cost must be lower than the cost of traditional fuels. Forty percent of the total cost of ethanol production is due to the raw material used. Similarly, if the use of microorganisms for their production arises, this method should be more efficient and more economical. Moreover, the choice of raw material depends on its availability, the type of enzymes used for treatment, and transportation to the place of processing. Some agroresidues are less preferred as they are used in cattle feed, though their use is possible in some cases because some wastes grand quantities that significantly exceeds demand occur.

2.2.4 SUGARCANE BAGASSE

Production of ethanol from sugar or starch that is extracted from sugarcane and cereals, respectively, impacts negatively on the economics of the process. Due to this, the technology development focus on the production of ethanol using other substrates, derived of food industries waste, as sugarcane bagasse, citrus peel, and even crop residues (stalks, husks, cobs, and other biomass unsuitable for direct human food), the main by-products in sugar, citrus juice, and grains production in several countries. In this sense, US agriculture could provide up to 155 million tons of residues for producing bioenergy in 2030 (UCS, 2012). The price of produced bioethanol in the United States was approximately $2.39 per gallon in 2013.

Sugarcane composition varies according to many factors the extraction process, sugarcane variety, and soil composition. In general, it consists of approximately 32–44% cellulose, 27–32% hemicellulose, and 19–24% lignin (Malherbe and Cloete, 2002). The lower content of another compounds (pectin, protein, mineral, and low molecular weight compounds) represent a great advantage for its bioconversion by microorganisms (Martins et al., 2011a). Sugarcane bagasse has the most positive net energy/balance of the cellulosic feedstock discussed today. Currently, 6000–7000 L of ethanol can be produced from 1 ha of sugarcane—not including bagasse. When bagasse can be utilized for ethanol production, the output is likely to 12,000–15,000 L/ha.

The economic viability of bioethanol production depends on four main factors: (1) cost of feedstock, (2) values of product (ethanol) and coproducts, (3) cost of processing, and (4) tax levels (Hossain et al., 2011). Studies have shown that the pretreatment step defines at which extension and cost the carbohydrates will be converted into ethanol (Chandel et al., 2007). In Brazil, the principal country in use sugarcane to ethanol production, 125–250 kg of bagasse per ton of sugarcane are obtained (Chandel et al., 2012; Vargas Betancur et al., 2010). The high carbohydrate content (60–70%) place it as a potential raw material for the production of bioenergetics as bioethanol. Moreover, with increase in high demand of global ethanol (66–125 million m³ between 2008 and 2020), more bagasse will be necessary (Gupta and Verma, 2015). However, the large-scale commercial production of bioethanol using lignocellulosic materials has not yet been implemented (Balat and Balat, 2009; Balat, 2011).

2.2.5 ANIMAL FEED

The increase in population has led to increased demand for animal products and, consequently, increased production of livestock and their necessary feed. The use of by-products as fruits residues shows nutritional and health-promoting features due to the presence of bioactive compounds. Citrus pulp is a common by-product in many countries; its price is relatively low and has a high percentage of pectin and soluble carbohydrates. This has meant that this product is used as livestock feed, replacing the cereals in ruminant diet (Scerra et al., 2001). However, it has antinutritional activity compounds, such as tannins. The fermentation process eliminates tannin-present compounds as an antinutritional effect. This process does not require complicated steps; the use of chemical compounds also increases the quality of the protein (Bostami et al., 2015).

In the same way, carrots by-products had great demand for feeding dairy cattle and possibly pigs. Most residues used for animals are by-products discarded from products of lesser importance such as sweet corn, lettuce, cabbage, cauliflower, and the leaves of other vegetables, mainly beets, turnips, and broccoli. Milk producers prefer to use the residual leaves of sweet corn, lettuce, cauliflower and, as a last resort, cabbage as feed for dairy cattle. The use of residues for animal feed has been proposed by a number of researchers in view of their high productive potential, as well as a way of eliminating an important source of contamination (Losada et al., 2000). However, a feature of the flow of these organic waste products from the commercialization centers to the livestock production unit is the fact that it costs nothing except for the cost of collection and transportation. Recently, some alternative uses have been explored by different research groups for the elaboration of composts and organic fertilizers (Cervantes et al., 2007; Saval, 2012).

On the other hand, one of the important areas for food supply is aquaculture. Cottonseed flour, soybeans, canola, corn, sorghum, maize-pulp, coffee, wheat, among others can be used to feed fish. In some cases, the heat treatment given to the extraction of oils can impair the quality of the protein present in these products. Even citrus peel is used to improve the food given to fish. The choice of raw material to be fed to fish depends heavily on the availability of the same. These products seek to replace commercial products used to feed fish; this food is mainly based on fish meal which considerably increases the cost (González-Salas et al., 2014).

Furthermore, but in this case in Mexico, in Mexico City, the production in this city can be characterized into three systems (urban, suburban, and periurban); in the east of Mexico City, there is the "Central de Abasto" (a local market), where there are sold more than 40% of the national agricultural production, and there are generated daily around 11,400 t of waste. From the 800 t/day of waste produced, mostly organic in origin, 100 t are used as a source of food-stuff for dairy cows in the market's zone of influence, that is, the eastern part of Mexico City. As well as forage, an important component of the organic waste is the fruit too ripe for human consumption and which is used in the cows' diet (Custardoy et al., 2015).

It is well known that to our planet the sustainability is the major challenge facing humanity in the 21st century where every sector of human activity will have to become sustainable. Since agriculture uses enormous land and water resources and contributes substantially to pollution, it has the most serious sustainability issues and needs to change its practices. To increase the eco-sustainability of food-processing industry, it is necessary to exploit by-products before they become wastes. Tons per year of waste from fruits, plants, grains, among others, occur. In most cases, they are discarded into the environment causing an impact. These wastes may be used for the production of fuels, primarily ethanol, and be used as cattle feed. In both cases, pretreatment is necessary and here that greater efforts should be made to improve performance in the case of biofuels, or in the case of feed for livestock and fish, improve the nutritional properties.

2.3   ENVIRONMENTAL BENEFITS

Currently, there is a growing trend toward the improvement, recovery, and recycling of wastes. This tendency has much potential in the food manufacturing industry, since the wastage, by-products, and effluents can be recovered or even improved to be high-quality and value-added products. The food industry is one of the largest manufacturing sectors in the global economy, in which the management and disposal of the wastes, as well as the production and utilization of by-products causes difficulties in the environment (Karaman et al., 2015).

Approximately, 1.6 billion tons of foods are annually wasted around the world which not only causes great economic losses (approximately USD 750 billion, almost the gross domestic product of Switzerland in 2011) but also terrible damage to natural resources. Even though 54% of the wastes are produced in the initial stages of food production, manipulation, and post-harvest storage, the remaining 46% occurs in the process, distribution, and the nonutilization of the entire product or its by-products. At a global level, these food wastes raise the greenhouse gas emission approximately 3300 million tons per year (FAO, 2013). Moreover, the production of plant foods requires a large amount of resources, such as water, energy, cropland, and agrochemicals, all of which are also thrown away when the plant products are wasted or not completely utilized.

There are different methodologies to decrease plant food wastes and to increase the utilization of products, such as improvement of the storage networks, development of the food's shelf-life, optimization of the transport and management of the products, and increase in the shelf-life, among others (Kummu et al., 2012). However, some of these procedures cannot be used due to the nature of the food product or unavailability of technologies to improve the use of the food without modifying the nutritional or sensorial characteristics; therefore, new strategies are necessary to recover by-products and ensure that the entire product is used. For this reason, the study of the use and recovery of by-products is indispensable to optimize the utilization of agricultural production worldwide, since reduction of food waste results in an immediate impact on our livelihood.

2.3.1 CONTRIBUTION OF GLOBAL FOOD WASTE TO ENVIRONMENTAL DAMAGE

As mentioned above, according to FAO, in 2007, almost 1.6 billion tons of food were wasted around the world, which is a quarter of the world's agricultural production per year, that is, 6.0 billion tons for food and nonfood uses. The contribution of food wastage from different regions of the world is shown in Table 2.1. It is observed that the food wasted from regions of industrialized Asia (China, Japan, and Republic of Korea) and South and Southeast Asia is almost half of the total food wasted in the world. On the other hand, Europe contributes with around 15% of the total waste, which is double than in the region of North America and Oceania (USA, Canada, New Zealand, and Australia). The remaining quarter is produced by Latin America, the Caribbean, Africa, and Western and Central Asia, the developing countries.

TABLE 2.1   Amount of Food Wastage and Carbon Footprint Contribution from Different Regions of the World.

	Food wastage (million tons)	Carbon footprint (% of total)
Industrialized Asia	456	34.0
South and Southeast Asia	352	21.0
Europe	248	15.0
Latin America and the Caribbean	152	8.5
Sub-Saharian Africa	152	4.5
North America and Oceania	128	10.0
North Africa, Western Asia, and Central Asia	112	7.0

Source: Adapted from FAO (2013).

From all the wasted food, 86% comes from plant sources (see Table 2.2). Cereals, starch roots, and vegetables correspond to almost 1 billion tons of the total food wastage, followed by wastage from fruits. Oil crops and legumes are better utilized plant foods with only 45.5 million tons wasted per year from more than 500 million tons produced per year. Moreover, 80% of the total food wastage is considered edible, meaning that it could still be consumed; however, the remaining 20% (approximately 300 million tons) corresponds to nonedible wastage. Nonedible wastage comes mostly from plant sources (approximately 260 million tons, 87% of the total edible food wastage), since wastage from animal products (meat, fish, seafood, milk, and eggs) is essentially edible (FAO, 2013). Plant food by-products are barely utilized as animal feed or by some industries such as cosmetics, paints, and varnishes, among others; however in most of the cases, they are disposed of in landfills or incinerated, generating different types of contaminations (O'Shea et al., 2012). Considering that the amount of nonedible disposals of plant foods is too high, and the current utilization of by-products is insufficient, the effect on the environment is clear. This information suggests that more research and development of effective techniques to reuse by-products from plant foods is required to reduce this environmental burden.

On the other hand, food wastage not only contaminates land and water but also has a great impact on the atmosphere. A food's carbon footprint is the amount of greenhouse gases emitted during its life cycle, expressed in CO_2 equivalents. In other words, carbon footprint is the impact of food on climate change (Virtanen et al., 2011). For example, the estimated amount of carbon footprint produced by food is about 3.3 billion tons of CO_2 equivalent. The carbon footprint of foods includes the greenhouse gases emissions during agricultural production. Table 2.1 shows the carbon foot-print contribution from different regions of the world. With the exception of the region of North America and Oceania, there is a direct correlation between the carbon footprint and the food wastage, which is explained by the food production rates of each region. However, the same correlation cannot be observed in the data shown in Table 2.2. The production of food from animal sources contributes to a higher carbon footprint than expected due to its wastage, mainly because of the energy used on farms and the methane (CH_4) emitted by livestock. Nevertheless, the carbon footprint of plant foods correlates with their wasted quantity. In general, plant food production emits low greenhouse gases. In most of the cases, the carbon footprint depends on the amount of fertilizers and diesel used in agricultural practices and heat production in greenhouses for the production of fruits and vegetables. However, at least two-thirds of the total carbon footprint is attributed to foods from plant sources, because of scale of the underutilized portion of its high production.

TABLE 2.2 Amount of Food Wastage and Carbon Footprint Contribution from Different Foods.

	Food wastage (million tons)	Edible part (million tons)	Nonedible part (million tons)	Food wastage (% of total)	Carbon footprint (% of total)
Cereals[a]	386.4	318.2	68.2	25	34
Starch roots	306.8	250.0	56.8	18.5	5
Oil crops and legumes	45.5	45.4	0.1	2.5	1.5
Fruits[a]	261.4	204.5	56.8	16	6
Vegetables	375.0	295.5	79.5	24	20.5
Meat	56.8	45.5	11.4	4	21
Fish and seafood	34.1	22.7	11.4	2.5	4.5
Milk and eggs[b]	113.6	113.5	0.1	7.5	7.5

Source: Adapted from FAO (2013).
[a]Excluding alcoholic beverages.
[b]Excluding butter.

Finally, the effect of food waste during the entire food supply chain on the environment has to be considered, specifically for uneaten food in the household. To produce the 1.6 billion tons of food wastage, 1.4 billion hectares of cropland is vainly utilized. Besides, the water resources used for agricultural purposes, that is, the consumption of water from surface and ground or the blue water footprint of food wastage, is approximately 250 km³. As expected, the vainly use of water and land has enormous impact on the biodiversity of natural ecosystems throughout the world.

2.3.2 POTENTIAL USE OF PLANT FOOD WASTE

By definition, food wastage is the part of the foods that are discarded in any part of the food chain. The wastage (culled or spoiled products) can occur within the agricultural production, postharvest manipulation, processing, distribution, and consumption. The processing of plant foods results in large amounts of peels, seeds, and pomace. Traditionally, these are used for feed and fertilizer because the global quantity of wastage produced in conjunction with the characteristic of the organic matter to decay limit the use of by-products to these purposes, resulting in inevitable contaminants in the environment. By-products (i.e., pits, pomace, and peels from fruit juice operations) that are obtained during these stages of plant food production cause different environmental problems that can be solved by different methods of their utilization (Gustavsson et al., 2012). For example, there is an increasing interest in foods with different functional properties, as well as consumption of foods with additional nutraceutical benefits, both of which can be obtained from plant by-products (Herrero et al., 2006). Moreover, even though most of the processes (i.e., bioethanol extraction and animal feed) have low production yields, the remarkable quantity of by-products would be sufficient for further utilization on an industrial scale. In Table 2.3, it is observed that the percentage of the nonusable portion from some plant foods, while considering the global production per year, is notable.

There are by-products that are of special interest because of both the great number of components that can be obtained and the large quantity of plant food produced and consumed. Some examples of both cases are tropical fruits, such as oranges, mangoes, and tomatoes. The orange juice industry is one of the largest producers of by-products around the world, in which the wastes resulting in the process are a serious environmental problem, since only approximately 50% of the fruit is used, while the other 50% is discarded (de Moraes Crizel et al., 2013). Similar is the case of mango, which is one of the main tropical fruits produced in the world. Its production per year is about 24 million tons (FAO and FOODS, 2004). It is used to generate a large number of products, such as puree, canned slices, nectar, among others. Through the different processes, a large amount of peel (15–20%) and seed (9–23%) is produced (Ajila et al., 2008). Some products, like tomato, are almost completely used to obtain processed foods; however, the amount of components of interest because their nutraceutical capacity, like their anti-oxidant activity, is important. In the production of tomato juice, about 3–7% is lost as waste. The pulp from this industry contains peel and seed. The oil from seed has a great fatty acids profile, rich in unsaturated fatty acids, the peel is rich in carotenoids, especially lycopene, and also have a large quantity of dietary fiber (Oreopoulou and Tzia, 2007).

TABLE 2.3   Percent of Waste from Some Plant Foods.

Plant product	Global production (million tons)[a]	Seed, peel, and pomace (% (w/w) of the plant)	Wastage per year (million tons)	Reference
Wine grapes	48	20.0	9.6	Schieber et al. (2001)
Oranges	65	50.0	32.5	de Moraes et al. (2013)
Mango	42	25–60	10.5–25.2	Ajila et al. (2008)
Banana	102	30.0	30.6	Schieber et al. (2001)
Kiwi	1.5	30.0	0.45	Schieber et al. (2001)
Tomato	162	3–7	5–11	Schieber et al. (2001)
Carrot	29	60.0	17.4	Schieber et al. (2001)
Potato	365	15–40	55–146	Schieber et al. (2001)

[a]Production in 2012 (FAO, 2013).

2.3.3   ENVIRONMENTAL BENEFITS OF CLEAN PRODUCTION OF PLANT FOOD PRODUCTS

Clean production is a manufacturing strategy for industry to reduce the wastes and effluents of the process and to optimize the use of energy and materials, while still obtaining the highest yields of productions (Paul and Ohlrogge, 1998). Plant food industries around the world have implemented numerous clean production practices in the different stages of the plant food life cycle to reduce wastage and produce additional incomes. Several examples are discussed in other chapters in this book; however, some of these practices not only contribute to decreasing the production of waste, but they also can be used to assist other clean production practices, or even improve products and processes of other plant food industries. Some of these practices are the recovery of multifunctional food ingredients with unusual characteristics from plant food residues to use in the juice, nectar, and/or bakery industries; such as the use of pectin from citrus by-products to enhance texture of nectar; the use of by-products as unique substrates to obtain fruit flavors by fermentation, such as pineapple, apple, and banana flavor (ethyl butyrate, ethyl pentanoate, and isoamyl acetate, respectively) from the fermentation of apple and carrot pomace and spent malt and hops by *Caralluma fimbriata*; and the production of bioadsorbents from plant food waste to treat industrial waste water, such as the use of dietary fiber obtained from apple and black currant to bind cadmium and lead to remove them from wastewater, among others (Laufenberg et al., 2003). There are at least three principal approaches in the plant food industry that must be taken into account to accomplish proper clean production and are described below as (a) modification of the processing plant, (b) waste treatment, and (c) obtaining and utilization of by-products.

2.3.3.1 MODIFICATION OF THE PROCESSING PLANT

The modification of manufacturing plants intends to optimize the processes to obtain products of high quality while reducing the use of energy and raw materials. The very fact that fewer resources can be used to obtain the same products has a great impact on the environment. However, more can be done in this area. Currently in the plant food industry, local and international environmental legislations have been upgraded to restrict or forbid the use of agrochemicals in plant food production or chemical reagents in the processing of products or by-products. In fact, some of the once highly used chemical agents are replaced by other harmless chemical agents, or even no longer used due to the changes of the processes where they were involved. Some examples are the almost complete replacement of elemental chlorine with sodium chlorite ($NaClO_2$) in bleaching and sanitization stages (Abraham et al., 2013), the use of enzymes to modify structures or facilitate extractions of compounds of interest, instead of acid or strong alkali, which even results in a more effective and cheaper process (Weightman et al., 1995), or the use of nonpathogenic microorganisms in the transformation of by-products in bioactive compounds, drugs, and ingredients, among others, instead the use of chemical reactions (Martins et al., 2011b).

2.3.3.2   WASTE TREATMENT

The treatment of wastes produced by the plant food industry is a corrective measure that does not avoid the introduction of matter to the ecosystem but intends to eliminate any potential danger to the environment and biodiversity. Although a great number of contaminants are released from these industries into the air, soil, and water, these are three situations of special concern. The first two are evident and there is enough legislation in most of the countries to solve them. There is soil contamination by agrochemicals, which may also lead to pollution of groundwater and problems related with the decay of organic matter. Therefore, the plant food industry should be aware of the bioactivity, bioavailability, and toxicology of their waste. Several studies have shown the health benefits of a great number of phytochemicals present in most of the plant by-products; however, there is also evidence to suggest that high doses of some of these compounds may possibly result in negative health effects (Schieber et al., 2001).

On the other hand, the use of plant food by-products has a great potential to treat industrial wastewater. Several compounds in such products have the inherent ability to adsorb and bind hazardous components that are usually soluble in polluted streams. Unfortunately, from an economic point of view, the use of plant food wastage as a real alternative to treat industrial wastewater is not possible yet, since the cost of the extraction and purification of these active compounds is still high (Laufenberg et al., 2003). Therefore, more research is necessary for efficient production of compounds from by-products, whose properties and functionality may be exploited for this purpose.

2.3.3.3   OBTAINING AND UTILIZATION OF BY-PRODUCTS

The recovery of by-products for the purpose of obtaining new value-added products or to recycle materials may be the most promising waste-management strategy. The quantity and characteristics of compounds or products that can be obtained from plant food wastage are numerous. Therefore, the food industry is becoming increasingly interested in this strategy to reduce the food wastage, to obtain novel products with improved or unique properties, and to obtain additional economic profits. However, the utilization of by-products is not enough; it is necessary to be complemented by environmental friendly procedures to obtain the by-products or to processes them. Currently, strategies and trends on the utilization of by-products from different sources and to several purposes are widely discussed in other chapters within this book.

2.4 CHALLENGES

Society and the food industry have realized the importance of utilizing waste and by-products for economic and environmental benefits; however, the implementation of such technologies and processes has not yet been fully embraced. Separation of waste materials for disposal, reuse or recycling is becoming more standard, but industry remains mostly concerned with how to dispose of these materials easily and cheaply within compliance of regulations (Chandrasekaran, 2012a). Herein presents the challenge of balancing the economic and environmental motivations for the safe disposal of enormous volumes of solid and liquid wastes without spending so much money, while also investing in new strategies for reusing or recycling these wastes or by-products. Industry environmental engineers and regulatory groups must collaborate to promote investment and research in addressing these obstacles.

2.4.1   ECONOMIC

Making business decisions require an extensive amount of research and analysis within an organization to assess the feasibility of undertaking a new opportunity. In the context of plant food wastes, the attractiveness of new uses must compete with the existing method of disposal. The biggest economic challenges in this regard surround the raw materials, industrial processes, and market opportunities.

Fruits and vegetables and their by-products contain protein, sugar, fat, waxes as well as aliphatic and aromatic compounds, which make them a valuable material for production of chemicals, food additives, energy, or other products. The types of waste streams that result from their processing include drifter materials (soil and extraneous plant materials); spoiled supply skins or hulls and trimmings; pits, seeds, and pulp or pomace (Murugan et al., 2013). These materials are characterized as having high variability, high organic load, and high moisture content, which contribute to microbial spoilage and high energy inputs for drying and storage. Therefore, the most traditional economical decision has been to send these materials as waste to the landfill, dried as animal feed or composted as fertilizer. The opportunity to utilize these materials for more valuable products is an economically enticing alternative but presents the challenge of maintaining or standardizing the quantity and quality of this raw material.

For example, phytochemicals used by the plant material for various functions—growth, reproduction, protection, and sensory attributes—include polyphenols which are of particular importance due to their utility as pigments, antioxidants, or additives in the food-processing industry. While these compounds are ubiquitous in plant material and particularly in outer layers of the skin, seed coats, and hulls, there is a wide variation in total phenolic content both between and within species of fruits and vegetables (Kammerer et al., 2014; Balasundram et al., 2006). Moreover, processing of these plant tissues may result in further uneven distribution of the compounds in the waste stream. To utilize by-products and waste streams with economy of scale principles for extraction and purification of such compounds, the quality of the waste must be evaluated prior to use as well as its establishment as a sufficiently available raw material. Due to the complexity of these compounds, this information may impact the methods of extraction or purification that will be used, the required investment in procurement and storage, and the break-even point for the organization (Frankel et al., 2013). Particularly in production of enzymes for industrial application or compounds derived via fermentation, the biggest challenge is the inability of a single waste stream to provide an adequate nutrient source to sustain the fermentation, thus requiring additions (Kao and Chen, 2012). The cost efficiency of manufacturing systems is given by the autonomous nature of the process, so interruptions or customized processes for different waste streams will be a challenge. The advent of technological standards for waste management or expectations regarding quality may help but is forthcoming.

Because of this aforementioned variability and complexity of the raw material, but rich nonetheless, the need for efficient recovery and purification procedures to transform them into valuable chemicals for commercial use like antioxidants, vitamins, concentrates, macromolecules (wax, cellulose, starch, lipids, enzymes, dyes) will be critical to the economic feasibility of this proposal. Research on such processing technologies, as supercritical carbon dioxide, pressurized liquid extraction, pulsed electric field, and ultrasound, has shown improved efficacy at the benchtop level with increasing specificity for compounds in comparison to conventional methods that rely on the power of solvents under application of heat. Readers are referred to several reviews for further information on these extraction technologies (Arancon et al., 2013; Azmir et al., 2013; Wijngaard et al., 2012).

While supercritical carbon dioxide methods are already used on an industrial scale—decaffeination of coffee and extraction of alpha acids from hops used to make beer—and modern chromatographic advances have improved extraction analysis, the significant challenge will be the implementation of these technologies for large-scale extraction and purification of commercial compounds of plant by-product origin. More experimentation is needed on the wide variety of plant materials and compounds to be utilized. In addition, depending on the process used and its cost, procurement of a sufficient volume of raw materials may require additional collection, transportation, and storage costs. For this reason, some researchers have proposed pretreatments and combination extraction steps to optimize the methods for larger scales or a variety of products. The lower energy inputs is a consideration for using these nonconventional processes, but the growing demand will continue to encourage the continuous improvement of this technology.

And, lastly with abundant waste streams and efficient processes for extraction and purification, the value-added product must remain competitive in the market. This competitiveness includes incentives for using wastes in these ways as well as promotion of suitable final markets to ensure profitability. The option to dispose waste in landfills is currently the most economical but is not environmentally sustainable in the long run as space is limited and it contributes to production of methane (greenhouse gas) (Arancon et al., 2013). Increased awareness about the environmental impacts has helped push industries to reconsider their waste management, but more incentives will bring value-addition of plant by-products into mainstream practices.

There are existing markets for raw materials to be used as alternative sources of energy, such as biofuels, and as animal feed. However, other potential profitable markets to be exploited are those of dietary supplements, food additives, and niche food products. The dietary supplement market already exists and polyphenols recovered from waste materials of fresh produce processing could contribute to the increased dietary intake of these compounds. The consumer push for more natural ingredients in food products has led to investigation in alternative sources for preservatives, colorants, thickeners, and flavorings. Moreover, spin-off industries for fad products and niche markets allow for diversification of products and entry into new business opportunities. These products could have a higher cost in the market place, thus yielding more profit on such a large scale compared to animal feed sales (Gowthaman and Poornima, 2012). What is necessary for the initial investment is creation of an enterprise that is acceptable and economically feasible.

2.4.2 ENVIRONMENTAL

In comparison to the composting runoff and landfill masses that have potential to contaminate ground and surface waters and contribute to greenhouse gas emissions, alternative uses of fruit and vegetable waste appear to be more environmentally sustainable. However, the dichotomy of defining the environmental benefits and challenges related to utilizing waste streams is highly dependable on the processing technology. Two main challenges for achieving a more environmentally friendly industry for plant by-products are the energy-intensive processes still required to utilize these by-products and the difficulty in conducting analyses to assess the true reduction in negative environmental impacts of such efforts.

The negative environmental impact of disposal of these plant by-products is due to their highly fermentable composition. And, this same composition of stable biopolymers, lignin, and pectin is what requires large energy inputs to degrade, usually extreme temperatures and pressures (Arancon et al., 2013). "Green" processing alternatives have been suggested and mentioned by previous reviews, such as enzyme and microorganism treatments, to reduce the energy requirements but are under continuous experimentation for industrial scale-up and variable composition of waste materials. Moreover, if the value-added processing does not occur at the same location as the by-product or waste stream, energy inputs for timely transportation and minimal storage will impart additional negative environmental impacts. While this environmental impact could be comparable to that of acquiring new raw materials; this is often a difficult analysis to conduct, as it will be discussed below.

International standards for making decisions based on environmental effects have established the life cycle assessment (LCA), in which all stages of a product, from inputs to outputs, are linked to environmental impacts. The goal of this assessment is to identify where in the production chain the largest environmental impacts or emissions are occurring in order for remediation. Another way LCA can be conducted is to identify the environmental consequences of a corrective action taken and has been used to analyze alternative waste handling systems. Readers are referred to Ohlsson (2004), for an example of LCA analysis of food production chains. Although it is useful, there are research gaps in LCA for the environmental sustainability potential of value-addition of waste and by-product streams in which the end-of-life for the products is ambiguous (Chandrasekaran, 2012a).. Moreover, the amount of data needed to include inputs of water, energy, raw materials, and subsequent outputs to air, land, and water for the alternative system presents a challenge often solved by making assumptions. In the end, the results of the analysis must be understandable and persuasive to those deciding on the implementation of new technologies and investment in alternative disposal methods.

2.4.3 REGULATORY/ETHICS

The purpose of the regulations in the food industry is to protect consumers by placing standards on equipment, tolerance levels, sanitary practices, qualifications, labels, and the like. The challenge in regulating the utilization of plant by-products is deciding what needs to be regulated (the process or the final product) and by which agencies (food, drug, public health, environmental). From the prior section on maintaining an adequate supply of these raw materials for economy of scale production, another regulatory and ethical issue of incentivizing waste reuse and recycling without promoting the production of waste is presented. It is unlikely that the waste streams will become a more profitable industry than the principle product, but promoting it as a business opportunity appears counterintuitive given the environmental considerations. For example, without regulations, dumping of large quantities of untreated waste in open land or undesignated areas would cause many environmental, economic, and societal issues. Currently, there are no separate or specific regulatory guidelines available for food waste or food processing waste but instead are included under general solid wastes (Chandrasekaran, 2012a). These policies vary but generally require waste disposal permits for facilities which can sometimes be discouraging. On the other hand, functional foods are on the boundary between foods and drugs making their regulation an important consumer safety issue, along with determining the validity of their health claims (Schieber et al., 2001). For regulating the processes, the challenge is making regulations with clear expectations, but not so tight that it becomes difficult for businesses to remain in compliance. Yet, if regulations are too flexible, ambiguities and inadequate recommendations may exist, making it difficult to know whether one is in compliance.

2.5 CONCLUSIONS AND FUTURE DIRECTIONS

As global food waste amounts to 1.3 billion tons per year and is a valuable source of materials in comparison to other industrial waste streams (sewage, municipal solids, construction waste), there is a great opportunity for diverting this disposal from landfills. Underutilization of available resources is a major impetus for the emphasis of valorization in the food industry, for example, the juice industry discards peels which are super valuable and packed with biomolecules of commercial importance (Chandrasekaran, 2012b). The economic and environmental benefits arise from the reduction of waste disposal volumes and the utilization of these as raw materials for profitable products. Potential markets of alternative energy sources, dietary supplements, and additives in the food industry offer incentives for the utilization of such waste streams. However, not only existing challenges of primarily minimizing waste production but also ensuring abundance of high-quality materials, implementing large-scale processing technologies, and clear regulations should be a priority in the near future.

It is clear that the current strategy of the plant food-processing industry is waste minimization. Current interdisciplinary research is also focused on optimizing food processing technology but is trending toward new methods for complete utilization of by-products on a large scale. Fermentation and enzyme processing are already used for industrial production of alcohol and yeast biomass from sugarcane by-products (molasses), demonstrating the efficiency of bioprocessing value-addition methods. Also, research on downstream technology for better extraction, separation, and purification of the desirable compounds from these complex by-products is on-going.

Future research is need in quality control systems to exclude toxins (solanin, patulin, ochratoxin, dioxins, and polycyclic aromatic hydrocarbons) and residual hazardous materials from useable waste streams; in assessment and profiling of bioactivity, bioavailability, and toxicology of phytochemicals in specific materials and their final products/functional compounds (Schieber et al., 2001), regulatory amendments for clear terminology and the inclusion of food processing by-product disposal or utilization; implementation of good agricultural practices and hazard analysis and critical control points principles; and market development. The reliable and increasing consumption of plant food products as well as the improving market for bioactive compounds and value-added products will continue to encourage this research and development to capitalize on such an economic and environmental opportunity.

KEYWORDS

- **waste management**
- **by-products**
- **plant materials**
- **thermal processing**
- **pomace**
- **treatment costs**

REFERENCES

Abraham, E.; Deepa, B.; Pothen, L.; Cintil, J.; Thomas, S.; John, M.; Anandjiwala, R.; Narine, S. Environmental Friendly Method for the Extraction of Coir fibre and Isolation of Nanofibre. *Carbohydr. Polym.* **2013,** *92* (2), 1477–1483.

Ajila, C.; Leelavathi, K.; Rao, U. P. Improvement of Dietary Fiber Content and Antioxidant Properties in Soft Dough Biscuits with the Incorporation of Mango Peel Powder. *J. Cer. Sci.* **2008,** *48* (2), 319–326.

Anal, A. K. Food Processing By-products. *Handbook of Plant Food Phytochemicals: Sources, Stability and Extraction*; John Wiley & Sons: Hoboken, NJ, 2013; pp 180–197.

Arancon, R. A. D.; Lin, C. S. K.; Chan, K. M.; Kwan, T. H.; Luque, R. Advances on Waste Valorization: New Horizons for a More Sustainable Society. *Energy Sci. Eng.* **2013,** *1* (2), 53–71.

Askar, A. Importance and Characteristics of Tropical Fruits. *Fruit Process.* **1998,** *8,* 273–276.

Ayala-Zavala, J.; Rosas-Domínguez, C.; Vega-Vega, V.; González-Aguilar, G. Antioxidant Enrichment and Antimicrobial Protection of Fresh-Cut Fruits Using their Own Byproducts: Looking for Integral Exploitation. *J. Food Sci.* **2010,** *75* (8), R175–R181.

Azmir, J.; Zaidul, I.; Rahman, M.; Sharif, K.; Mohamed, A.; Sahena, F.; Jahurul, M.; Ghafoor, K.; Norulaini, N.; Omar, A. Techniques for Extraction of Bioactive Compounds from Plant Materials: A Review. *J. Food Eng.* **2013,** *117* (4), 426–436.

Baiano, A. Recovery of Biomolecules from Food Wastes—A Review. *Molecules* **2014,** *19* (9), 14821–14842.

Balasundram, N.; Sundram, K.; Samman, S. Phenolic Compounds in Plants and Agri-industrial By-products: Antioxidant Activity, Occurrence, and Potential Uses. *Food Chem.* **2006,** *99* (1), 191–203.

Balat, M. Production of Bioethanol from Lignocellulosic Materials via the Biochemical Pathway: A Review. *Energy Convers. Manage.* **2011,** *52* (2), 858–875.

Balat, M.; Balat, H. Recent Trends in Global Production and Utilization of Bio-ethanol fuel. *Appl. Energy* **2009,** *86* (11), 2273–2282.

Barbulova, A.; Colucci, G.; Apone, F. New Trends in Cosmetics: By-products of Plant Origin and their Potential Use as Cosmetic Active Ingredients. *Cosmetics* **2015,** *2* (2), 82–92.

Bhatia, L.; Paliwal, S. Banana Peel Waste as Substrate for Ethanol Production. *Int. J. Biotechnol. Bioeng. Res.* **2010,** *1* (2), 213–218.

Bostami, A.; Ahmed, S.; Islam, M.; Mun, H.; Ko, S.; Kim, S.; Yang, C. Growth Performance, Fecal Noxious Gas Emission and Economic Efficacy in Broilers Fed Fermented Pomegranate Byproducts as Residue of Fruit Industry. *Int. J. Adv. Res.* **2015,** *3* (3), 102–114.

Cervantes, F. J.; Saldívar-Cabrales, J.; Yescas, J. F. Estrategias para el aprovechamiento de desechos porcinos en la agricultura. *Rev. Latinoam. Recur. Nat.* **2007,** *3* (1), 3–12.

Chandel, A. K.; Chan, E.; Rudravaram, R.; Narasu, M. L.; Rao, L. V.; Ravindra, P. Economics and Environmental Impact of Bioethanol Production Technologies: An Appraisal. *Biotechnol. Mol. Biol. Rev.* **2007,** *2* (1), 14–32.

Chandel, A. K.; da Silva, S. S.; Carvalho, W.; Singh, O. V. Sugarcane Bagasse and Leaves: Foreseeable Biomass of Biofuel and Bio-products. *J. Chem. Technol. Biotechnol.* **2012,** *87* (1), 11–20.

Chandrasekaran, M. Future Prospects and the Need for Research. *Valorization of Food Processing By-Products*; CRC Press: Boca Raton, FL, 2012a; pp 757–772.

Chandrasekaran, M. Regulatory Issues and Concerns of Valorization of Food Processing By-products. *Valorization of Food Processing By-Products* **2012b,** *3,* 63–90.

Chandrasekaran, M. *Valorization of Food Processing By-products*. CRC Press: Boca Raton, FL, 2012c.

Choi, I. S.; Lee, Y. G.; Khanal, S. K.; Park, B. J.; Bae, H.-J. A Low-Energy, Cost-Effective Approach to Fruit and Citrus Peel Waste Processing for Bioethanol Production. *Appl. Energy* **2015**, *140*, 65–74.

Custardoy, H. R. L.; Rodríguez, L. L.; Zorrilla, J. C.; Romero, J. M. V. The Use of Organic Waste from Animals and Plants as Important Input to Urban Agriculture in México City. *Int. J. Appl. Sci. Technol.* **2015**, *5* (1), 38–44.

de Moraes Crizel, T.; Jablonski, A.; de Oliveira Rios, A.; Rech, R.; Flôres, S. H. Dietary Fiber from Orange Byproducts as a Potential Fat Replacer. *LWT—Food Sci. Technol.* **2013**, *53* (1), 9–14.

DeSA, U. *World Population Prospects: The 2012 Revision*. Population Division of the Department of Economic and Social Affairs of the United Nations Secretariat: New York, 2013.

FAO, J.; FOODS, M. H. I. http://faostat.fao.org. Food and Agriculture Organization of the United Nations: Rome, 2004.

FAO. Food Wastage Footprint. Impacts on Natural Resources. In *Summary Report*. FAO: Rome, 2013.

Ferreira, D. Antioxidant, Antimalarial and Antimicrobial Activities of Tannin-Rich Fractions, Ellagitannins and Phenolic Acids from *Punica granatum* L. *Planta Med.* **2007**, *73* (5), 461.

Frankel, E.; Bakhouche, A.; Lozano-Sánchez, J.; Segura-Carretero, A.; Fernández-Gutiérrez, A. Literature Review on Production Process to Obtain Extra Virgin Olive Oil Enriched in Bioactive Compounds. Potential Use of Byproducts as Alternative Sources of Polyphenols. *J. Agric. Food Chem.* **2013**, *61* (22), 5179–5188.

Gil, M. I.; Tomás-Barberán, F. A.; Hess-Pierce, B.; Holcroft, D. M.; Kader, A. A. Antioxidant Activity of Pomegranate Juice and Its Relationship with Phenolic Composition and Processing. *J. Agric. Food Chem.* **2000**, *48* (10), 4581–4589.

Goñi, I.; Hervert-Hernández, D. *By-Products from Plant Foods are Sources of Dietary Fibre and Antioxidants*; INTECH Open Access Publisher, 2011.

González-Salas, R.; Romero-Cruz, O.; Valdivié-Navarro, M.; Ponce-Palafox, J. Los productos y subproductos vegetales, animales y agroindustriales: Una alternativa para la alimentación de la tilapia. *Rev. Bio Cienc.* **2014**, *2* (4), 240–251.

Gowthaman, M. K.; Poornima, G. Process Engineering and Economics. In *Valorization of Food Processing By-Products*; CRC Press: Boca Raton, FL, 2012; pp 147–166.

Gupta, A.; Verma, J. P. Sustainable Bio-ethanol Production from Agro-residues: A Review. *Renew. Sustain. Energy Rev.* **2015**, *41*, 550–567.

Gustavsson, J.; Cederberg, C.; Alimentación, O. d. l. N. U. p. l. A. y. l. *Pérdidas y desperdicio de alimentos en el mundo: alcance, causas y prevención*. In Congreso Internacional Save Food, 2012.

Gustavsson, J.; Cederberg, C.; Sonesson, U.; Van Otterdijk, R.; Meybeck, A. *Global Food Losses and Food Waste*; Food and Agriculture Organization of the United Nations: Rome, 2011.

Herrero, M.; Cifuentes, A.; Ibanez, E. Sub- and Supercritical Fluid Extraction of Functional Ingredients from Different Natural Sources: Plants, Food-By-products, Algae and Microalgae: A Review. *Food Chem.* **2006**, *98* (1), 136–148.

Hossain, A.; Ahmed, S.; Alshammari, A. M.; Adnan, F.; Annuar, M.; Mustafa, H.; Hammad, N. Bioethanol Fuel Production from Rotten Banana as an Environmental Waste Management and Sustainable Energy. *Afr. J. Microbiol. Res.* **2011**, *5* (6), 586–598.

Kammerer, D. R.; Kammerer, J.; Valet, R.; Carle, R. Recovery of Polyphenols from the By-products of Plant Food Processing and Application as Valuable Food Ingredients. *Food Res. Int.* **2014,** *65,* 2–12.

Kao, T. H.; Chen, B. H. Fruits and Vegetables. In *Valorization of Food Processing By-Products*, CRC Press: Boca Raton, FL, 2012; pp 517–558.

Karaman, S.; Karasu, S.; Tornuk, F.; Toker, O. S.; Geçgel, Ü.; Sagdic, O.; Ozcan, N.; Gül, O. Recovery Potential of Cold Press Byproducts Obtained from the Edible Oil Industry: Physicochemical, Bioactive, and Antimicrobial Properties. *J. Agric. Food Chem.* **2015,** *63* (8), 2305–2313.

Kulkarni, A. P.; Aradhya, S. M. Chemical Changes and Antioxidant Activity in Pomegranate Arils during Fruit Development. *Food Chem.* **2005,** *93* (2), 319–324.

Kummu, M.; De Moel, H.; Porkka, M.; Siebert, S.; Varis, O.; Ward, P. Lost Food, Wasted Resources: Global Food Supply Chain Losses and their Impacts on Freshwater, Cropland, and Fertiliser Use. *Sci. Tot. Environ.* **2012,** *438,* 477–489.

Laufenberg, G.; Kunz, B.; Nystroem, M. Transformation of Vegetable Waste into Value Added Products: (A) The Upgrading Concept; (B) Practical Implementations. *Bioresour. Technol.* **2003,** *87* (2), 167–198.

Limtong, S.; Sringiew, C.; Yongmanitchai, W. Production of Fuel Ethanol at High Temperature from Sugar Cane Juice by a Newly Isolated *Kluyveromyces marxianus. Bioresour. Technol.* **2007,** *98* (17), 3367–3374.

Liu, Y.; Heying, E.; Tanumihardjo, S. A. History, Global Distribution, and Nutritional Importance of Citrus Fruits. *Compr. Rev. Food Sci. Food Saf.* **2012,** *11* (6), 530–545.

Losada, H.; Bennett, R.; Soriano, R.; Vieyra, J.; Cortes, J. Urban Agriculture in Mexico City: Functions Provided by the Use of Space for Dairy Based Livelihoods. *Cities* **2000,** *17* (6), 419–431.

Malherbe, S.; Cloete, T. E. Lignocellulose Biodegradation: Fundamentals and Applications. *Rev. Environ. Sci. Biotechnol.* **2002,** *1* (2), 105–114.

Martins, D. A. B.; Gomes, E.; do Prado, H. F. A.; Ferreira, H.; de Souza Moretti, M. M.; da Silva, R.; Leite, R. S. R. *Agroindustrial Wastes as Substrates for Microbial Enzymes Production and Source of Sugar for Bioethanol Production*; INTECH Open Access Publisher, 2011a.

Martins, S.; Mussatto, S. I.; Martínez-Avila, G.; Montañez-Saenz, J.; Aguilar, C. N.; Teixeira, J. A. Bioactive Phenolic Compounds: Production and Extraction by Solid-State Fermentation. A Review. *Biotechnol. Adv.* **2011b,** *29* (3), 365–373.

Mazza, G.; Miniati, E. *Anthocyanins in Fruits, Vegetables, and Grains.* CRC Press: Boca Raton, FL, 1993.

Monier, V.; Mudgal, S.; Escalon, V.; O'Connor, C.; Anderson, G.; Montoux, H.; Reisinger, H.; Dolley, P.; Oglivie, S.; Morton, G. *Preparatory Study on Food Waste across EU 27.* European Commission, 2011.

Mtui, G. Y. Recent Advances in Pretreatment of Lignocellulosic Wastes and Production of Value Added Products. *Afr. J. Biotechnol.* **2009,** *8* (8).

Murugan, K.; Chandrasekaran, S.; Karthikeyan, P.; Al-Sohaibani, S.; Chandrasekaran, M. *Current State of the Art of Food Processing by Products*; CRC Press: Boca Raton, FL, 2013.

O'Shea, N.; Arendt, E. K.; Gallagher, E. Dietary Fibre and Phytochemical Characteristics of Fruit and Vegetable By-products and their Recent Applications as Novel Ingredients in Food Products. *Innov. Food Sci. Emerg. Technol.* **2012,** *16,* 1–10.

Ohlsson, T. Food Waste Management by Life Cycle Assessment of the Food Chain. *J. Food Sci.* **2004,** *69* (3), CRH107–CRH109.

Oreopoulou, V.; Tzia, C. Utilization of Plant By-products for the Recovery of Proteins, Dietary Fibers, Antioxidants, and Colorants. In *Utilization of By-products and Treatment of Waste in the Food Industry*; Springer: Berlin-Heidelberg, 2007; pp 209–232.

Ozaki, Y.; Miyake, M.; Inaba, N.; Ayano, S.; Ifuku, Y.; Hasegawa, S. Limonoid Glucosides of Satsuma Mandarin (*Citrus unshiu* Marcov.) and its Processing Products. *ACS Symp. Ser.* **2000,** *758,* 107–119.

Paul, D.; Ohlrogge, K. Membrane Separation Processes for Clean Production. *Environ. Progr.* **1998,** *17* (3), 137–141.

Reinhardt, A.; Rodriguez, L. *Industrial Processing of Pineapple-Trends and Perspectives.* In VI International Pineapple Symposium 822, 2007; pp 323–328.

Salazar-Ordóñez, M.; Pérez-Hernández, P. P.; Martín-Lozano, J. M. Sugar Beet for Bioethanol Production: An Approach Based on Environmental Agricultural Outputs. *Energy Policy* **2013,** *55,* 662–668.

Sánchez Orozco, R.; Balderas Hernández, P.; Flores Ramírez, N.; Roa Morales, G.; Saucedo Luna, J.; Castro Montoya, A. J. Gamma Irradiation Induced Degradation of Orange Peels. *Energies* **2012,** *5* (8), 3051–3063.

Sandhu, S. K.; Oberoi, H. S.; Dhaliwal, S. S.; Babbar, N.; Kaur, U.; Nanda, D.; Kumar, D. Ethanol Production from Kinnow Mandarin (*Citrus reticulata*) Peels via Simultaneous Saccharification and Fermentation Using Crude Enzyme Produced by *Aspergillus oryzae* and the Thermotolerant *Pichia kudriavzevii* Strain. *Ann. Microbiol.* **2012,** *62* (2), 655–666.

Saval, S. Aprovechamiento de residuos agroindustriales: Pasado, presente y futuro. *Rev. Soc. Mex. Biotecnol. Bioing., AC* **2012,** 14–46.

Scerra, V.; Caparra, P.; Foti, F.; Lanza, M.; Priolo, A. Citrus Pulp and Wheat Straw Silage as an Ingredient in Lamb Diets: Effects on Growth and Carcass and Meat Quality. *Small Rumin. Res.* **2001,** *40* (1), 51–56.

Schieber, A.; Hilt, P.; Streker, P.; Endreß, H.-U.; Rentschler, C.; Carle, R. A New Process for the Combined Recovery of Pectin and Phenolic Compounds from Apple Pomace. *Innov. Food Sci. Emerg. Technol.* **2003,** *4* (1), 99–107.

Schieber, A.; Stintzing, F.; Carle, R. By-products of Plant Food Processing as a Source of Functional Compounds—Recent Developments. *Trends Food Sci. Technol.* **2001,** *12* (11), 401–413.

Seeram, N. P.; Adams, L. S.; Henning, S. M.; Niu, Y.; Zhang, Y.; Nair, M. G.; Heber, D. In Vitro Antiproliferative, Apoptotic and Antioxidant Activities of Punicalagin, Ellagic Acid and a Total Pomegranate Tannin Extract Are Enhanced in Combination with Other Polyphenols as Found in Pomegranate Juice. *J. Nutr. Biochem.* **2005,** *16* (6), 360–367.

Spatafora, C.; Tringali, C. Valorization of Vegetable Waste: Identification of Bioactive Compounds and their Chemo-Enzymatic Optimization. *Open Agric. J.* **2012,** *6,* 9–16.

Tang, Y.-J.; Zhang, W.; Liu, R.-S.; Zhu, L.-W.; Zhong, J.-J. Scale-up Study on the Fed-Batch Fermentation of *Ganoderma lucidum* for the Hyperproduction of Ganoderic Acid and Ganoderma Polysaccharides. *Process Biochem.* **2011,** *46* (1), 404–408.

Ubalua, A. Cassava Wastes: Treatment Options and Value Addition Alternatives. *Afr. J. Biotechnol.* **2007,** *6* (18), 2065–2073.

UCS, U. o. C. S. *The Promise of Biomass: Clean Power and Fuel—If Handked Right,* 2012 [Online] (accessed August, 2015).

Vargas Betancur, G. J.; Pereira Jr., N. Sugar Cane Bagasse as Feedstock for Second Generation Ethanol Production: Part I: Diluted Acid Pretreatment Optimization. *Electr. J. Biotechnol.* **2010,** *13* (3), 10–11.

Virtanen, Y.; Kurppa, S.; Saarinen, M.; Katajajuuri, J.-M.; Usva, K.; Mäenpää, I.; Mäkelä, J.; Grönroos, J.; Nissinen, A. Carbon Footprint of Food—Approaches from National Input–Output Statistics and a LCA of a Food Portion. *J. Clean. Prod.* **2011**, *19* (16), 1849–1856.

Weightman, R.; Renard, C.; Gallant, D.; Thibault, J.-F. Structure and Properties of the Polysaccharides from Pea Hulls—II. Modification of the Composition and Physico-chemical Properties of Pea Hulls by Chemical Extraction of the Constituent Polysaccharides. *Carbohydr. Polym.* **1995**, *26* (2), 121–128.

Wijngaard, H.; Hossain, M. B.; Rai, D. K.; Brunton, N. Techniques to Extract Bioactive Compounds from Food By-products of Plant Origin. *Food Res. Int.* **2012**, *46* (2), 505–513.

Wu, Z. S.; Zhang, M.; Wang, S. J. Effects of High-Pressure Argon and Nitrogen Treatments on Respiration, Browning and Antioxidant Potential of Minimally Processed Pineapples during Shelf Life. *J. Sci. Food Agric.* **2012**, *92* (11), 2250–2259.

Zhou, W.; Widmer, W.; Grohmann, K. *Developments in Ethanol Production from Citrus Peel Waste.* In Proc. Fla State Hortic. Soc. 2008; pp 307–310.

Zhou, W.; Widmer, W.; Grohmann, K. *Economic Analysis of Ethanol Production from Citrus Peel Waste.* In Proc. Fla State Hortic. Soc. 2007; pp 310–315.

CHAPTER 3

EXTRACTION TECHNOLOGIES FOR THE PRODUCTION OF BIOACTIVE COMPOUNDS FROM PLANT FOOD BY-PRODUCTS

L. A. ORTEGA-RAMIREZ, G. A. GONZÁLEZ-AGUILAR, J. F. AYALA-ZAVALA, and M. R. CRUZ-VALENZUELA*

Centro de Investigacion en Alimentacion y Desarrollo, A.C. (CIAD, AC), Carretera a la Victoria Km 0.6, La Victoria, Hermosillo, Sonora 83000, Mexico

Corresponding author. E-mail: reynaldo@ciad.mx

CONTENTS

ABSTRACT

Processing of foods of plant origin generates vast quantities of by-products. Disposal of these by-products represents both a cost to the food processor and a potential negative impact on the environment. Different extraction methods have been used to obtain bioactive compounds from plants as solvent extraction, pressurized liquid extraction, subcritical fluid extraction, supercritical extraction, microwave-assisted extraction, pulsed electric fields, and ultrasonic-assisted extraction. These technologies can be used to take advantage of the bioactive compounds present in agro-industrial wastes and give them added value, although the efficiency of the conventional and nonconventional extraction method will depend mainly on the nature of the plant material, the type of bioactive compounds, and the selectivity of the method. The application of these technologies for obtaining of active compounds could provide an innovative approach to increase the production of specific compounds and their potential use as nutraceutical agents, antimicrobials, antioxidants, or as ingredients in the design of functional foods.

3.1   INTRODUCTION

The high content of bioactive compounds present in fruit by-products can be used as natural food additives. If this approach is realized, it would be feasible to fulfill the requirements of consumers for natural and preserved healthy food (Ayala-Zavala et al., 2011). In addition, the full utilization of fruits could lead the industry to a lower waste agribusiness, increasing industrial profitability to highlight the agro-industrial potential of exotic fruit by-products as a source of natural antioxidants, antimicrobials, flavoring, colorants and texturizer additives, and their possible uses in the food industry (Ayala-Zavala et al., 2011). The search for environmentally friendly and low-cost raw materials and technologies is forcing the food industry to develop new methods to guarantee the sustainability of the food chain. As industrialization continues, food production will become more concentrated, creating greater quantities of waste at a given location. While this can create greater environmental problems, the concentrated waste can often be more easily re-assimilated into the food cycle (Corrales et al., 2008). Thus, the use of novel technologies that are able to enhance the high content of compounds may represent not only an economical but also an environmental alternative, since by-products could be recycled in the food industry in the form of value-added ingredients or additives (Toepfl et al., 2006; Corrales et al.,

2008). The aim of this chapter is to summarize and give an overview of the technologies and methods that have been developed to improve the production and isolation of bioactive compounds, with special attention to antioxidants with their possible application in the design of nutraceuticals and functional food products.

3.2   ANTIMICROBIAL AND ANTIOXIDANT USES OF BY-PRODUCTS OF FOOD PLANTS

Phenolic compounds are some of the most important bioactive compounds in mango fruit, mainly found in peels and seeds (Ribeiro et al., 2008). Moreover, catechin, chlorogenic acid, and phloridzin, three phenolic compounds that are abundant in apple processing by-products exhibited varying degree of inhibitory action toward the growth of tested food pathogenic and spoilage bacteria, fungi, and yeasts (Muthuswamy and Rupasinghe, 2007). Phenolic compounds can be found in different tissues of fruits, for example, (1) peels are rich in phenolic compounds with antioxidant, antimicrobial, and colorant properties, some of these are natural defenses against pathogens attack and environmental conditions; (2) the pulp possess a lower content of phenolic compounds; however, it is the main source of dietary antioxidants for humans; and finally (3) the seed can be also rich in phenolics, mainly tannins with antioxidant an antimicrobial properties that protect this tissue to perpetuate the regeneration of the species (Ayala-Zavala et al., 2011).

In other study (Vega-Vega et al., 2013), different methods to extraction of phenolics from seed mango Haden were used, such as methanolic-polar, methanolic-nonpolar, ethanolic-polar, ethanolic-nonpolar, and water infusion. The total phenolic content of the ethanolic-nonpolar extract showed 875.06 mg/g, DPPH EC_{50}: 0.04 mg/mL, causing a 100% inhibition of bacteria pathogens applying 25 mg/mL and inhibition of 89.78% against *Alternaria* sp. applying 6.25 mg/mL. In other study, the antimicrobial properties of mango seed kernel phenolic extracts were investigated. Minimum inhibitory concentrations of mango kernel extract against 18 species of 43 strains, containing food-borne pathogenic bacteria, were determined using the agar dilution method. The mango kernel extracts had a broad antimicrobial spectrum and was more active against Gram-positive than Gram-negative bacteria with a few exceptions. These results also indicated that the active component of the mango kernel extract was a type of polyphenol (Kabuki et al., 2000).

On the other hand, the antimicrobial and antioxidant potential of pomegranate peel and seed extract were investigated in chicken products (Li et al., 2006a; Kanatt et al., 2010). Pomegranate peel extract showed excellent antioxidant activity, while the seed extract did not have any significant activity, probably to the difference in the type and amount of bioactive compounds present in both tissues. Pomegranate peel extract showed good antimicrobial activity against *Staphylococcus aureus* and *Bacillus cereus*. In general, addition of pomegranate peel extract to popular chicken and meat products enhanced its shelf life by 2–3 weeks, during chilling temperature storage. Pomegranate peel extract was also effective in controlling oxidative rancidity in these chicken products (Kanatt et al., 2010). Water infusion of *Cocos nucifera* L. husk fiber has been used in northeastern Brazil traditional medicine for treatment of diarrhea and arthritis. The crude extract rich in catechin revealed antimicrobial and antiviral activities. Catechin and epicatechin together with condensed tannins (B-type procyanidins) were demonstrated to be the components of the water extract from *Cocos* by-products (Esquenazi et al., 2002).

Mandalari et al. (2007) evaluated a flavonoid-rich extract from the peel of Bergamot citrus fruit, an important by-product in the processing industry, against different bacteria and yeast. The enzyme preparation pectinase 62 L efficiently converted common glycosides into their aglycones from bergamot extracts, and this deglycosylation increased the antimicrobial potency of Citrus flavonoids. Pair-wise combinations of eriodictyol, naringenin, and hesperidins showed both synergistic and indifferent interactions that were dependent on the test indicator organism. This study concluded that Bergamot peel is a potential source of natural antimicrobials that are active against Gram-negative bacteria. In other study (Cruz-Valenzuela et al., 2013), orange peel and seed extracts were used as additives to preserve quality of fresh-cut orange using ethanolic extraction from seeds fractionated in polar (PE) and nonpolar (NPE) solvents, and essential oils from peels (PEO) were obtained. NPE showed the best response in phenolic content, antioxidant, and antifungal activities, followed by the PE, while the PEO showed the lowest scores. NPE was selected and applied to fresh-cut orange. Treated fruit showed the highest values of total phenolic content (32.6 mg GAE/g), flavonoids (26.3 mg QE/g) and antioxidant activity (8270 μmol TE/g), and the lowest mesophilic bacteria (1854 CFU/g) and total molds and yeast (3435 CFU/g) growth compared to controls (6.7 mg GAE/g; 5.4 mg QE/g; 1836 μmol TE/g; 2848 CFU/g; 19,030 CFU/g, respectively). This work demonstrates the potential of seed by-products rich in antioxidant and antimicrobial compounds that can be used to preserve quality of fresh-cut oranges.

3.3 EXTRACTION TECHNOLOGIES OF BIOACTIVE COMPOUNDS OF FOOD PLANTS

Processing of foods of plant origin generates vast quantities of by-products. Disposal of these by-products represents both a cost to the food processor and a potential negative impact on the environment. Research over the past 20 years has revealed that many of these by-products could serve as a source of potentially valuable bioactive compounds. Despite this, the vast majority of by-products are currently not exploited as sources of these compounds. This is in part due to the lack of appropriate techniques for extraction of these compounds (Table 3.1). In recent times, a number of novel extraction techniques have been used to optimize extraction of bioactive compounds from by-products (Herrero et al., 2006; Wijngaard et al., 2012).

3.3.1 SOLVENT EXTRACTION

Bioactive compounds from plant materials can be extracted by various classical extraction techniques. Most of these techniques are based on the extracting power of different solvents in use and the application of heat and/or mixing. To obtain bioactive compounds from plants, the existing classical techniques as Soxhlet extraction, maceration, and hydrodistillation (Azmir et al., 2013). Some of the most widely used solvents in the extraction procedures are hexane, ether, chloroform, acetonitrile, benzene, and ethanol and are commonly used in different ratios with water (Starmans and Nijhuis, 1996). A large number of natural bioactive compounds have been traditionally extracted from natural sources with organic solvents representing the most important step in optimizing recovery of desirable components by this technique (Li et al., 2006b).

Lapornik et al. (2005) studied the effect of solvent and extraction time on the yield of extracted antioxidants from grape, black and redcurrant by-products. Results showed that ethanol and methanol extracts of red and blackcurrant contain twice more anthocyanins and polyphenols than water extracts, extracts made from grape marc had seven times higher values than water extracts. In water extracts, the yields of polyphenols decreased, while in methanol and ethanol extracts their content increased with the time of extraction. However, various solvents must be used with care as they are toxic for humans and dangerous for the environment; moreover, the extraction conditions are sometimes laborious (Li et al., 2006b; Miron et al., 2011). The solvent must be separated from the final extract, especially if the product is to be used in food applications (Starmans and Nijhuis, 1996).

TABLE 3.1 Obtaining Bioactive Compounds from By-products of the Food Industry Using Different Extraction Methods.

Method of extraction	By-product	Compounds extracted	Research	Bioactivity	References
SE methanol, acetone, acidified waters, or mixtures (1:1, v/v), 1 or 120 in, pH 3 or 8, 22 or 55°C	Banana peels from 2 cultivars (Grande Naine and Gruesa)	Total phenols and anthocyanins	3.1–3.8 GAE/100 g and 434 µg c-3-gE/100 g of dry material	Antioxidant/compounds with food applications and health benefits	González-Montelongo et al. (2010)
Supercritical fluids CO_2 313.15 and 323.15 K; 180, 200, and 220 bar 1.7×10^{-4} kg^{-1}	Grape seed oil	Triacylglycerides	Oil composition and antioxidant capacity of the extracts using different operating conditions	Antioxidants	Passos et al. (2010)
Supercritical fluids CO_2; 40–100°C; 20–40 MPa 1.0–2.0 mL/min	Tomato skins	Lycopene	Influence of the operating conditions on yield and antioxidant activity of the extracts	Protective effect against cardiovascular, coronary heart diseases, and cancer	Yi et al. (2009)
Supercritical fluids 500 mL of CO_2, 86°C, and 34.4 MPa	Tomato seeds	Carotenoids (lycopene)	Optimization of extraction of lycopene with SFE using CO_2 without the use of cosolvents. The information obtained in this study can be used for scale-up of the extraction process	Anticarcinogenic properties: lower incidence of digestive tract cancer, decreased risk of prostate cancer	Rozzi et al. (2002)
Supercritical fluids CO_2 40–60°C, 148–602 bar, 10–60 min	Roasted wheat germ	Phenolic compounds and tocopherols	Recovery of phenolic compounds and tocopherols, yields, and measurement of antioxidant activity by TPC, TTC, and DPPH	Pharmaceutical, food, and cosmetic formulation. Biological control agents against insects	Gelmez et al. (2009)
Supercritical fluids CO_2 311–331 K, 116–180 bar, 0.60 g/mL CO_2	Coriander seeds	Antioxidant fractions	Antioxidant activity (DPPH), yield, and effect of the operating conditions. Fractions were compared with commercial antioxidants	Diet supplements in nutraceuticals industries	Yepez et al. (2002)
Supercritical fluids CO_2, 50°C, and 30 MPa 0, 2, and 3% EtOH)	Mangosteen pericarp	Xanthones	Characterization of the extracts by HPLC/LC-ESI-MS and antioxidant activity by DPPH	Inhibition of lipid peroxidation, antioxidant activity, neuroprotective, and inhibitor of HIV-1 protease	Zarena and Sankar (2009)

TABLE 3.1 (Continued)

Method of extraction	By-product	Compounds extracted	Research	Bioactivity	References
Supercritical fluids CO_2, 30 and 50°C, 250, 275, and 300 bar, 5 g/min CO_2 5–20% EtOH	Grape seed	Proanthocyanidins	Effect of different pressure, temperature, and ethanol percentage. Characterization of the extracts by HPLC	Anticarcinogenic, antiviral, anticancer, and also may act against oxidation of low density lipoproteins	Yilmaz et al. (2011)
Supercritical fluids CO_2, 40 MPa, 35°C, using 5% v/v ethanol	Grape (seed, stem, skin, and pomace)	Resveratrol (19.2 mg/100 g DW)	Grape pomace is a potential source of resveratrol. SC-CO_2 extraction under optimized conditions ensures its extraction, thus allowing the final by-product to be reused for other activities	Antioxidant activity and may reduce oxidant-induced apoptosis and LDL oxidation, decreased risk of coronary heart disease. The cardioprotective activity; chemoprevention agent, inhibit tumor initiation, promotion, and progression, inhibit the growth of cancerous cells; reduce inflammation via inhibition of prostaglandin production, estrogenic activity of resveratrol	King et al. (2006), Casas et al. (2010)
Supercritical fluids CO_2, 40, 60, and 80°C; 200, 400, and 600 bar; 25 g/min; 150 min	*Hibiscus cannabinus* L. seed	Edible oil	Comparison with Soxhlet extraction and ultrasonic-assisted extraction. Antioxidant activity of extract was compared with 7 commercial edible oils	Functional foods	Chan and Ismail (2009)
UAE MetOH, 40°C, 60 kHz	Penggan peel	Hesperidin	The effect of ultrasonic on extraction the yield of hesperidin depends on many ultrasonic parameters that produce physical, chemical, or mechanical effects, which will play an important role for isolation bioactive compounds from Penggan peels by individually or in combination, which is difficult to determinate the interaction of many parameters	Food and pharmaceutical industry; Posse's antioxidant, anti-inflammatory, and antiallergic activities	Ma et al. (2008)

TABLE 3.1 *(Continued)*

Method of extraction	By-product	Compounds extracted	Research	Bioactivity	References
UAE EtOH 72%, 65°C, 37 min, 40 W	Hawthorn seeds	Flavonoids	UAE was reliable and feasible for extracting the flavonoids compounds from hawthorn seed and gave better results than the water-bath extraction techniques for the extraction of flavonoids	Health-promoting compounds negatively associated with coronary heart diseases	Pan et al. (2012)
UAE water, 45 min, 222	Litchi seeds	Polysaccharides (arabinose, fructose, galactose, glucose, and mannose)	The optimal conditions to obtain the highest yield of Litchi seeds polysaccharides were determined as follows: 15.0 mL/g of solvent/material ratio, 45.0 min of ultrasonic time and 222 W of ultrasonic power	Food and biomedical applications. Posse's antitumoral, antioxidant, and hypoglycemic properties	Chen et al. (2011)
MAE EtOH 30%, 1.5 g skins, 30 s at 855 W	Peanut skins	Phenolic compounds	Peanut skin extracts obtained using MAE, exhibited relatively high ORAC activity, especially when compared to the ORAC values reported for some common fruits and spices known to have high antioxidant activities	Pharmaceutical applications of health-promoting compounds including cancer prevention	Ballard et al. (2010)
MAE EtOH 47.2%, 60°C, 4.6 min at 150 W	Grape seeds of cultivars Cabernet Sauvignon, Shoraz, Sauvignon Blanc and Chardonnay	Polyphenols	Sequential application of the optimal conditions to one sample revealed that approximately 92% of the total polyphenols were extracted in the first instance. One key finding was that varying the applied power to the extraction was essentially irrelevant; inspection of the applied power profile during extraction revealed that the power was strictly modulated to maintain a constant temperature in the reaction cell	Pharmaceutical, cosmetic, and food industry	Li et al. (2011)

TABLE 3.1 *(Continued)*

Method of extraction	By-product	Compounds extracted	Research	Bioactivity	References
Subcritical water extraction 50–200°C; 5 min 20–260, 5 min and 200 and 260°C, 5–120 min	Rice bran biomass	Phenolic compounds	Phenolic compounds, total phenolic content, antioxidant activity, total soluble sugar, pH, and electrical conductivity was performed to see the effects of temperature and exposure time in this parameters	Beneficial effect against cancer and diabetes, among others. Application in food industries and health and cosmetic markets	Pourali et al. (2010)
Subcritical water extraction 100–190°C; 5–30 min; 90–131 bar	Onion skin	Quercetin	Effect of temperature and extraction time on yield extraction. Subcritical water was compared with conventional methods such as ethanol, methanol, and water at boiling point. Chemical composition by HPLC	Anticancer, antivirus, and anti-inflammatory activities	Ko et al. (2011)

3.3.2   PRESSURIZED LIQUID EXTRACTION

Traditionally, the nature and amounts of target compounds in natural products are determined after exhaustive extraction of the sample using solid–liquid extraction techniques. Pressurized solvents use elevated pressures and sometimes temperatures which drastically improve the speed of the extraction process. In ideal solid–liquid extractions, the desired compound should have high solubility in the solvent employed while other compounds from the solid matrix should not be solubilized during extraction (Pronyk and Mazza, 2009). In reality however, this is rarely achieved, and therefore much research has been carried out on optimizing conditions, such as solvent-to-feed ratio, particle size, modifier concentration extraction temperature, pressure and time, and flow rate, to enhance the recovery of bioactive compounds from food by-products of plant origin (Kaur et al., 2008; Wijngaard and Brunton, 2010). However, conventional solid–liquid extraction techniques, such as Soxhlet extraction and maceration, are time consuming and use high amounts of solvents (Wang and Weller, 2006) emphasizing the need for more sustainable techniques, including techniques based on pressurized fluids.

The use of pressurized solvent techniques also offers the advantage of enhanced target-molecule specificity and speed due to physicochemical properties of the solvent, including density, diffusivity, viscosity, and dielectric constant, which can be controlled by varying pressure and temperature of the extraction system. Two pressurized fluid extraction methods are very popular: supercritical fluid extraction (SFE), called supercritical CO_2 extraction (SC-CO_2) when CO_2 is used, and pressurized liquid extraction, where 100% water is used; this technique is frequently called subcritical water extraction (SWE) (Pronyk and Mazza, 2009).

3.3.3   SUBCRITICAL FLUID EXTRACTION

SWE, that is, extraction using hot water under pressure, has recently emerged as a useful tool to replace the traditional extraction methods. SWE is an environmentally friendly technique that can provide higher extraction yields from solid samples. SWE is carried out using hot water (from 100 to 374°C, the latter being the water critical temperature) under high pressure (usually from 10 to 60 bar) to maintain water in the liquid state (Herrero et al., 2006).

This technique has emerged as a useful tool to replace traditional extraction methods. SWE presents a series of important advantages over the traditional extraction techniques; it is faster, produces high yields, and the use of solvents can be greatly reduced (Plaza et al., 2010). Therefore, this novel extraction technique is gaining increasing attention due to the advantages it provides compared to other traditional extraction approaches and because it is environmentally friendly compared to conventional organic liquid (Herrero et al., 2006). The most important factor to consider in this type of extraction procedure is the variability of the dielectric constant with temperature. Water at room temperature is a very polar solvent, with a dielectric constant close to 80. However, this value can be significantly decreased to values close to 27 when water is heated up to 250°C, while maintaining its liquid state by applying the appropriate pressure (Miller and Hawthorne, 2000).

SWE has been widely used to extract different compounds from several vegetable by-product matrices. SWE was applied to extract essential oils from *Trachyspermum ammi* seeds and obtain thymol as the main component of this oil. Finding that the most suitable condition for extraction of essential oils was found to be at temperature 175°C, water flow rate 4 mL/min, mean particle size 1 mm and pressure 2 MPa. At these optimum extraction parameters, the maximum yield of thymol obtained experimentally was 12.9634 mg/g dry sample. Thymol is a very important compound because it has antimicrobial and antioxidant activity and provides characteristic flavor and odor in foods (Rodriguez-Garcia et al., 2016). Another study, decomposition and conversion of rice bran into valuable chemical compounds were successfully conducted using SWE, showing that the degradation of the lignin/phenolics–carbohydrates complexes of rice bran were achieved (up to 92% of rice bran) in the water without using organic solvent, acid, base, and/or enzyme. Decomposition of rice bran and defatted rice bran have resulted almost the same amount of phenolic compounds; it was understood that phenolic compounds were mainly produced from decomposition of bounds between lignin, carbohydrate, and phenolic compounds, and not from rice-bran oil. Protocatechuic and vanillic acids were the major ones among identified phenolic compounds (Pourali et al., 2010).

Some advantages SWE over conventional methods are shorter extraction time, higher quality extract, lower cost of water, and lack of using of organic solvent. This method is developed as a strong possibility for plant sample extraction (Khajenoori et al., 2015). Therefore, the bioactive properties of the individual compounds obtained by SWE must be evaluated to obtain a better knowledge of the potential application of these extracts.

3.3.4 SUPERCRITICAL EXTRACTION

SFE has also been widely investigated for the recovery of high-added-value compounds and this technology is of paramount importance mainly due to the purity of the extracts provided (Barba et al., 2016). SFE is based on the use of solvents at conditions above their critical point and enables the continuous modulation of the solvation power and the selectivity of the solvent. Carbon dioxide (CO_2) is commonly used in SFE because, in contrast with organic solvents, it is nontoxic, inexpensive and volatile, with moderate critical conditions. This allows the use of moderate temperatures, which allied with other advantages of this method, such as the absence of light and oxygen during the extraction process, avoids degradation reactions of bioactive substances, and extracts are solvent-free (Cruz et al., 2017).

Frequently, the processes that take place in the food industry generate products (the so-called by-products) that are discarded, causing subsequent environmental problems. In recent years, companies have devoted effort to find value-added application for these food by-products. Thus, several studies have been developed to extract β-carotene and lycopene from by-products of tomato industry (Kalogeropoulos et al., 2012). These compounds are natural pigments, belonging to the carotenoid group, and their antioxidant properties are well known (Kalogeropoulos et al., 2012). Other interesting by-products are those from the wine industry and their interest is related to the type and amount of phenolic compounds that are found in grape seeds and skins. Isolation of phenolic compounds from grape seeds has been attempted using supercritical carbon dioxide (Murga et al., 2000). Palma and Taylor (1999) observed that the recovery of catechin, gallic acid, and epicatechin from grape seeds was higher when using SC-CO_2, with methanol as modifier, rather than when using traditional solid–liquid extraction. On the other hand, by-products from the olive oil industry to extract tocopherols was suggested by Ibanez et al. (2000), where the separation of tocopherols from the olive pomace was achieved by means of a supercritical carbon dioxide extraction with two-step fractionation. In the second fractionation step, where complete depressurization took place, the CO_2 density was very low and enrichment of the extract with tocopherols was observed Ibanez et al. (2000). The extraction of natural pigments from food by-products is of major importance, not only because of the increasing demand for natural ingredients by the food industry but also because some of these pigments may also have associated antioxidant activity that can even more, increase their added-value (Herrero et al., 2006). However, sometimes pigments themselves are the target compounds. SFE is considered the most suitable method for producing fractions with high biological activities, for this reason, the isolation and separation of bioactive compounds from by-products applying supercritical fluid technology provides an interesting approach to exploit such by-products in an environmentally friendly way.

3.3.5 MICROWAVE-ASSISTED EXTRACTION

Microwaves are electromagnetic waves, which are usually operated at a frequency of 2.45 GHz. Microwaves can access biological matrices and interact with polar molecules, such as water, and generate heat. The temperature will rise, which generally leads to enhanced extraction efficiency (Wang and Weller, 2006). By using microwave-assisted extraction (MAE), the extraction time can be reduced in comparison to conventional extraction methods (Ballard et al., 2010; Pérez-Serradilla and de Castro, 2011). MAE extraction efficiency principally depends on microwave energy, treatment time, and temperature used. It should be noted that in general temperature increase enhances extraction rates of solutes. For example, Tsubaki et al. (2010) reported that the proportion of polyphenol in the extract of tea residue increased from 25.3% to 74.4% when the temperature was increased from 110°C to 230°C using MAE. However, 5-(hydroxymethyl) furfural, a potentially harmful compound was also formed at 230°C. Therefore, caution should be taken while increasing the temperature in MAE.

3.3.6 PULSED ELECTRIC FIELDS

In pulsed electric fields (PEFs), material located between two electrodes is exposed to a strong electrical field. If the stress caused by the electrical field on the membrane is large enough, pore formation occurs. The pore formation can be reversible or irreversible and depends on the conditions of PEF treatment, such as electric field strength, pulse duration, and the number of pulses. Pore formation enhances cell permeability (Angersbach et al., 2000). PEF has been extensively investigated as a nonthermal preservation technique. On the other hand, the use of PEF in the recovery of bioactive compounds from by-products is not well studied up to now. The technique has been mainly used to extract polyphenols from grape by-products.

In red grape by-product, the level of anthocyanins was 60% increased when PEF was applied as a pretreatment of 1 min at 25°C in combination with a conventional thermal extraction at 70°C for 1 h (Corrales et al., 2008).

When white grape skins were treated with PEF at a temperature of 20°C, 10% more polyphenols than nontreated samples were extracted (Boussetta et al., 2009). PEF seems a potential technique to use as pretreatment in the extraction of polyphenols from by-products, although industrial scale equipment is still under development and the technique does not apply to solid products (Han, 2008).

3.3.7 ULTRASONIC-ASSISTED EXTRACTION

Ultrasound-assisted extraction can also be used in the food industry to perform extractions. The technique uses high frequency sound waves (higher than 20 kHz). When the ultrasound waves are strong enough, bubbles are formed in the liquid. Eventually the formed bubbles cannot absorb the energy any longer and will collapse: "cavitation" takes place. This collapse causes a change in temperature and pressure within the bubble and hence energy for chemical reactions is generated. Extremely high temperatures of 5000°C and pressures of 1000 bar have been measured. The process is affected by mechanical forces surrounding the bubble when a solid matrix is present (Luque-García and De Castro, 2003). Due to the cavitation process, plant cell walls can be penetrated, which provides an easier cell access. In addition, ultrasound can result in swelling of the plant material, which in turn can enhance extraction (Vinatoru, 2001). Various studies have tested ultrasound and its effect on the extraction of polyphenols. Ultrasound has been shown to enhance the extraction of bioactive compounds such as polyphenols from different plant by-products in several studies (Vilkhu et al., 2008; Khan et al., 2010; Virot et al., 2010).

Orange peels were treated with ultrasound to extract the flavanones hesperidin and naringin. Temperature, ethanol:water ratio, and sonication power were optimized using response surface methodology. A temperature of 40°C, a 4:1 (v/v) ethanol:water ratio, and a sonication power of 150 W were determined as optimal. These conditions resulted in an enhanced extraction of 38% for naringin and 41% for hesperidin when compared to nonsonicated samples (Khan et al., 2010). When ultrasound was applied to apple pomace, a by-product of the apple cider industry, catechins were 20% better extracted than when a conventional extraction was used. In addition, a scale-up test was performed with an ultrasonic bath of a volume of 30 L (Virot et al., 2010). Ultrasound has also been successfully applied to enhance the extraction of carotenoids from different by-products of plant origin.

Sun et al. (2011) investigated the effect of various factors (particle size, solvent type, solid:solvent ratio, temperature, extraction time, acoustic intensity, height of liquid, and duty cycle of ultrasound exposure) on the ultrasound assisted extraction yield of all-trans-β-carotene from citrus peel. All tested ultrasound-assisted extraction (UAE) conditions yielded higher all-trans-β-carotene than that of conventional solid/liquid extraction, with a maximum of ca. 11 μg all-trans-β-carotene/g DW at an extraction time of 120 min. The height of the liquid present in the sample beaker showed an inverse relationship with extraction yield in UAE. The authors reported that this may be explained by the fact that the cavitation intensity decreases with increasing height due to the attenuation of the waves caused by absorption and scattering.

3.4 CONCLUDING REMARKS

It is clear that bioactive compounds are the future in the field of functional foods and nutraceuticals that can be found in plant food by-products. Functional foods are gaining importance due to changes in eating habits and concern about health. Nevertheless, some topics must be addressed prior to successful applications in the food and pharmaceutical industries to replace the common "synthetic pharmaceuticals" by "natural nutraceuticals."

KEYWORDS

- **food additives**
- **food safety**
- **phenolic compounds**
- **antimicrobial activity**
- **antioxidants**

REFERENCES

Angersbach, A.; Heinz, V.; Knorr, D. Effects of Pulsed Electric Fields on Cell Membranes in Real Food Systems. *Innov. Food Sci. Emerg. Technol.* **2000,** *1* (2), 135–149.

Ayala-Zavala, J.; Vega-Vega, V.; Rosas-Domínguez, C.; Palafox-Carlos, H.; Villa-Rodriguez, J.; Siddiqui, M. W.; Dávila-Aviña, J.; González-Aguilar, G. Agro-industrial Potential of Exotic Fruit Byproducts as a Source of Food Additives. *Food Res. Int.* **2011,** *44* (7), 1866–1874.

Azmir, J.; Zaidul, I. S. M.; Rahman, M. M.; Sharif, K. M.; Mohamed, A.; Sahena, F.; Jahurul, M. H. A.; Ghafoor, K.; Norulaini, N. A. N.; Omar, A. K. M. Techniques for Extraction of Bioactive Compounds from Plant Materials: A Review. *J. Food Eng.* **2013,** *117* (4), 426–436.

Ballard, T. S.; Mallikarjunan, P.; Zhou, K.; O'Keefe, S. Microwave-Assisted Extraction of Phenolic Antioxidant Compounds from Peanut Skins. *Food Chem.* **2010,** *120* (4), 1185–1192.

Barba, F. J.; Zhu, Z.; Koubaa, M.; Sant'Ana, A. S.; Orlien, V. Green Alternative Methods for the Extraction of Antioxidant Bioactive Compounds from Winery Wastes and By-products: A Review. *Trends Food Sci. Technol.* **2016,** *49*, 96–109.

Boussetta, N.; Lebovka, N.; Vorobiev, E.; Adenier, H.; Bedel-Cloutour, C.; Lanoiselle, J.-L. Electrically Assisted Extraction of Soluble Matter from Chardonnay Grape Skins for Polyphenol Recovery. *J. Agric. Food Chem.* **2009,** *57* (4), 1491–1497.

Casas, L.; Mantell, C.; Rodríguez, M.; de la Ossa, E. M.; Roldán, A.; De Ory, I.; Caro, I.; Blandino, A. Extraction of Resveratrol from the Pomace of *Palomino fino* Grapes by Supercritical Carbon Dioxide. *J. Food Eng.* **2010,** *96* (2), 304–308.

Corrales, M.; Toepfl, S.; Butz, P.; Knorr, D.; Tauscher, B. Extraction of Anthocyanins from Grape By-products Assisted by Ultrasonics, High Hydrostatic Pressure or Pulsed Electric Fields: A Comparison. *Innov. Food Sci. Emerg. Technol.* **2008,** *9* (1), 85–91.

Cruz-Valenzuela, M. R.; Carrazco-Lugo, D. K.; Vega-Vega, V.; Gonzalez-Aguilar, G. A.; Ayala-Zavala, J. F. Fresh-Cut Orange Treated with Its Own Seed By-products Presented Higher Antioxidant Capacity and Lower Microbial Growth. *Int. J. Postharv. Technol. Innov.* **2013,** *3* (1), 13–27.

Cruz, P. N.; Pereira, T. C. S.; Guindani, C.; Oliveira, D. A.; Rossi, M. J.; Ferreira, S. R. S. Antioxidant and Antibacterial Potential of Butia (*Butia catarinensis*) Seed Extracts Obtained by Supercritical Fluid Extraction. *J. Supercrit. Fluids.* **2017,** *119*, 229–237.

Chan, K. W.; Ismail, M. Supercritical Carbon Dioxide Fluid Extraction of *Hibiscus cannabinus* L. Seed Oil: A Potential Solvent-Free and High Antioxidative Edible Oil. *Food Chem.* **2009,** *114* (3), 970–975.

Chen, Y.; Luo, H.; Gao, A.; Zhu, M. Ultrasound-Assisted Extraction of Polysaccharides from Litchi (*Litchi chinensis* Sonn.) Seed by Response Surface Methodology and their Structural Characteristics. *Innov. Food Sci. Emerg. Technol.* **2011,** *12* (3), 305–309.

Esquenazi, D.; Wigg, M. D.; Miranda, M. M.; Rodrigues, H. M.; Tostes, J. B.; Rozental, S.; da Silva, A. J.; Alviano, C. S. Antimicrobial and Antiviral Activities of Polyphenolics from *Cocos nucifera* Linn. (Palmae) Husk Fiber Extract. *Res. Microbiol.* **2002,** *153* (10), 647–652.

Gelmez, N.; Kıncal, N. S.; Yener, M. E. Optimization of Supercritical Carbon Dioxide Extraction of Antioxidants from Roasted Wheat Germ Based on Yield, Total Phenolic and Tocopherol Contents, and Antioxidant Activities of the Extracts. *J. Supercrit. Fluids* **2009,** *48* (3), 217–224.

González-Montelongo, R.; Lobo, M. G.; González, M. Antioxidant Activity in Banana Peel Extracts: Testing Extraction Conditions and Related Bioactive Compounds. *Food Chem.* **2010,** *119* (3), 1030–1039.

Han, J. H. *Packaging for Nonthermal Processing of Food*; John Wiley & Sons: Hoboken, NJ, 2008.

Herrero, M.; Cifuentes, A.; Ibanez, E. Sub- and Supercritical Fluid Extraction of Functional Ingredients from Different Natural Sources: Plants, Food-By-Products, Algae and Microalgae: A Review. *Food Chem.* **2006,** *98* (1), 136–148.

Ibanez, E.; Palacios, J.; Senorans, F.; Santa-Maria, G.; Tabera, J.; Reglero, G. Isolation and Separation of Tocopherols from Olive By-products with Supercritical Fluids. *J. Am. Oil Chem. Soc.* **2000,** *77* (2), 187–190.

Kabuki, T.; Nakajima, H.; Arai, M.; Ueda, S.; Kuwabara, Y.; Dosako, S. I. Characterization of Novel Antimicrobial Compounds from Mango (*Mangifera indica* L.) Kernel Seeds. *Food Chem.* **2000,** *71* (1), 61–66.

Kalogeropoulos, N.; Chiou, A.; Pyriochou, V.; Peristeraki, A.; Karathanos, V. T. Bioactive Phytochemicals in Industrial Tomatoes and their Processing Byproducts. *LWT—Food Sci. Technol.* **2012,** *49* (2), 213–216.

Kanatt, S. R.; Chander, R.; Sharma, A. Antioxidant and Antimicrobial Activity of Pomegranate Peel Extract Improves the Shelf Life of Chicken Products. *Int. J. Food Sci. Technol.* **2010,** *45* (2), 216–222.

Kaur, D.; Wani, A. A.; Oberoi, D.; Sogi, D. Effect of Extraction Conditions on Lycopene Extractions from Tomato Processing Waste Skin Using Response Surface Methodology. *Food Chem.* **2008,** *108* (2), 711–718.

Khajenoori, M.; Asl, A. H.; Eikani, M. H. Subcritical Water Extraction of Essential Oils from *Trachyspermum ammi* Seeds. *J. Essen. Oil Bear. Plants* **2015,** *18* (5), 1165–1173.

Khan, M. K.; Abert-Vian, M.; Fabiano-Tixier, A.-S.; Dangles, O.; Chemat, F. Ultrasound-Assisted Extraction of Polyphenols (Flavanone Glycosides) from Orange (*Citrus sinensis* L.) Peel. *Food Chem.* **2010,** *119* (2), 851–858.

King, R. E.; Bomser, J. A.; Min, D. B. Bioactivity of Resveratrol. *Comprehens. Rev. Food Sci Food Saf.* **2006,** *5* (3), 65–70.

Ko, M.-J.; Cheigh, C.-I.; Cho, S.-W.; Chung, M.-S. Subcritical Water Extraction of Flavonol Quercetin from Onion Skin. *J. Food Eng.* **2011,** *102* (4), 327–333.

Lapornik, B.; Prošek, M.; Golc Wondra, A. Comparison of Extracts Prepared from Plant By-products Using Different Solvents and Extraction Time. *J. Food Eng.* **2005,** *71* (2), 214–222.

Li, Y.; Guo, C.; Yang, J.; Wei, J.; Xu, J.; Cheng, S. Evaluation of Antioxidant Properties of Pomegranate Peel Extract in Comparison with Pomegranate Pulp Extract. *Food Chem.* **2006a,** *96* (2), 254–260.

Li, B.; Smith, B.; Hossain, M. M. Extraction of Phenolics from Citrus Peels: II. Enzyme-Assisted Extraction Method. *Sep. Purif. Technol.* **2006b,** *48* (2), 189–196.

Li, Y.; Skouroumounis, G. K.; Elsey, G. M.; Taylor, D. K. Microwave-Assistance Provides Very Rapid and Efficient Extraction of Grape Seed Polyphenols. *Food Chem.* **2011,** *129* (2), 570–576.

Luque-Garcia, J.; De Castro, M. L. Ultrasound: A Powerful Tool for Leaching. *TrAC—Trends Anal. Chem.* **2003,** *22* (1), 41–47.

Ma, Y.; Ye, X.; Hao, Y.; Xu, G.; Xu, G.; Liu, D. Ultrasound-Assisted Extraction of Hesperidin from Penggan (*Citrus reticulata*) Peel. *Ultrasonics Sonochem.* **2008,** *15* (3), 227–232.

Mandalari, G.; Bennett, R.; Bisignano, G.; Trombetta, D.; Saija, A.; Faulds, C.; Gasson, M.; Narbad, A. Antimicrobial Activity of Flavonoids Extracted from Bergamot (*Citrus bergamia* Risso) Peel, a Byproduct of the Essential Oil Industry. *J. Appl. Microbiol.* **2007,** *103* (6), 2056–2064.

Miller, D. J.; Hawthorne, S. B. Solubility of Liquid Organic Flavor and Fragrance Compounds in Subcritical (Hot/Liquid) Water from 298 K to 473 K. *J. Chem. Eng. Data* **2000,** *45* (2), 315–318.

Miron, T.; Plaza, M.; Bahrim, G.; Ibáñez, E.; Herrero, M. Chemical Composition of Bioactive Pressurized Extracts of Romanian Aromatic Plants. *J. Chromatogr. A* **2011,** *1218* (30), 4918–4927.

Murga, R.; Ruiz, R.; Beltrán, S.; Cabezas, J. L. Extraction of Natural Complex Phenols and Tannins from Grape Seeds by Using Supercritical Mixtures of Carbon Dioxide and Alcohol. *J. Agric. Food Chem.* **2000,** *48* (8), 3408–3412.

Muthuswamy, S.; Rupasinghe, H. V. Fruit Phenolics as Natural Antimicrobial Agents: Selective Antimicrobial Activity of Catechin, Chlorogenic Acid and Phloridzin. *J. Food Agric. Environ.* **2007,** *5* (3/4), 81.

Palma, M.; Taylor, L. T. Extraction of Polyphenolic Compounds from Grape Seeds with near Critical Carbon Dioxide. *J. Chromatogr. A* **1999,** *849* (1), 117–124.

Pan, G.; Yu, G.; Zhu, C.; Qiao, J. Optimization of Ultrasound-Assisted Extraction (UAE) of Flavonoids Compounds (FC) from Hawthorn Seed (HS). *Ultrasonics Sonochem.* **2012,** *19* (3), 486–490.

Passos, C. P.; Silva, R. M.; Da Silva, F. A.; Coimbra, M. A.; Silva, C. M. Supercritical Fluid Extraction of Grape Seed (*Vitis vinifera* L.) Oil. Effect of the Operating Conditions upon Oil Composition and Antioxidant Capacity. *Chem. Eng. J.* **2010,** *160* (2), 634–640.

Pérez-Serradilla, J.; de Castro, M. L. Microwave-Assisted Extraction of Phenolic Compounds from Wine Lees and Spray-Drying of the Extract. *Food Chem.* **2011,** *124* (4), 1652–1659.

Plaza, M.; Amigo-Benavent, M.; Del Castillo, M. D.; Ibáñez, E.; Herrero, M. Facts about the Formation of New Antioxidants in Natural Samples after Subcritical Water Extraction. *Food Res. Int.* **2010,** *43* (10), 2341–2348.

Pourali, O.; Asghari, F. S.; Yoshida, H. Production of Phenolic Compounds from Rice Bran Biomass under Subcritical Water Conditions. *Chem. Eng. J.* **2010,** *160* (1), 259–266.

Pronyk, C.; Mazza, G. Design and Scale-Up of Pressurized Fluid Extractors for Food and Bioproducts. *J. Food Eng.* **2009,** *95* (2), 215–226.

Ribeiro, S.; Barbosa, L.; Queiroz, J.; Knödler, M.; Schieber, A. Phenolic Compounds and Antioxidant Capacity of Brazilian Mango (*Mangifera indica* L.) Varieties. *Food Chem.* **2008,** *110* (3), 620–626.

Rodriguez-Garcia, I.; Silva-Espinoza, B.; Ortega-Ramirez, L.; Leyva, J.; Siddiqui, M. W.; Cruz-Valenzuela, M.; Gonzalez-Aguilar, G.; Ayala-Zavala, J. Oregano Essential Oil as an Antimicrobial and Antioxidant Additive in Food Products. *Crit. Rev. Food Sci. Nutr.* **2016,** *56* (10), 1717–1727.

Rozzi, N.; Singh, R.; Vierling, R.; Watkins, B. Supercritical Fluid Extraction of Lycopene from Tomato Processing Byproducts. *J. Agric. Food Chem.* **2002,** *50* (9), 2638–2643.

Starmans, D. A.; Nijhuis, H. H. Extraction of Secondary Metabolites from Plant Material: A Review. *Trends Food Sci. Technol.* **1996,** *7* (6), 191–197.

Sun, Y.; Liu, D.; Chen, J.; Ye, X.; Yu, D. Effects of Different Factors of Ultrasound Treatment on the Extraction Yield of the All-Trans-β-Carotene from Citrus Peels. *Ultrasonics Sonochem.* **2011,** *18* (1), 243–249.

Toepfl, S.; Mathys, A.; Heinz, V.; Knorr, D. Review: Potential of High Hydrostatic Pressure and Pulsed Electric Fields for Energy Efficient and Environmentally Friendly Food Processing. *Food Rev. Int.* **2006,** *22* (4), 405–423.

Tsubaki, S.; Sakamoto, M.; Azuma, J.-I. Microwave-Assisted Extraction of Phenolic Compounds from Tea Residues under Autohydrolytic Conditions. *Food Chem.* **2010,** *123* (4), 1255–1258.

Vega-Vega, V.; Silva-Espinoza, B. A.; Cruz-Valenzuela, M. R.; Bernal-Mercado, A. T.; Gonzalez-Aguilar, G. A.; Ruiz-Cruz, S.; Moctezuma, E.; Siddiqui, M. W.; Ayala-Zavala, J. F. Antimicrobial and Antioxidant Properties of Byproduct Extracts of Mango Fruit. *J. Appl. Bot. Food Quality* **2013,** *86* (1). DOI:10.5073/JABFQ.2013.086.028.

Vilkhu, K.; Mawson, R.; Simons, L.; Bates, D. Applications and Opportunities for Ultrasound Assisted Extraction in the Food Industry—A Review. *Innov. Food Sci. Emerg. Technol.* **2008,** *9* (2), 161–169.

Vinatoru, M. An Overview of the Ultrasonically Assisted Extraction of Bioactive Principles from Herbs. *Ultrasonics Sonochem.* **2001,** *8* (3), 303–313.

Virot, M.; Tomao, V.; Le Bourvellec, C.; Renard, C. M.; Chemat, F. Towards the Industrial Production of Antioxidants from Food Processing By-products with Ultrasound-Assisted Extraction. *Ultrasonics Sonochem.* **2010,** *17* (6), 1066–1074.

Wang, L.; Weller, C. L. Recent Advances in Extraction of Nutraceuticals from Plants. *Trends Food Sci. Technol.* **2006,** *17* (6), 300–312.

Wijngaard, H.; Hossain, M. B.; Rai, D. K.; Brunton, N. Techniques to Extract Bioactive Compounds from Food By-products of Plant Origin. *Food Res. Int.* **2012,** *46* (2), 505–513.

Wijngaard, H. H.; Brunton, N. The Optimisation of Solid–Liquid Extraction of Antioxidants from Apple Pomace by Response Surface Methodology. *J. Food Eng.* **2010,** *96* (1), 134–140.

Yepez, B.; Espinosa, M.; López, S.; Bolanos, G. Producing Antioxidant Fractions from Herbaceous Matrices by Supercritical Fluid Extraction. *Fluid Phase Equilibr.* **2002,** *194*, 879–884.

Yi, C.; Shi, J.; Xue, S. J.; Jiang, Y.; Li, D. Effects of Supercritical Fluid Extraction Parameters on Lycopene Yield and Antioxidant Activity. *Food Chem.* **2009,** *113* (4), 1088–1094.

Yilmaz, E. E.; Özvural, E. B.; Vural, H. Extraction and Identification of Proanthocyanidins from Grape Seed (*Vitis vinifera*) Using Supercritical Carbon Dioxide. *J. Supercrit. Fluids* **2011,** *55* (3), 924–928.

Zarena, A.; Sankar, K. U. Supercritical Carbon Dioxide Extraction of Xanthones with Antioxidant Activity from *Garcinia mangostana*: Characterization by HPLC/LC-ESI-MS. *J. Supercrit. Fluids* **2009,** *49* (3), 330–337.

CHAPTER 4

PLANT TISSUES AS A SOURCE OF NUTRACEUTICAL COMPOUNDS: FRUIT SEEDS, LEAVES, FLOWERS, AND STEMS

G. R. VELDERRAIN-RODRIGUEZ[1], R. PACHECO-ORDAZ[1],
M. G. GOÑI[2], L. SIQUEIRA-OLIVEIRA[3], A. WALL-MEDRANO[4],
J. F. AYALA-ZAVALA[1], and G. A. GONZÁLEZ-AGUILAR[1*]

[1]Centro de Investigación en Alimentación y Desarrollo, AC
(CIAD, AC), Carretera a la Victoria Km 0.6, La Victoria CP 83000,
Hermosillo, Sonora, México

[2]Grupo de Investigación en Ingeniería en Alimentos, Facultad de
Ingeniería, Universidad Nacional de Mar del Plata, Juan B,
Justo 4302, 7600 Mar del Plata, Buenos Aires, Argentina

[3]Department of Biochemistry and Molecular Biology, Federal
University of Ceará, Av. Mr. Hull 2297 Bl. 907, Campus do Pici,
60455-760 Fortaleza, CE, Brazil

[4]Instituto de Ciencias Biomédicas, Universidad Autónoma de Ciudad
Juárez, Chihuahua, México

*Corresponding author. E-mail: gustavo@ciad.mx

CONTENTS

ABSTRACT

Plants have a huge role in human health and nutrition since ancient time, when man started to build sedentary societies. Man started to domesticate plants to use them as food, and eventually realized that some plant tissues had beneficial properties against diverse illness and problems they faced at the time. Nowadays, scientific research had exhaustively studied thousands of different plant tissues, from fruit seeds to flowers and stems. Scientific data now have identified that most of the chemical compounds in plants, also known as phytochemicals, involved in the color and texture of diverse plant tissues are responsible for many of the benefits that its consumption may bring to humans. In that sense, the diverse colors, shapes, and textures of plant tissues may be key points to consider at the time of looking for the right phytochemical to use in food products. The growing demand on economic and healthy food products leads the food industry to the exploitation and development of novel functional foods and nutraceutical products. In order that as the content of phytochemicals such as polyphenols, dietary fiber, carotenoids, free fatty acids, and vitamins in plant tissues may vary from one species to another, it is essential to analyze the common origin of each phytochemical compound from one plant tissue to another. Hence, this chapter summarizes the most common phytochemicals considered as nutraceutical compounds, its related health benefits, and the diverse plant tissues where we could find a specific type of phytochemical.

4.1 INTRODUCTION

Scientific research has confirmed and explained through the pass of time the huge role of plant tissues to human health. Even when ancient civilizations already had found cures and relief in plants, nowadays, science has established and identified the main chemical compounds (phytochemicals) responsible for its benefits within to manipulate or even extract these compounds (Brekhman, 2013). With the upswing of the functional foods, and the growing demand that are getting these days, food industry must be at the edge of food science limits, trying to surpass the frontier of conventional foods to develop new food products able to relief illness or contribute to enhance human health as possible. Ever since man realized that plant tissues contain beneficial compounds to human health, it started looking for ways to extract and isolate these compounds (Wang and Weller, 2006). Nowadays, science has optimized some of the phytochemical extraction methods, and the novel techniques includes ultrasound-assisted extraction (UAE), micro-wave-assisted extraction (MAE), supercritical fluid extraction (SFE), and accelerated solvent extraction (Gupta et al., 2012).

Likewise, scientific data reveal specific plant tissues where we can find abundance of specific types of phytochemicals, either as fiber in grains or polyphenols in flowers and fruits. The plant tissue where a phytochemical or nutraceutical compound is embedded, especially the components comprised over the different plant tissues, is first barrier to consider before its extraction and isolation. The chemistry nature of the desired phytochemical compound is also essential prior to the selection of the most adequate method for its extraction. In that sense, the current chapter describes the most common phytochemical compounds used as nutraceuticals, as well as their health-related benefits, their occurrence in plant tissues, and examples of phyto-chemical extraction methods.

4.2 NUTRACEUTICALS POTENTIAL IN HEALTH PROMOTION

Belief in the medicinal power of foods is not a recent, it has been widely accepted for years; almost 2500 years ago, Hippocrates said: "let food be thy medicine and medicine be thy food." Herbal medicine, along with all other plant components, has been used for healing since the beginning of human civilization; however, with modern medicine, most herbal products were left unused and many of them were forgotten by western or traditional medicine. Nowadays, natural medicine is steadily gaining interest all over the world again. Moreover, consumers are aware of the health benefits of selected foods and their components. However, if the consumers recognize the benefits of plant components for their health, it is essential that they received factual and reliable information to make informed decisions about the merits or risks associated with changing dietary habits.

In certain countries, functional foods and nutraceuticals are used inter-changeably; however, in all cases, the main focus is on improving health and reducing disease risk through, mainly, prevention (Shahidi, 2009). The term "Nutraceuticals" is a hybrid word with an intended double implication of "nutrition" and "pharmaceutical" (Chaturvedi et al., 2011; Das et al., 2012). It commonly refers to a nontoxic extract supplement that has scientifically proven health benefits for both disease treatment and prevention. It should not be mistaken with the term functional foods, which refers to foods with relevant effects on well-being and health or result in a reduction in disease (Dillard and German, 2000). Nutraceuticals has now been amalgamated in a new category under natural health products that promote health, that not only include nutraceuticals but also encompass herbal and other natural products.

The short-term goal of functional foods, nutraceuticals, and dietary supplements is to improve the quality of life and enhance health status, while its long-term goal is to increase lifespan while maintaining health. Due to the high and increasing incidents of cancer and heart disease in industrialized countries, governments have been making concerted efforts to raise public awareness about the advantages of eating a healthy diet. Indeed, numerous epidemiological studies have already documented an inverse association between fruit and vegetable consumption and chronic diseases. Hence, in some countries as a measure taken by the government, public programs to promote healthy eating are trying to increase people awareness in the diverse benefits of healthy eating, and in some cases, even promoting the intake of food with special purpose (functional foods).

However, the current lifestyle of the modern man, difficult the administration of their own time, which in most cases results in reducing the time spent for eating or food preparation. Therefore, as food industry realized of the economic impact of these modern lifestyles, the production of ready-to-eat foods comes to solve this problem, and likewise, the development of nutraceutical products seems like an attractive choice to most consumers. Nutraceuticals covers most of the therapeutics areas such as anti-arthritic, cold and cough, sleeping disorders, digestion and prevention of certain cancers, osteoporosis, blood pressure, cholesterol control, pain killers, depression, and diabetes (Chaturvedi et al., 2011; Das et al., 2012; Delzenne et al., 2013). A new paradigm has been presented by researchers in the field, with more emphasis in the nutritional aspects of the diet; and in agreement with the new lifestyle adopted today, which has changed the basic food habits (Das et al., 2012). Because prevention is a more effective strategy than treatment for chronic diseases, a constant supply of phytochemical-containing products with desirable health benefits beyond basic nutrition is essential to furnish the defensive mechanism to reduce the risk of chronic diseases in humans (Chu et al., 2002). The importance of this approach on health-care cost is enormous (Shahidi, 2009).

However, over the years, people's diet habits are based on a high consumption of fast food and processed meals (ready-to-eat), leading to a number of diseases caused due to improper nutrition. The prevalence of obesity and type 2 diabetes is rapidly increasing, becoming a major challenge for health-care professionals (Devalaraja et al., 2011). Obesity is now recognized as a global issue. It is a major problem affluent societies are facing as well as the rest of the world, since reduced activity and lack of exercise may lead to obesity with the consequence of a host of diseases and the so-called metabolic syndrome (Shahidi, 2009). Likewise, heart disease continues to be a primary cause of death in most of the developing countries worldwide, followed by cancer, osteoporosis, arthritis, and many others (Das et al., 2012). It is well established that a correction of lifestyle, such as the intake of a healthy and low energy diet along with increased physical activity is the most effective preventive therapy to ameliorate the prevalence of obesity and diabetes in society. It was suggested that in addition to standard prescribed therapies, the addition of dietary supplements to the diet could be used to manage obesity and diabetes (Devalaraja et al., 2011).

There has been a boom in the sale of nutraceuticals due to the adverse effects of traditional pharmaceuticals, increase tendency of patients to self-medication, and the growing aging of the population associated to increase in certain chronic diseases (Chaturvedi et al., 2011). Pharmaceutical and nutritional industries are conscious of the monetary success taking advantage of the more health-seeking consumers. Phytochemicals are likely to form the basis of nutraceuticals and the uses of plants as a source of them represents an enormous opportunity for growth and expansion, see Table 4.1. Leaves, stems, seeds, flowers, and fruits of plants have numerous natural compounds that have been associated to reducing the risk of coronary heart disease, diabetes, tumor incidence, cancer, blood pressure, reduces the rate of cholesterol and fat absorption, delaying gastrointestinal emptying, and providing gastrointestinal health. Recent studies have demonstrated that the phytochemicals in plants are primarily located in the outermost layers and/or skin and thus, in some cases, their extraction would lead to products that are less beneficial to health (Shahidi, 2009).

Food components used as nutraceuticals are often dietary fiber (DF), prebiotics, probiotics, polyunsaturated fatty acids (PUFAs), antioxidants, and other different types of herbal/natural foods (Das et al., 2012). There are several nutraceuticals available in the market from the food or the pharmaceutical industry, the herbal and dietary supplement market, or from the newly pharmaceutical/agribusiness/nutrition associations. Several commercial nutraceuticals are available, some as isolated nutrients, herbal-based products, genetically engineered or "designer" foods and processed products such as cereals, soups, and beverage (Fig. 4.1). Hence, it is essential to know the type of phytochemical that we can extract from the different plant tissue, as well as the health benefits that we can get from their diverse chemical structures. Likewise, nutraceutical products must be exhaustively studied, with a complete characterization of the phytochemicals comprised in the product, as well as a series of research about the effects of the phytochemical

TABLE 4.1 Phytochemical Compounds of Common Use as Nutraceuticals.

Plant tissue	Name	Terpenoids	Phenolic compounds	N- and S-based compounds	Carbohydrates	Fatty acids	Minerals	References
Fruits	*Malus domestica*	NR	Epicatechin Catechin Clorogenic acid Quercetin	NR	Starch Rafinose Sucrose Xylose	NR	NR	Krawitzky et al. (2014), Feng et al. (2014a)
	Hylocereus polyrhizus	NR	NR	Beta-cyarins	Pectin	NR	NR	Gengatharan et al. (2015), Muhammad et al. (2014)
	Punica granatum	NR	Punicalagins Rutin quercetin, gallic acid Ellagic acid	NR	NR	NR	NR	Middha et al. (2013)
	Ribes nigrum and *R. rubrum*	NR	Cyanidin Delphinidin *p*-Coumaroylquinic acid Myricetin, quercetin Kaempferol	NR	NR	NR	NR	Aneta et al. (2013)
	Mangifera indica L.	Lupeol β-Carotene, *cis-* and *trans-* violaxanthin	Maguiferin Quercetin Gallic acid Vanillic acid Clorogenic acid Protocatechuic acid Ellagic acid	NR	NR	NR	NR	Dorta et al. (2014), Low et al. (2015), Palafox-Carlos et al. (2012), Ruiz-Montañez et al. (2014)

TABLE 4.1 *(Continued)*

Plant tissue	Name	Terpenoids	Phenolic compounds	N- and S-based compounds	Carbohydrates	Fatty acids	Minerals	References
Seeds	*Hylocereus* spp.	NR	NR	NR	NR	Palmitic Oleic Linoleic	NR	Liaotrakoon et al. (2013)
	Phoenix dactylifera L.	β-Carotene Leuteina Echinenone	Coumaric, sinapic and dihydrocinamic acids, apigenin, naringenin and catechin	NR	NR	NR	NR	Habib et al. (2013), Messaoudi et al. (2013)
	Paulllinia cupana	NR	Procyanidin A, B, C Epicatechin Catechin	Caffeine and theobromine	NR	Linoleic Linolenic Oleic Palmitic Stearic	Ca P Mg	Hamerski et al. (2013), Schimpl et al. (2014)
	Vitis vinifera L.	NR	Gallic acid Catechin Epicatechin Quercetin *Trans*-resveratrol	NR	NR	Linoleic Oleic Palmitic Stearic	NR	Dang et al. (2014), Fernandes et al. (2013)
	Psidium guajava L.	NR	NR	NR	Starch Pectin Fructose	Lauric Myristic Palmitic Oleic Stearic Linoleic	Ca Mg Fe Zn Na K P Mn	Uchôa-Thomaz et al. (2014)

TABLE 4.1 (Continued)

Plant tissue	Name	Terpenoids	Phenolic compounds	N- and S-based compounds	Carbohydrates	Fatty acids	Minerals	References
Leaves	Vitis vinifera L.	NR	Catechin, epicatechin, quercetin, kampferol, cis- and trans-resveratrol, apigenin	NR	NR	NR	NR	Katalinic et al. (2013)
	Paullinia cupana	NR	NR	Theobromine Theophylline Caffeine	NR	NR	NR	Schimpl et al. (2014)
	Olea europaea	NR	Hydroxytyroso Oleuropein Luteolin apigenin	NR	Arabinose Galactose Glucose Mannose xylose	NR	NR	Romero-Garcia et al. (2016)
	Diospyros kaki L.	NR	Gallic acid Hydroxytyroso Hydroxybenzoic acid Vanillic acid Caffeic acid p-Coumaric acid Sinapic acid Catechin Gallocatechin Naringenin Apigein Kaempferol Quercetin	NR	NR	NR	NR	Martinez-Las Heras et al. (2016)

TABLE 4.1 (Continued)

Plant tissue	Name	Terpenoids	Phenolic compounds	N- and S-based compounds	Carbohydrates	Fatty acids	Minerals	References
	Rosmarinus officinalis	Rosmanol Carnosol Carnosic acid Rosmadial Epirosmanol	Siringic acid Gallocatechin Nepetrin Hesperidin Rosmarinic acid	NR	NR	NR	NR	Borrás-Linares et al. (2014)
Flowers	*Calluna vulgaris*	α-Amyrin Lupeol Uvaol Ursolic acid Friedelin Taraxasterol	NR	NR	NR	NR	NR	Szakiel et al. (2013)
	Musa × paradisiaca	NR	Gallic acid Quinic acid Shikimic acid	NR	Glycerol D-Mannitol D-Sorbitol Sucrose Xylitol	Lauric Linoleic Myristic Oleic Palmitic Stearic	NR	Acharya et al. (2016)
	Nymphaea spp.	NR	Quercetin 3-O-rhamnoside Quercetin 7-O-galactoside Naringenin 7-O-galactoside Kaempferol 3-O-glucosyl	NR	NR	NR	NR	Yin et al. (2015)

TABLE 4.1 (Continued)

Plant tissue	Name	Terpenoids	Phenolic compounds	N- and S-based compounds	Carbohydrates	Fatty acids	Minerals	References
Stems	*Paullinia cupana*	NR	NR	Theobromine and theophylline	NR	NR	NR	Schimpl et al. (2014)
	Rhus verniciflua	NR	Gallic acid, Fustin, Fisetin, Sulfuretin, Quercetin	NR	NR	NR	NR	Kim et al. (2013)
	Panax ginseng	Saponins	NR	NR	NR	NR	NR	Ma et al. (2015)
	Vitis vinifera L.	β-Myrcene, β-Ocimene, α-Pinene, Limonene	Gallic acid, Catechin, Epicatechin, Procyanidin b3 and b2, Epicatechin gallate, *Trans*-resveratrol, Quercetin, Kaempferol, Caffeic acid, Syringic acid, Rutin	NR	NR	NR	NR	Apostolou et al. (2013), Matarese et al. (2014)
	Pinus ponderosa	Limonene, α-Pinene, Myrcene, Camphene, Linalool	NR	NR	NR	NR	NR	Keefover-Ring et al. (2016)

NR, Not reported.

intake to human health. In that sense, once characterized the type and the amount of phytochemicals added to a nutraceutical product, the in vitro and in vivo evaluation of isolated and combined compounds must assure that the nutraceutical product is safe for human consumption.

Nutraceuticals: product obtained from foods that have extra health benefits added to their nutritional value.	
Dietary supplements	*Functional Food Ingredients*
Vitamins *Minerals* *Amino acids* *Herbal extracts*	*Vitamins* *Fats* *Carbohydrates* *Amino acids* *Fiber*
Examples	Fortified cereals (containing minerals, vitamins or extra fiber) Supplemented beverages (containing vitamins, antioxidants, soluble fiber and minerals) Energy drinks (with herbal extracts) Reducing cholesterol beverages (containing fiber, antioxidants)

FIGURE 4.1   Nutraceuticals, their classification, and commercial presentation.

4.3   PHYTOCHEMICALS AND ITS ROLE IN HUMAN HEALTH

Plants are an important source of phytochemicals (biologically active substances) that present antioxidant, anticarcinogenic, and antimutagenic properties (Ajila et al., 2008; Cavalcanti et al., 2011; Chiva-Blanch and Visioli, 2012). It has been suggested that including fruits and vegetables in the diet has a beneficial effect on health due to the reduction of incidence of heart disease, cancer, and some neurodegenerative disorders (Cavalcanti et al., 2011). Phytochemicals quantity and quality can vary significantly according to different intrinsic and extrinsic factors, such as plant genetics and cultivar, soil composition and growing conditions, maturity state, and part of the plant, among others (Faller and Fialho, 2010). For example, a stressful environment for plants can lead to the induction of secondary metabolism and the subsequent increment in the production of phytochemicals, especially those modulated by enzymatic activity (Reyes et al., 2007). Moreover, those phytochemicals can be used immediately to counteract the cause of the stress (as a free-radical scavenging agent against an oxidative stressful environment) or relocated elsewhere in the plant for later use.

Oxidative stress results from either a decrease of natural cell antioxidant capacity or an increased amount of reactive oxygen species (ROS) in organisms. If the balance between oxidants and antioxidants in the body is shifted by the overproduction of free radicals, it will lead to oxidative stress and DNA damage. When left unrepaired, it can cause base mutation, single- and double-strand breaks, DNA cross-linking, and chromosomal breakage and rearrangement (Contreras-Calderón et al., 2011; Chu et al., 2002; Chun et al., 2005). The development of chronic diseases, like metabolic syndrome, rheumatoide arthritis, cardiovascular diseases, cancer, arteriosclerosis, hypertension, neurodegenerative diseases, and aging process and type II diabetes, among others, involves the production of high quantities of free radicals which leads to oxidative stress in the tissues (Cavalcanti et al., 2011; Contreras-Calderón et al., 2011; Chu et al., 2002; Moure et al., 2001). An increment in the consumption of fruits and vegetables has been associated with lower risk of chronic diseases, because in addition to its vitamin and mineral composition, it will also contain other compounds with protective effects, in particular antioxidants (Contreras-Calderón et al., 2011; Chu et al., 2002; Chun et al., 2005; Faller and Fialho, 2010).

The protection that phytochemicals provided against these illnesses and syndromes has been attributed to the presence of several antioxidant compounds, like vitamin C, vitamin E, tocopherols, carotenes, and polyphenolic compounds (Chu et al., 2002). In vivo evidence of the formation of oxidized low-density lipoprotein includes the presence of excessive ROS production in the bloodstream (where they circulate as negatively charged). Moreover, their circulating levels have been positively correlated with the progression of carotid lesions and many epidemiological studies have correlated a high dietary intake of antioxidants (e.g., tocopherols, carotenoids, flavonoids, and polyphenols) with a lower incidence of cardiovascular diseases (Chiva-Blanch and Visioli, 2012; Chun et al., 2005).

Antioxidants have several roles on lipid oxidation, such as binding metal ions, scavenging radicals, and decomposing peroxides. Often, more than one mechanism is involved, which in some cases may lead to synergism or anta-gonic effects. Even if this is the most accepted theory on the action mecha-nism of fruits and vegetables in health, it is necessary take into consideration that including more vegetables in the diet can also be healthy due to the reduction in the intake of other potentially noxious foods, such as those rich in animal protein and saturated fats (Chiva-Blanch and Visioli, 2012). Plants also contain polyunsaturated fats, fiber, and vitamins that can interact with antioxidants to increases its health benefits, resulting in an overall effect which is difficult to ascertain to individual components. It is also suggested that through overlapping or complementary effects, the complex mixture of phytochemicals in fruits and vegetables provides a better protective effect on health than single phytochemicals (Contreras-Calderón et al., 2011; Chu et al., 2002).

4.3.1 NITROGEN AND SULFUR-BASED COMPOUNDS

Among nitrogen-containing compounds, glucosinolates and S-methylcystine sulfoxide are the major sulfur compounds present in vegetables (Manchali et al., 2012). Glucosinolates (thioglucoside-N-hydroxysulphates) constitute a homogeneous class of naturally occurring thiosaccharidic compounds mainly found in the botanical order *Brassicales* (Prakash and Gupta, 2012). More than 120 glucosinolates and its precursors have been identified in plants, which are known for their fungicidal, bactericidal, nematocidal, and allelopathic properties. Many recent studies are focusing in the iso-thiocya-nates, one of the hydrolyzed products of glucosinolates, due to their role in the organism and its anticarcinogenic activity (Prakash and Gupta, 2012).

4.3.2 ANTIOXIDANT VITAMINS

Vitamins like vitamin C, vitamin E, and carotenoids are collectively known as antioxidant vitamins, due to their biological activity associated to the prevention of oxidative stress, which leads to several degenerative diseases and most of their protective action is related to their free-radical scavenging activity (Das et al., 2012). These vitamins are present in fresh fruits and vege-tables. The human body cannot synthesize lipid- and water-soluble vitamins, such as vitamins E and C and must be derived from food (Chiva-Blanch and Visioli, 2012). The carotenoids are also present in fruits and vegetables; they are usually responsible for the color of the tissues (yellow, red, and orange in particular). Several carotenes have vitamin A activity; however, these molecules are now being studied intensively from a different point of view as phytochemical compounds with anticancer properties (Moure et al., 2001). Vitamin C or ascorbic acid can exert its antioxidant action by donating hydrogen atoms to lipid radicals, quenching singlet oxygen radicals and removing molecular oxygen from the medium.

Vitamin E is an essential, fat soluble nutrient that functions as an antioxidant in the human body, and foods and supplements must provide it (Sen et al., 2006). At present, vitamin E represents a generic term for all chemical compounds having the biological activity of RRR-α-tocopherol, which includes α-, β-, γ-, and δ-tocopherol; and α-, β-, γ-, and δ-tocotrienol. The tocopherols are saturated forms of vitamin E, whereas the tocotrienols are unsaturated and possess an isoprenoid side chain (Aggarwal et al., 2010). Some evidence suggests that human tissues can convert tocotrienols to tocopherols. Tocotrienols are the primary form of vitamin E in the seed endosperm of most monocots, including agronomically important cereal grains such as wheat, rice, and barley; it is especially found in palm oil (up to 800 mg/kg), mainly consisting of γ-tocotrienol and α-tocotrienol (Sen et al., 2010). Tocopherols occur ubiquitously in plant tissues and are the exclusive form of vitamin E in leaves of plants and seeds of most plants. In contrast, tocotrienols are considerably less widespread in the plant kingdom, being only found in significant amounts in nonphotosynthetic tissues and organs, like in seeds and fruits.

In the human body, vitamin E may exert functions beyond its antioxidant property (Sen et al., 2010). Deficiency of this vitamin is known to cause severe degenerative diseases such as ataxia, Duchenne muscular dystrophy-like muscle degeneration, and infertility (Aggarwal et al., 2010). Several studies showed that vitamin E has a neuroprotective effect, reduces Apo B levels in hypercholesterolemic subjects, modulates the normal growth of the mammary glands, has anticancer, anti-inflamatory, antihypertensile, anti-aging, anti-angiogenic, and antioxidant effect, increases immune function, and reduces serum triglycerides (Sen et al., 2006). Moreover, vitamin E and selenium has a synergistic role against lipid peroxidation (Das et al., 2012).

Carotenoids are widely distributed in plants, especially in their leaves, flowers, and fruits and are responsible for the yellow, orange, and red color of their tissues. Carotenoids are intracellular products and are usually located in the membranes of chloroplasts, mitochondria, or endoplasmatic reticulum due to their hydrophobic nature. Their function in the plant is strongly related to photosynthesis and regulation of the membranes fluidity, and they play an essential role in the protection of the tissues against excessive light and photooxidative stress (Chun et al., 2005; Jaswir et al., 2011). Carotenoids are structurally classified as carotenes (like α-carotene, β-carotene, and lycopene) and xanthophylls (like lutein, zeaxanthin, fucoxanthin, and astaxanthin).

They are essential nutrients; animals cannot produce carotenoids and must obtain them from the ingestion of vegetables. There are more than 700 known carotenoids from plants, lycopene, α-carotene, β-carotene, lutein, and zeaxanthin are the most abundant in human plasma (Aizawa and Inakuma, 2007). One of the main characteristics of carotenoids that made them important for the human diet is their provitamin A activity (Yeum and Russell, 2002). However, only three of them have a significant provitamin A activity in humans: α-carotene, β-carotene, and β-cryptoxanthin which are converted to vitamin A or retinol in the body (Jaswir et al., 2011). In turn, vitamin A is responsible of preventing eye diseases, such as night blindness. Carotenoids in general, can also act as antioxidants and reduce oxidative stress by scavenging ROS generated during photooxidative stress, via electron transfer, hydrogen transfer, or forming resonance-stabilized carbon-centered radicals (against lipid peroxyl radicals for example). Several studies have suggested the positive effect of carotenoids in cancer prevention, risk reduction of cardiovascular diseases, age-related macular degeneration and cataract formation, enhancement of the immune system, maintaining bone health, and preventing osteoporosis (Chang et al., 2010; Das et al., 2012; Jaswir et al., 2011).

4.3.3 DIETARY FIBER

Increasing attention has been given to the beneficial effects on humans of the consumption of the nondigestible components of vegetable foods, commonly known as DF. The term "dietary fiber" is used to refer to the edible parts of plants that are resistant to hydrolysis by digestive enzymes in humans (Nawirska and Kwaśniewska, 2005). It is a mixture of carbohydrate polymers, composed of resistant starch, cellulose, hemicellulose, pectin, and inulin, also associated with lignin, gums, waxes, cutin, polyphenols, resistant protein, and alginates, among others (Elleuch et al., 2011). It is widely known that DF obtained from different sources behaves differently during their transit through the gastrointestinal tract, depending on their chemical composition and physicochemical characteristics (Figuerola et al., 2005).

Although these components are not as active biologically as polyphenolic compounds or vitamins, they have a significant effect in the metabolism (Contreras-Calderón et al., 2011; Chun et al., 2005). Published reports indicate numerous health benefits associated with an increased intake of DF, including reduced risk of coronary heart disease, diabetes, obesity, and some forms of cancer (Elleuch et al., 2011; González-Centeno et al., 2010). DF exerts a buffering effect in the gastrointestinal tract, increases the fecal bulk, and stimulates the intestinal activity, stimulates colonic fermentation, improvement of gastrointestinal function, regulates postprandial insulin response, and reduces total and low-density cholesterol (Elleuch et al., 2011; Mildner-Szkudlarz et al., 2013).

DF also has a positive effect on the desirable microbiota of the intestine by creating a favorable environment for their growth. DF in general but hemicellulose and pectins in particular, have the ability to sorb harmful substances or to bind mineral components and heavy metals. Polyphenolic compounds are often associated with the food matrix and, as consequence, only a small part of them can be absorbed in the small intestine, reducing their bioavailability (Hervert-Hernández et al., 2011). These polyphenolic compounds can exert their biological activity once they reach the colon by interaction with the colonic microbiota as fermentable substrates (Saura-Calixto et al., 2007).

DF is a complex group of phytochemicals with different chemical and physical properties, often related to the source of the mix or the extraction methodology and are usually classified as water soluble (SDF) or water insoluble (IDF). The average daily requirement of total DF is 25 g/day for women younger than 50, 21 g/day for women older than 50; 38 g/day for men younger than 50, and 30 g/day for men older than 50 (Elleuch et al., 2011). In addition, most nutritionists suggest that 20–30% of daily DF intake should come from SDF, with a suggested proportion of SDF to IDF of 1:3 or 1:4. Maintaining that proportion is helpful in reducing the risk of heart diseases, level of cholesterol, triglycerides, and glucose in blood (Mildner-Szkudlarz et al., 2013). Moreover, the SDF/IDF ratio is also important for maintaining its functional properties (Figuerola et al., 2005). Therefore, the content of DF in food products and nutraceuticals can also have desirable effects on the texture for its gelling, emulsifying, and stabilizing properties (Abdul-Hamid and Luan, 2000).

DF derived from fruits and vegetables have a considerably higher proportion of SDF than those obtained from cereals (González-Centeno et al., 2010). The ideal DF extract should be as concentrated as possible, be bland in taste, color, and odor; have a balanced composition and adequate amount of associated bioactive compounds; have a good shelf life; be compatible with food processing; and have the expected physiological effects (Figuerola et al., 2005). DF concentrates can be used in various applications in the food industry with excellent results, such as nutraceuticals.

Due to all these characteristics, DF has an important role in both prevention and treatment of obesity, atherosclerosis, coronary heart diseases, large intestine cancer, and diabetes (Nawirska and Kwaśniewska, 2005). Dietary supplementation with DF could be effective in protecting against oxidative damage in the tissues due to lipid peroxidation by improving the antioxidant defense systems (Mildner-Szkudlarz et al., 2013). An increase in the level of DF in the daily diet has been recommended and because of this, it would be interesting to investigate different foods that can supply DF, like fruits and different parts of the plants. Fiber extraction and concentration could help to overcome the fiber deficit in the diet. Supplementation with DF can result in fitness-promoting foods, low in calories, cholesterol, and fat.

4.3.4 POLYUNSATURATED FATTY ACIDS

PUFAs are very important in the human diet for several body's function and need to be included in the diet since mammals are unable to synthetize them. There are two major classes of PUFAs: omega-3 and omega-6. Among omega-3, α-linolenic acid (ALA), eicosapentanoic acid (EPA), docosahexanoic acid (DHA) are the most abundant and therefore the most popular ones for the nutraceuticals industry. While EPA and DHA are found mainly in fatty fishes, ALA is often obtained from soybean, canola, walnuts, or other types of seeds. For the other major class, omega-6 the most common ones are linoleic acid (LA), γ-linolenic acid and arachidonic acid. LA is often obtained from vegetable oils like corn, safflower, soybean, and sunflower, among others. FDA recommends a maximum of 3 g/day intake of EPA and DHA omega-3 fatty acids, with no more than 2 g/day from a dietary supplement (Das et al., 2012). Several clinical studies have focused in the health benefits of PUFAs, suggesting that their main effects are in the prevention of cardiovascular diseases. Omega-3-oils in particular, have also shown positive effects in asthma, depression, diabetes, and also to be beneficial at various stages of life. It has been stated that a minimum intake of selenium per day or u3 fatty acids from marine oils would reduce the burden on health care tremendously as nearly one-third of all diseases are life-style related (Shahidi, 2009).

4.3.5 PREBIOTICS

Prebiotics are food components that beneficially affect the host by selectively altering the composition or metabolism of the gut microbiota (Das et al., 2012). Most of them are indigestible polysaccharides, in particular fructose-based oligosaccharides. These components can be naturally present in food or added. Vegetables like chicory roots, banana, tomato, alliums are rich in fructo-oligosaccharides (Das et al., 2012). Other oligosaccharides often considered as prebiotic are raffinose and stachyose, commonly found in beans and peas.

In recent years, there has been increased attention focused on the bacteria that colonize the gut, which in ideal conditions live in symbiosis with the host. Gut microbiota is complex system, with self-regulating mechanisms that can offer protection to the host by acting as a barrier against pathogens. However, these activities could be reduced during illness, stress, or aging, or by antibiotic treatment. It can also be diminished if disbiosis (change in the microbiota composition) occurs due to changes in the host diet (Macfarlane and Macfarlane, 2013). Experiments performed in a murine system colonized with the human gut microbiota reveals that changes in the diet composition (from high carbohydrates to western diet) allows a rapid switch of the microbial community. Those data suggest that the gut microbiota composition/activity associated with nutritional imbalance might contribute to obesity and related disorders (Delzenne et al., 2013). The prebiotic consumption generally promotes the *Lactobacillus* and *Bifidobacteria* growth in the gut, thus helping in metabolism. The health benefits of the prebiotics include improved lactose tolerance, antitumor properties, neutralization of toxins, and stimulation of intestinal immune system, reduction of constipation, blood lipids, and blood cholesterol levels.

4.3.6 PHENOLIC COMPOUNDS

One of the most studied phytochemicals are phenolic compounds (PC), also known as polyphenols, a group of secondary metabolites with diverse chemical structures and many functions. These compounds are usually produced during plant growth and development and/or as a response to various forms of environmental stress (Anastasiadi et al., 2010), via the shikimic acid pathway and/or phenylpropanoid metabolism (Anastasiadi et al., 2012). Phenolic compounds are one of the most abundant phytochemicals, with consumption of about a 1 g/day, which is 10 times higher than average vitamin C intake (Chun et al., 2005; Scalbert et al., 2005). PC are widely distributed, therefore they are present in fruits, vegetables, cereals, legumes, leaves, grains, seeds, flowers, and stems foods, being present not only in fruits and vegetables but also in foods (Cavalcanti et al., 2011; Faller and Fialho, 2010; Scalbert et al., 2005). Despite their wide distribution in plants, the health benefits of including PC in the diet have been taken into consideration in the mid-1990s. Before that, the most widely studied antioxidants were vitamins, carotenoids, and minerals. Currently, the contribution of PC to the prevention of cardiovascular diseases, cancers, osteoporosis, neurodegenerative diseases, and diabetes mellitus has been repeatedly proved (Anastasiadi et al., 2010; Cavalcanti et al., 2011; Scalbert et al., 2005).

Several studies on animals have shown that including PC in their diet reduces the incidence of cancer, cardiovascular diseases, neurodegenerative diseases, diabetes, and osteoporosis (Anastasiadi et al., 2010). However, most of these studies made on the prevention of diseases by PC that are carried out in vitro or with animal testing, often use doses much higher than those present on a normal diet (Scalbert et al., 2005). To elucidate the significance of polyphenols in human health, it is essential to know the amount of polyphenols consumed in the diet and their bioavailability (Saura-Calixto et al., 2007). Most of the studies advocating the use of antioxidants suggest that intake should be well above the general levels of consumption, shifting the focus from dietary consumption to pharmacological treatment or nutraceuticals production (Chiva-Blanch and Visioli, 2012).

Other major difficulty in the study of the effects of PC in human health is the large number of them found in plants and the differences in their chemical structures. It is also necessary to consider the differences in their bioaccessibility and bioavailability, mainly due to the interaction with other components in the matrix of the tissue plant and the digestion process (Saura-Calixto et al., 2007). Food going through the human gastrointestinal tract is digested in the stomach (strong acid environment), small intestine (mild basic environment), and then colon (neutral pH), all with different enzymes present (Chu et al., 2002). Phenolic compounds are present in vegetables in two forms: free or bound, with a different resistance to digestion. Bound PC may resist the stomach and small intestine conditions and reach the colon intact, where they are released and exert bioactivity. In most published research, only free phenolics are determined on the basis of the solvent-soluble extraction, as consequence total phenolic contents and their antioxidant activities are often underestimated by not including bound phenolics (Saura-Calixto et al., 2007). Moreover, their health benefit resides in their ability to reach the colon unaffected by the gastric and small intestine digestion.

While there are several studies showing increased plasma antioxidant capacity following the intake of polyphenol-rich food items, many investigators suggest that polyphenols' bioavailability is too low to produce a significant effect on the antioxidant metabolism (Chiva-Blanch and Visioli, 2012). As a consequence, research on phytochemicals in general has expanded and is now focusing on different bioactive compounds instead of only paying attention to antioxidants. Several products are presented as a result, to complement the increase on the diet of fruits and vegetables, including functional foods, designer foods, and nutraceuticals.

Synthetic antioxidants are being replaced for natural ones, not only based on its functionality (such as solubility both in oil and water solvents, emulsification properties) but also associated to its additional health benefits. Many of them are obtained from spices or herbs and have some limitations in its practical applications in spite of their high antioxidant capacity due to its sensorial impact, especially for their flavor (Moure et al., 2001). More research is needed to find new sources of natural antioxidants that can be freely used in the food, pharmacological and nutraceuticals industries with reduced sensorial impact. However, natural phytochemical substances also need safety testing; its application should be thoroughly confirmed before nutraceuticals are developed from them. Moreover, it should also be taken into account that in food-related systems, antioxidant activity means chain-breaking inhibition of lipid peroxidation, whereas in in vivo systems, free radicals can damage proteins, DNA, and other small molecules (Moure et al., 2001). As PC are one of the most studied nutraceutical compounds due to their high abundance in most of the plant tissues, and besides that these molecules due to their diverse chemical structures has been related to several health benefits, the following section describes the most common methods of PC extraction from plants.

4.4 TECHNIQUES FOR EXTRACTION OF PHENOLIC COMPOUNDS FROM PLANTS

One of the most important step for the analysis of PC derived from plant materials is extraction, which is influenced by the PC chemical nature (molecular structure, polarity, concentration, number of aromatic rings, and hydroxyl groups), the extraction method employed, sample particle size, time and condition of storage, as well as presence of interfering substances. According to the chemical nature, plant phenolics vary from simple to highly polymerized substances that may also be present as complexes with carbohydrates, protein, and other types of cell plant components. Therefore, phenolic extracts from plant materials are always a mixture of different classes of phenolics that include varying proportions of flavonoids, phenolic acid, tannins, lignins, and simple phenols.

The analysis of polyphenols can be realized through determination of total phenolic content or quantifying an individual phenolic or a specific class of phenolic using spectrophotometry, gas chromatography, high-performance liquid chromatography (HPLC), or capillary electrophoresis methods. However, PC must first be extracted from different plant tissues and, the ideal extraction method must provide high extraction rates and should be nondestructive and time saving (Rombaut et al., 2014). In addition, optimization and standardization of extraction parameters for these phytochemical are important to retain their antioxidant properties.

Several extraction techniques have previously been reported to extract PC from plant materials, generally, are carried out using the conventional method, employ solvents assisted by mechanical agitation, pressing, or heating system, although, more recently, modern methods have been used, including SFE, UAE, MAE, subcritical water extraction (SWE), and high hydrostatic pressure processing.

4.4.1 CONVENTIONAL

The most common technique to extract phenolics from plant materials involves the use of organic and inorganic solvents. Isolation and identification analysis of plants polyphenols are mostly dependent on the extraction solvent and technique used and, generally, solvent-type effect is related with polarity of the solvents and the solubility of target compounds in them, though solubility of PC is governed by the type of solvent, degree of polymerization of phenolics, as well as, interaction of this compounds with other food compounds and formation of insoluble complexes (Naczk and Shahidi, 2004; Turkmen et al., 2006).

Complete extraction of phenolic compounds is a crucial step which is specific to the food matrix. Several factors may contribute to the optimization of phenolics extraction including solvent-to-sample ratio, extraction time and temperature, number of extraction, solvent type, pH of the aqueous solvent, ratio of water in the solvent system, besides state, cultivar, and particle size of the material (González-Montelongo et al., 2010; Khoddami et al., 2013; Tuncel and Yılmaz, 2015). The most frequently used solvents for the extraction of phenolics in plant material are methanol, ethanol, acetone, water, ethyl acetate, propanol, and their combinations. However, the choice of extraction solvent will influence the yields of phenolics extracted, seeing that the extraction of PC from a sample is directly related to the compatibility of the compounds with the solvent system according to the "like dissolves like" principle (Tuncel and Yılmaz, 2015).

A high yield of phenolics can be extracted from sorghum leaf using water (Agbangnan et al., 2012), although the use of water as the only solvent yields to an extract with a high content of impurities (e.g., organic acids, sugars, soluble proteins) along with polar compounds which could interfere in the identification and quantification (Hijazi et al., 2013). From lychee flowers, to the highest extraction yield of antioxidant components (phenols, flavonoids, and condensed tannins), acetone was required (Liu et al., 2009). In another example, an investigation into the effect of different solvents on extraction of phenolics from *Limnophila aromatic* showed that 50% aqueous acetone extract was more efficient than the pure solvent which may be facilitated the extraction of all PC soluble in both water and organic solvent (Do et al., 2014). Study reports that the application of water combined with other organic solvents makes it a moderately polar medium ensuring the optimal conditions for extraction. Besides, using water in combination with alcohols leads to an increase in swelling of plant materials and the contact surface area between the plant matrix and the solvent finally improves the extraction yield (Hijazi et al., 2013).

In general, the different properties of the phenolic components of the plant materials could concern these differences. For extraction of the total phenolics from black currant leaves, aqueous acetone was found to be more effective than methanol and water (Tabart et al., 2007). According to the authors, acetone and methanol have distinct specificities in the extraction of PC and this is due to polarity of the solvent and solubility of phenolics in them. In contrast, no significant differences between ethanol and methanol were observed in terms of extraction efficiency ($p > 0.05$) in research aiming to optimize the extraction of PC in feijoa fruit (Tuncel and Yılmaz, 2015).

Two other important parameters that affect the extraction yield of PC, in addition to selecting of extraction solvent, are temperature and time of extraction. It has been reported that high temperatures improve the efficiency of extraction, due to the enhanced diffusion rate and solubility of phytochemicals in solvents, and thus reduce the extraction time to reach maximum polyphenol content recovery. Dorta et al. (2012) demonstrated that the optimum extraction of phytochemical (flavonoids, tannins, and proanthocyanins) from mango peel and seed was between 50°C and 70°C for 60 min. Investigation into the effects of the phenolics extraction conditions from *Morinda citrifolia* showed that the optimized condition was at 65°C for 80 min (Thoo et al., 2010). Extraction of phenolics from grape stems was most efficient at 84.2°C during 23.4 min for red variety and at 95°C during 23 min for white variety (Dominguez-Perles et al., 2014). Tuncel and Yılmaz (2015) demonstrated that increasing temperature from 25°C to 40°C significantly increased the amount of PC extracted from feijoa fruit. Increased temperature could also negatively affect the phenolic extraction since this condition may not be suitable for all kinds of PC leading to their degradation or loss by volatilization or reaction with other components of plant material. Thus, only samples with higher proportions of thermally stable polyphenols are more appropriate to extract under elevated temperature (Thoo et al., 2010).

The solvent-to-sample ratio and the number of replicate extractions performed for each sample also affect the recovery of phenolics (Khoddami et al., 2013). Extraction amount of total phenolics increased almost linearly by increasing solvent ratio, so the higher the solvent ratio higher is the total amount of solids obtained due to the mass transfer principles (Al-Farsi and Lee, 2008). These authors reported that a 60:1 ratio of solvent to sample in a two-stage procedure is sufficient to extract most phenolics from plant tissues. Tuncel and Yılmaz (2015) founded that a 60:1 ratio of solvent to sample was most efficient to extraction of PC in feijoa fruit. Particle size also can influence phenolic extraction from plant materials. Tabart et al. (2007) showed that the use of lyophilized material (black currant leaves and buds) allowed a better extraction of phenolics antioxidants due to a better grinding of the tissues and thus a reduced particle size in the simple and degradation of some phenolics and antioxidants in undried plant material.

Conventional extraction is usually performed using reflux, cold maceration, Soxhlet, and simple distillation techniques. These methods have been used for many decades. However, conventional extraction techniques use large quantities of toxic organic solvents, are labor intense, require long extraction times, possess low selectivity and/or low extraction yields, and can result in the exposure of extracts to excessive heat, light, and oxygen (Hossain et al., 2011).

4.4.2 NONCONVENTIONAL

An efficient extraction process should maximize the recovery of target compounds with minimal degradation, resulting in an extract with high antioxidant activity using environmentally friendly technologies and low-cost raw materials (Santos et al., 2010). In this context, recently novel methods for phenolic antioxidant extraction have been employed including UAE, MAE, SFE, and SWE. These extraction methods presenting advantages when compared to conventional methods because they decreasing solvent consumption, shortening extraction time, due to the possibility of working at elevated temperatures or pressures in inert atmosphere, and giving higher yield than the conventional methods of extraction; furthermore, they can be carried out in the absence of light and oxygen (Nayak et al., 2015; Wang and Weller, 2006).

4.4.2.1 ULTRASOUND-ASSISTED EXTRACTION

UAE is an efficient extraction method and might be a potential means to extract bioactive compounds. The sonication is characterized by the production of sound waves that create cavitation bubbles near the sample tissue which releasing cell contents and extraction efficiency can be enhanced through acoustic cavitation and mechanical effects. Acoustic cavitations can disrupt cell walls facilitating solvent penetration into the plant material and allowing the intracellular content to be released, whereas mechanical effects caused by ultrasound could also be the agitation of the solvent used for extraction, thus increasing the surface contact area between the solvent and the targeted compounds by allowing greater penetration of the solvent into the sample matrix (Corbin et al., 2015).

Probe and bath systems are the two most common ways of applying ultrasound waves to the sample, and the PC extraction from plant material can be carried out using both static and dynamic modes. Whereas in a static system, there is a closed-vessel extraction for which no continuous transfer of solvent occurs, in a dynamic system extraction, solvent is supplied continuously, which allows efficient adsorption of analytes and their effective transfer from the extraction vessel (Khoddami et al., 2013). It is important to regard that in the UAE method, extract recovery is influenced not only by sonication time, temperature, and solvent selection but also by wave frequency and ultrasonic-wave distribution (Wang and Weller, 2006).

Compared to conventional methods, UAE is one of the most simple, inexpensive extraction systems and can be operated rapidly using a range of solvents. As a method to extract PC of flaxseed, UAE showed to be very efficient for the reduction of mucilage entrapment of these phenolics, thus, allowing a high extraction yield (Corbin et al., 2015). Good yield was also possible to reach using UAE of bioactive compounds (anthocyanins and phenolics) from jaboticaba peels which was shown to be a more efficient extraction method than the sophisticated extraction systems such as high pressure CO_2-assisted extraction (Rodrigues et al., 2015).

4.4.2.2 MICROWAVE-ASSISTED EXTRACTION

MAE is an alternative process that uses microwave energy to extract compounds from materials and appears to be one of the best nonconventional methods to extract PC from plant due to the special microwave/matter interactions and the vary rapid extraction time (Setyaningsih et al., 2015). Microwaves are nonionizing radiation ranging in frequency from 300 MHz to 300 GHz. The effect of microwaves is strictly related to the conversion of electromagnetic energy to heat, which is based on the direct effect of microwaves by ionic polarization and dipolar rotation on molecules with dipoles. Therefore, when microwave radiation is applied in materials or solvents, the time variation of the wave electric field leads to dipolar rotation that is due to the alignment on the electric field of molecules possessing a dipole moment and this produces molecular friction and collisions with consequent liberation of thermal energy into the medium results in fast dielectric heating.

The efficiency of microwave heating at a given frequency and temperature depends on the ability of the material to absorb eletromagnetic energy and to dissipate heat (Flórez et al., 2015). In addition, an increase in the extraction yield of solutes from food matrix is reached when solvents are used with a high dielectric constant and a high dissipation factor which also facilitates distribution of heat throughout the matrix. Then, polar solvents have a higher dielectric constant than nonpolar solvents and can absorb more microwave energy, which can result in a higher yield of phenolics. Application of microwave radiation induces changes at cellular level. During MAE, polar molecules absorbed efficiently the energy and the sudden heating generated cause liquid vaporization and pressure built up within the cells that drastically change the physical properties of the cell walls. In addition, the cell structure is broken down and improves the capillary–porous structure of the tissues, facilitating faster diffusion out of the solid (Flórez et al., 2015).

Seeing that different chemical substances absorb microwaves to different extents, the MAE is an efficient method for extractions and this behavior makes it possible to selectively extract target compounds from complex food matrix at a higher rate than in conventional extraction, since it is induced by highly localized heating caused by microwaves, which could be due to a synergy combination of mass and heat transfer phenomena acting in the same direction (Flórez et al., 2015; Setyaningsih et al., 2015). Furthermore, the increase of movement and collision efficiency of the molecules induced by the change in the electric field direction weakens hydrogen bonds which facilitates disruption of the solute–matrix interactions and release of target compounds.

MAE has recently received much attention due to its many advantages, including shorter extraction time, lower organic solvent requirement, and increased extraction yields. In addition, this system rapidly generates heat and this result in good quality extracts with better target compound recovery. However, the efficiency of this extraction method depends on extraction time, extraction temperature, solid–liquid ratio, and the type and composition of solvent used (Setyaningsih et al., 2015). MAE can be used for the extraction of polyphenols from plant material since these compounds are dipoles and can absorb microwave energy due to their hydroxyl groups which are distributed along its molecular structure (Khoddami et al., 2013). In addition, this alternative extraction method can be applied to extraction of heat-sensitive bioactive compounds from plant materials where rapid heating and hence shorter extraction time is desired.

Researches have been performed to determine the best operating conditions to extract different PC using this method. MAE method was used for the extraction of PC from rice grains and the highest yield was obtained applying extraction temperature 185°C, microwave power 1000 W, extraction time 20 min, solvent 100% methanol, and solvent-to-sample ratio 10:1 (Setyaningsih et al., 2015). MAE method induced high extraction selectivity of bioactive compounds from *Urtica dioica* leaves and stems and high yield of extracted compounds was obtained with microwave power 750°C, extraction time 2 min, and solvent 100% ethanol (Hijazi et al., 2013).

The MAE under optimum conditions can be considered as a powerful tool for the extraction of PC from variety of plant material.

4.4.2.3   ENZYME-ASSISTED EXTRACTION

Enzyme-assisted extraction (EAE) of bioactive compounds from plants have been widely investigated due to its advantages that including easy operation and environment friendship, becoming a potential alternative to conventional solvent based extraction methods. This method also has shown faster extraction, higher recovery, reduced solvent usage, and lower energy consumption when compared to nonenzymatic methods (Puri et al., 2012). In recent years, EAE has also been gaining attention as efficient method to enhance the release and recovery of PC from plants since enzymes can effectively work catalyzing the degradation of vegetables cell wall, favoring the release of bioactive components contained inside the cell as well as those reported to be linked to cell wall polysaccharides by hydrophobic interactions and hydrogen bonds.

It is known that the primary cell wall of plants is mainly composed of cellulose, hemicellulose, and pectin; thus cellulases, hemicellulases, and pectinases as well as others enzymes can be used to catalyze and hydrolyze the cell wall polysaccharides and therefore enabling a better release and a more efficient extraction of PC (Miron et al., 2013). Different enzymes have been employed (alone or in combination) to enhance and accelerate phenolics extraction of plant material. Pectinases were employed to extraction and release anthocyanins from saffron tepals and showed high efficiency as compared to conventional ethanol extraction (Lotfi et al., 2015). PC from thymus leaves were efficiently extracted by enzyme-assisted method with cellulase and polygalacturonase which induced increase in antioxidant capacity of the extracts (Cerda et al., 2013). In another study, extraction of antioxidant phenolics from ginger was studied, and the enzymes' pretreatment had a significant influence on the yield of 6-gingerol and the total polyphenols.

For some studies, enzyme pretreatment of raw material normally results in a reduction in extraction time, minimizes usage of solvents, and provides increased yield and quality of product. However, compared to prior research evaluating the use of enzymatic treatment for enhancement of antioxidant compounds extraction from thymus, better results were observed in this investigation when commercial enzymes were incorporated during the solvent extraction of PC and not when applied as a pretreatment process. This study suggests that the difference can be due to a long extraction time in the second case, once the enzyme was in contact with the thymus sample during the pretreatment time and during the extraction time (21 h total), which decreases the concentration of phenols in the medium because of their degradation by the same enzyme activity still present in the extract (Cerda et al., 2013).

In most studies, the enzyme incorporation increases the presence of PC and antioxidant activity. The enzymes required for extraction process can be derived from bacteria, fungi, animal organs, or vegetable/fruit extracts. However, to use enzymes most effectively for extraction applications, it is important to understand their catalytic property and mode of action, optimal operational conditions, and which enzyme or enzymes combination is appropriate for the plant material selected. In addition, the parameters impacting enzyme-assisted release of bioactives (pH, time, temperature, and concentration) need to be optimized for each specific process: enzymes normally function at an optimal temperature, however, they can still be used over a range of temperatures, providing flexibility for both cost and product quality; substrate particle size reduction prior to enzymatic treatment provides better accessibility of the enzyme to the cell to increase extraction yields significantly; in aqueous extraction, the enzymes can rupture the polysaccharide–protein colloid in the cell wall creating an emulsion that interferes with extraction, so nonaqueous systems are preferable for some materials (Puri et al., 2012).

The EAE is an attractive proposition to enhance the yields of bioactive compounds. Although further investigations are needed, in particular to synthesize of new enzymes and purification of enzymatic mixtures, helping to improve the level of released bioactives making it a viable extraction process at commercial scale.

4.4.2.4 SUBCRITICAL WATER EXTRACTION

Recently, SWE, also referred as pressurized or low-polarity water extraction, has been used as an alternative technique for the extraction of bioactive compounds. Subcritical water (SW) is defined as water at a temperature between its boiling and critical point where the pressure is regulated in such a way that water remains in its liquid (Herrero et al., 2012). The technique has been receiving much attention in the field of natural compounds extraction because SW is an efficient solvent for both polar and nonpolar compounds and its versatility as a solvent is related to the tunable polarity of water, which is directly dependent upon the temperature; when the temperature of water is increased, its polarity decreased, so the solubility of nonpolar organic increases, whereas the solubility of polar organics decreases (Carr et al., 2011). Under subcritical condition, the change in the water temperature can lead to changes in the water's dielectric constant and consequently the water polarity: under standard temperature and pressure (25°C and 101 kPa) water is a polar compound with dielectric constant of about 80, but when the temperature is increased to about 200–350°C, the dielectric constant drops to around 20–30, which is similar to the range of dielectric constants of conventional solvents usually applying in the conventional bioactive extraction process like methanol, ethanol, and acetone at room temperature (Carr et al., 2011; Duba et al., 2015; Herrero et al., 2012).

However, these solvents are often toxic and rigorous organic solvent removal is necessary due to the extract can be ingested as an ingredient food or pharmaceutical. Therefore, SW is an ideal candidate for use as solvent for the natural compounds extraction from plant material because water is ubiquitous, nontoxic, and has low disposal costs. SWE process based on the thermodynamic properties of water which are described in terms of hydrogen bonding strength and its structure. Changes in hydrogen bonding strength are reflected in the dielectric constant; at higher temperatures, the increased thermal agitation reduces the hydrogen bond strength in water and leads to a reduction in dielectric constant value which makes the water a solvent of less polarity which in turn increases the solubility of some organic compounds as polyphenols. However, the solubility of organic compounds in SW depends on several factors like chain length, type, and position of side groups, molecular weight, position of hydrogen bonding, etc. of the solute being solubilized (Carr et al., 2011).

Treatment with SWE has been shown to be sufficiently powerful to extract polyphenols from grape skins and defatted grape seeds (Duba et al., 2015). The study shown that high yields of total polyphenols were obtained for both skins and seeds. For extraction of flavonols from black tea, celery, and ginseng leaf, the effectiveness of SWE compared to that of other extraction solvents, such as ethanol, methanol, and hot water has been studied and the results indicated that SWE is a highly selective and rapid method for extraction of flavonols from plants. Therefore, SW could be an excellent alternative to organic solvent as a medium for extracting flavonols (Cheigh et al., 2015). SWE has been demonstrated to be an effective extraction method for a wide range of PC providing higher yield and reduction on extraction time by up to 50% of conventional method extraction time. In addition, SWE could be a good alternative industrial method to use for extraction of large amounts of PC without toxic organic solvent residues.

4.4.2.5   SUPERCRITICAL FLUID EXTRACTION

SFE is another environmentally friendly extraction technique which in recent years has received greater attention as an important alternative to traditional solvent extraction methods since degradation and decomposition of the active compounds is avoided operating at reduced temperatures, in absence of light and oxygen (Meneses et al., 2015). In addition, the extracts obtained by SFE are natural origin, present absence of residual organic solvent, and composition controlled by process selectivity (Paes et al., 2014). SFE method extracts soluble components from a raw material exploiting the unique properties of gases above their critical points (Mallikarjun Gouda et al., 2015). Thus, the relatively low viscosity (near to the gas) and the high diffusivity of supercritical fluids (SCF) help to penetrate the porous solid materials more efficiently than liquid solvents, thus resulting in faster and more efficient extractions (Otero-Pareja et al., 2015). SFE is usually performed in inert atmospheres, absence of light, at moderate temperatures and short time contributing to avoid oxidation, thermal degradation, and others chemical changes in bioactive compounds (Nyam et al., 2010).

In SFE method, methane, carbon dioxide, ethane, propane, ammonia, ethanol, benzene, and water are the usual SCF applied. Among them, carbon dioxide (CO_2) is the most used because is nontoxic, nonflammable, and noncorrosive, in addition to be inert to most materials, cheap, and readily available in bulk quantity with satisfied purity (Nyam et al., 2010). Thus, CO_2 is an ideal solvent which has been used to extracting of phenolics from asparagus, peach leaves, and myrtle leaves (Kazan et al., 2014; Pujol et al., 2013; Solana et al., 2015). However, CO_2 has a very limited capacity to dissolve polar and high molecular weight compounds being the addition of polar cosolvent such as methanol, ethanol, and acetone recommended to modify its polarity (Meneses et al., 2015).

4.5 PLANT TISSUE SELECTION

To date, there are at least 14 groups of plant secondary metabolites with nutraceutical potential. Alkaloids, amines, cyanogenic glycosides, terpenes, isoprenoids, tocopherols, phenylpropanoids, glucosinolates, polyacetylenes, polyketides, saponins, steroids, small peptides, and nonprotein amino acids are just few examples (Karim and Azlan, 2012). Despite that, these phytochemicals are distributed along the diverse plant tissues and may vary upon species, some of these phytochemicals are more common to occur in a specific plant tissue according to their physiological role in plants, see Figure 4.2. The selection of the most suitable source of bioactives to design effective nutraceuticals implies the selection of high-quality raw materials and a strict control of by-side contaminants (Lockwood, 2011), among other factors. This represents an everyday dilemma for agronomists, food chemists, and pharmaceutical developers, which not only have to face the problem of obtaining highly pure ingredients but also must be sensitive to the rapid changes of the functional/nutraceutical market (Siro et al., 2008). A comprehensive path to obtain nutraceutical bioactives (*NB*) is depicted in Figure 4.3, summarizing some important considerations at each stage. Anticipating that the selection of plant/tissue source (Fig. 4.3, **Stage 1**) turns out to be the critical starting point in the research and development process, in the following paragraphs, this aspect is addressed in detail.

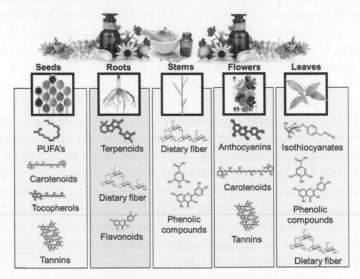

FIGURE 4.2   Common phytochemical distribution in plant tissues.

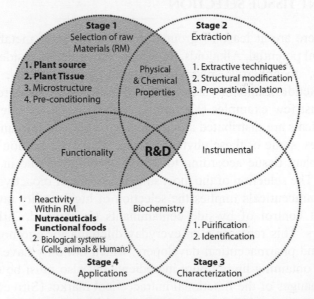

FIGURE 4.3   Research and development (R&D) of functional foods and nutraceuticals.

First, the specific chemical fingerprint of a particular plant, along with its health-promoting properties, far outstrips its taxonomy. The natural distribution of a particular *NB* in fruits, the edible usually fleshy and sweet smelling part of a plant that may or may not contain seeds, is primarily committed by their taxa-specific richness. For example, *lycopene* a linear carotenoid (Saini et al., 2015) that plays a modest role on prostate cancer (Chen et al., 2013) and cardiovascular disease (Müller et al., 2015) is found in a higher amount (>1000 times) in mature fruits of watermelon (*Cucurbitaceae*) and tomato (*Solanaceae*) as compared to mango (*Anacardiaceae*) or carrot (*Apiaceae*), though the later are richer in another carotenoid: *β-carotene*.

As if this were not enough, phylogenetically related plants also have marked differences in *NB* content. For example, the *Solanaceae* family compromises 98 genera and approximately 2700 species; *Capsicum annum* belongs to it and includes a wide variety of hot peppers with graded levels of pungency (Scoville units) associated with their *capsaicin* content, an alkaloid with a protective antioxidant effect against nonalcoholic fatty liver disease and hyperglycemic-induced endothelial disorders (McCarty et al., 2015), adipogenesis (Alcalá-Hernández et al., 2015), and cancer (Swain and Kumar Mishra, 2015). Fortunately, information on structure–function as well as plant sources for specific *NB*, is continuously deposited in several open access databases (e.g., USDA-NDB, phenol-explorer, PhytAMP, SOFA, and GMD) in such way that the selection of the richest source of a particular *NB* seems to be no longer the problem. The systematization within these databases also helps to bring together several plants and even specific tissues, with the same functional/nutraceutical action (Andrade-Cetto and Heinrich, 2005).

Mother Nature also distributes several *NB* in other tissues during the plant's lifetime, according to their physiological role. For example, leaves have the highest photosynthetic activity and so they are rich in *chlorophyll*, a porphyrin pigment with antioxidant and anticancer activity (İnanç, 2011). Straws, stalks, and stems, whose main function is to support for and the elevation of leaves, flowers, and fruits, are rich in *xylo-oligosaccharides* (Carvalho et al., 2013) useful to improve gut's health, while seeds (the ripened ovule of a plant, containing the embryo and the endosperm, wrapped in a protective coat) have the highest content of *fatty acids*, *phytosterols*, and *proteins*, all having a pivotal role on preventing cardiovascular diseases. Edible flowers are excellent sources of volatile *NB* secreted during specialized ant pollination (de Vega et al., 2014), with several medical applications such as antiseptic, antispasmodic, and antiparasitic actions (Mlcek and Rop, 2011). Lastly, from woody barks and fruit exocarps (e.g., shells, peels, and pods) is possible to get many *alkaloids* an PC such as *condensed* and *hydrolyzed tannins*, which protect the plant from predators' attack but also have and hypoglycemic effect (Olivas-Aguirre et al., 2014; Shan et al., 2005). The development of sophisticated techniques to isolate and identify *NB* from all of the above-mentioned plant tissues (Fig. 4.3, **stage 3**) has been used in *Plant Molecular Systematics* to consolidate three research lines:

Molecular and nutraceutical characterization of a single NB obtained from different plant sources.

New tools for the molecular identification of *NB* (HPLC–MS, MS, MALDI-TOF, etc.) has led to the conclusion that many nutraceutical actions are restricted to just one metabolite. For example, *Butein* and *chalconoid* isolated from stems, barks, flowers, fruits, heartwoods, and leaves of at least 30 different plants (Semwal et al., 2015) has many medical applications as an analgesic, antibiotic, antithrombotic, anticancer, and anti-inflamma- tory agent. Also, apigenin-7-*O*-glycoside (flavan-3-ol) isolated from olive leaves, chamomile flowers, salvia stems is either a strong hematopoietic (Samet et al., 2015) or an anti-anxiolitic (Kumar and Bhat, 2014). *Lunasin* (Fig. 4.4), a 43 amino acid peptide found in soybean, wheat, barley, rice, rye, triticale, and amaranth, has anticancer and anti-inflamatory capacity (Malaguti et al., 2014).

Identification of different NBs from distinct plant sources with a same nutraceutical action.

From folk medicine, it has been possible to gain knowledge on several plant sources (and tissues) for a specific metabolic condition or disease. For example, Mexican diabetic patients use several "phytoremedies" (Table 4.2) including those known as *tizanas* which are herbal infusions prepared with flowers, fruits, or roots from different plants but with the same medicinal purpose (Johnson et al., 2006). However, despite the accumulated empirical knowledge, very little is known about their pharmacological effectiveness and their safe intake/toxicity level. In this sense, the isolation and purifica- tion of the associated *NB* to these *tizanas* is a crucial step (Fig. 4.3, **stage 3**) to develop culturally accepted nutraceuticals, either alone or combined.

To further complicate this matter, the natural distribution of *NB* within a single plant, can differ from one tissue to another. For example, it is well known that several noncommunicable chronic diseases are associated with pro-inflammatory conditions along with an altered redox status. These condi- tions could be reversed by consuming plant antioxidants such as PCs. PCs are widely distributed in the plant kingdom, grouping a plethora of chemical compounds with radical scavenging capacity (RSC). Fresh-cut aromatic herbs such as coriander, mint, and parsley are believed to provide the organism with extra *NB* with RSC and anti-inflammatory capacity (Santos et al., 2014). However, a graphical adaptation of how their RSC is differently distributed from stem to leaves, as described by Al-Juhaimi et al. (2011), is shown in Figure 4.5. This same phenomenon extends to the comparison of peels and seeds vs. pulp (common edible part) of tropical fruits such as avocado, guava, mango, and pomegranate (Ayala-Zavala et al., 2011).

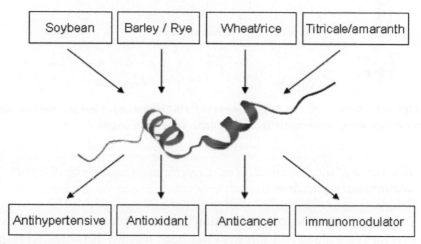

FIGURE 4.4 Lunasin: sources and nutraceutical actions.

TABLE 4.2 Mexican Plants with Hypoglycemic Effect and Related *NB*.

Common name	Family	Plant part	Associated *NB*
Maguey	*Agavaceae*	Stem	Sapogenins
Onion	*Liliaceae*	Bulb	Sulfuric derivates
Angel grass	*Asteraceae*	Leaf/stem	Terpenes
Sabila	*Liliaceae*	Stem (roasted)	Polysaccharides/flavonoids
Pineapple	*Bromeliaceae*	Fruit (juice)	Monoterpenoids/lactones
Peanut	*Fabaceae*	Seed (oil)	Sterols/flavonoids
Pingüica	*Ericaceae*	Leaves	Alkaloid/flavonoids
Beet	*Chenopodiaceae*	Stem	Alkaloids/flavonoids
Prodigiosa	*Asteraceae*	Leaf/stem	Sesquiterpenes/lactones
White zapote	*Rutaceae*	Bark	Alkaloids
Tejocote	*Rosaceae*	Root	Tannins/flavonoids
Guayacan flower	*Solanaceae*	Flower	Alkaloids

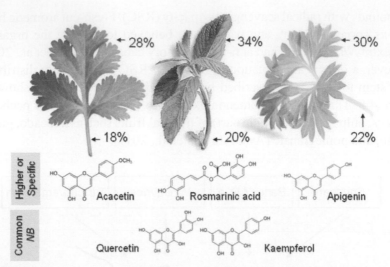

FIGURE 4.5 Selected PC and RSC activity (% DPPH inhibition) of leaf and stem extracts from culinary herbs. Coriander (left), mint (center), and parsley (right).

Generation of metabolomic data on a specific plant source with different nutraceutical applications.

The concept of "integral exploitation" commonly used in the agribusiness sector (agriculture and forestry) has been recently benefited with the second *MSP* research line. Since the 1970s, this sector realized the need to diversify the applications of several plant resources, building up a strategic partnership between primary producers and academics. As an example, several patents have been authorized on the use of tropical, semi-arid, and arid crops as potential sources of *NB* (Souto et al., 2014). Also, researchers worldwide have focused their efforts on elucidating the biological control (metabolomics) and the tissue-specific richness in *NB* of a single plant source (Ayala-Zavala et al., 2011; Patel, 2015; Sharma et al., 2015). A few examples of specific plant tissues rich in *NB*, as well as their potential use to enhance human health are summarized in Table 4.3.

For example, *bitter gourd* (*Momordica charantia* L.), has been traditionally used as a food in tropical regions such as India, Malaya, China, tropical Africa, Middle East, America. As a medicinal plant, it has been reported to possess antilipolytic, hyploglycemic, analgesic, abortifacient, antiviral, anticytotoxic, and antimutagenic properties. These preventive effects are related to the differential antioxidant capacity (as assayed FRAP) and PC content (phenolic acids) of its parts.

TABLE 4.3 Phenolic Compounds Screening and Nutraceutical Potential from Selected Plant Sources.

Plant	Tissue	NB	Structure	Action	References
Bitter gourd *Momordica charantia* L	Leaf	Caffeic acid		Hepatoprotective Neuroprotective Anti-inflammatory Cardioprotective	Kubola and Siriamornpun (2008), Tolba et al. (2013)
	Stem	Ferulic acid		Anticytotoxic Cardioprotective Hepatoprotective	Alam et al. (2013), Kubola and Siriamornpun (2008)
	Fruit (ripe)	Gallic acid		Neuroprotective Hepatoprotective Anti-inflammatory Anti-allergic Antiulcerogenic	Kubola and Siriamornpun (2008), Sen et al. (2013)
		Cucurbitane-type triterpene glycosides		Hypoglycemic Hepatoprotective Antiproliferative	Hsiao et al. (2013), Torres-Moreno et al. (2015)
Guamuchil/manila tamarind *Pithecellobium dulce* (Roxb.) *Benth*	Seed	Oleanolic acid		Antidiabetic Hypolipidemic Anti-inflammatory	Nagmoti et al. (2015), Wang et al. (2013)

TABLE 4.3 *(Continued)*

	Leaves	Quercetin	Anti-inflammatory Antibacterial Neuroprotective Hypolipidemic	Chandran and Balaji (2008), Ghosh et al. (2013), Jung et al. (2013)
	Bark	Lupenol	Anticancer Antiparasitic Antidislipidemic	Monroy and Colin (2004), Pitchai et al. (2014), Srivastava et al. (2013)
Flamboyant *Delonix regia*	Fruit	Afzelechin	Elastase (neutrophyl) inhibitory activity	Feng et al. (2014b), Huang et al. (2013), Siddiqui et al. (2014)
	Stem bark	Catechin	Antityrosinase activity Cytoprotective Antidiabetic Anti-inflammatory	Feng et al. (2014b), Gawlik-Dziki et al. (2016), Liu et al. (2014), Morel et al. (1993)
	Leaf	(Epi)-gallo-chatechin	Antioxidant Antiproliferative Anti-inflammatory	Feng et al. (2014b)

In summary, the careful selection of a primary plant source of *NB*s will determine the success or failure of the final manufactured nutraceutical. In this sense, it is crucial to select the most suitable plant tissue for a particular purpose, as well as other important factors such as ripening stage and proper extractive methods and technologies, which could hinder the quality and performance of the final *NB*.

4.6 NUTRACEUTICAL PRODUCTS IN MARKET

Nutraceuticals market is growing rapidly, the global profits for 2013 was approximately 175 billion US dollars and is expected to grow to 424 billion by 2017 (Daliri and Lee, 2015). There has been an increase in the consumption of nutraceuticals in the last decade, the United States being the largest consumer of nutraceuticals and functional foods followed by Japan and Europe, and other countries such as India, China, Russia, and Canada. The rise in the incidence of obesity, cardiovascular disease, cancer, and diabetes, along with the knowledge of reduce risk of several diseases and maintain a state of health through a diet rich in fruits and vegetables, and the need for products that offer higher bioavailability of one or more bioactive ingredients, among other factors have contributed to the growth of this industry (Zawistowski and Debasis, 2008). People's awareness in the health-related benefits of dietary supplements, and the practical presentation of most of the nutraceutical products and functional foods are key to their successful growth.

Most of the nutraceuticals products are targeting weight control, vascular health, general nutrition, and sports nutrition mainly. Nutraceuticals also focus in other health areas such as eye health, diabetes, mental health, cancer, arthritis, ageing, and sexual health and performance. Most common nutraceuticals in the market are those that contain ingredients like DF, vitamin E, PUFAs, inulin, probiotics, conjugated LA, soy and plant antioxidant (polyphenols). Nutraceuticals normally are consumed as pills, capsules, soft gels, and tinctures and can be categorized as nutrients (vitamins and minerals), botanicals (nutraceuticals made from plant parts), and dietary supplements (Dureja et al., 2003). However, nutraceuticals can be taken either as a food additive or a powder concentrate and it depends on the consumer's preference and needs.

The growing demand of consumers for nutraceuticals has increased the release of new functional products every year. Between 2008 and 2009, the United States launched 881 healthy products, followed by Italy and Japan with 325 and 314, respectively (Valls et al., 2013). There are organizations from different countries such as the Food and Drug Administration (FDA), the European Authority of Food Safety, or the Ministry of Health, Labour, and Welfare (MHLW) that are in charge to regulate this type of products and to assure that its claims are not misleading consumers. FDA is responsible for the regulation of nutraceuticals under the authority of the Federal Food, Drug, and Cosmetic Act and evaluates the safety and labeling of dietary supplements fulfill the requirements of Dietary Supplement Health and Education and FDA before marketing. The FDA does not recognize nutraceuticals as they are not legally defined, they are regulated under the same statutes as food products and not like drugs and must not have claims that says the product use is for the treatment or prevention of a specific disease. The manufacturer must notify and give information to the FDA that the new product is safe and it is under the conditions of use as stipulated in its labeling. Until now, most of the dietary supplements that are selling in the United States possess one of the three claims (nutrient content, structure/function, and health claims) (Hasler, 2008).

In Europe, the situation is similar than in the United States, nutraceuticals products are regulated by the same legislation than foods. The European Union distinguishes two types of claims. Nutrition claims: it suggests that food has particular beneficial nutritional properties due to the energy it provides or the nutrients it contains. Health claims: it implies that there exists a relationship between food or one of food constituents and health. Inside this claim exist other claims such as reduction disease risk claims which state that the consumption of a food or food constituent significantly reduces a risk factor in the development of human disease (Hasler, 2008; Verhagen et al., 2010). The MHLW enacted functional foods or Food for Specified Health Uses (FOSHU) as a regulatory system for approval of food with health claims that can be used on a label to inform consumers about their functionality (Regulations, 1996). Japan is the only country that has a specific food category besides food and medicine. Also, FOSHU is the only category of functional foods that are qualified to carry health function claims in Japan; their health benefits are aim especially to gastrointestinal health, blood pressure, dental hygiene, bone health, serum cholesterol, mineral absorption, and blood glucose (Bagchi, 2008).

To assure the future of the nutraceuticals industry, it is necessary to legally define the term nutraceutical and try to uniform the legislations of the different countries for these products. Companies should perform long-term clinical trials in animals and humans to scientifically validate the health benefit they are selling and to identify the presence of many ingredients that can harmful to the consumers. Also, health professionals and academics should work together along with manufacturers to provide scientific evidence for the development of new functional products.

4.7   CONCLUDING REMARKS

The scientific breakthroughs and constant innovation in food science and technologies are pushed by the desire of people to get healthier choices of food products that they may add to their diet at low cost. The growing demand for novel functional foods and nutraceutical products are leading scientist around the world to the continuous search of new natural sources of bioactive compounds with optimum extraction yield to the food industry, to get low cost of product manufacture. Plant tissues as product of their secondary metabolism has a great variety of phytochemicals, from PC to DF, or even carotenoids, vitamins, and amino acids. Hence, it depends on the desirable health effect of the plant tissue that should be selected for the extraction of the appropriate phytochemical to incorporate it in the nutraceutical product. In that sense, plant tissues represent a great source of bioactive compounds due to their great amount and diversity of phytochemicals. Moreover, once the phytochemical desired is on target, and the plant tissue source of this compound is found, it is essential to select a proper extraction method to achieve the optimum extraction yield of the desirable compound. However, there are plenty of plant tissues yet uncharacterized in many countries around the world, and scientists around the world must continue the contribution at least with information about the local plant tissues of their own region.

KEYWORDS

- **functional foods**
- **phytochemicals**
- **design foods**
- **functional ingredients**
- **extraction methods**

REFERENCES

Abdul-Hamid, A.; Luan, Y. S. Functional Properties of Dietary Fibre Prepared from Defatted Rice Bran. *Food Chem.* **2000,** *68* (1), 15–19.

Acharya, J.; Karak, S.; De, B. Metabolite Profile and Bioactivity of *Musa* × *paradisiaca* L. Flower Extracts. *J. Food Biochem.* **2016** *40* (6), 724–730.

Agbangnan, P.; Tachon, D.; Dangou, J.; Chrostowska, A.; Fouquet, E.; Sohounhloue, D. Optimization of the Extraction of Sorghum's Polyphenols for Industrial Production by Membrane Processes. *Res. J. Res. Sci.* **2012,** *1,* 1–8.

Aggarwal, B. B.; Sundaram, C.; Prasad, S.; Kannappan, R. Tocotrienols, the Vitamin E of the 21st Century: Its Potential against Cancer and Other Chronic Diseases. *Biochem. Pharmacol.* **2010,** *80* (11), 1613–1631.

Aizawa, K.; Inakuma, T. Quantitation of Carotenoids in Commonly Consumed Vegetables in Japan. *Food Sci. Technol. Res.* **2007,** *13* (3), 247–252.

Ajila, C.; Leelavathi, K.; Rao, U. P. Improvement of Dietary Fiber Content and Antioxidant Properties in Soft Dough Biscuits with the Incorporation of Mango Peel Powder. *J. Cer. Sci.* **2008,** *48* (2), 319–326.

Al-Farsi, M. A.; Lee, C. Y. Nutritional and Functional Properties of Dates: A Review. *Crit. Rev. Food Sci. Nutr.* **2008,** *48* (10), 877–887.

Al-Juhaimi, F.; Ghafoor, K. Total Phenols and Antioxidant Activities of Leaf and Stem Extracts from Coriander, Mint and Parsley Grown in Saudi Arabia. *Pak. J. Bot.* **2011,** *43* (4), 2235–2237.

Alam, M. A.; Sernia, C.; Brown, L. Ferulic Acid Improves Cardiovascular and Kidney Structure and Function in Hypertensive Rats. *J. Cardiovasc. Pharmacol.* **2013,** *61* (3), 240–249.

Alcalá-Hernández, C. F.; de la Rosa, L. A. A.; Wall-Medrano, A.; López-Díaz, J. A.; Álvarez-Parrilla, E. Avances en Terapia Farmacológica y Fitoquímica de la Adipogénesis. *Nutr. Hosp.* **2015,** *32* (n02).

Anastasiadi, M.; Pratsinis, H.; Kletsas, D.; Skaltsounis, A.-L.; Haroutounian, S. A. Bioactive Non-coloured Polyphenols Content of Grapes, Wines and Vinification By-products: Evaluation of the Antioxidant Activities of their Extracts. *Food Res. Int.* **2010,** *43* (3), 805–813.

Anastasiadi, M.; Pratsinis, H.; Kletsas, D.; Skaltsounis, A.-L.; Haroutounian, S. A. Grape Stem Extracts: Polyphenolic Content and Assessment of their In Vitro Antioxidant Properties. *LWT—Food Sci. Technol.* **2012,** *48* (2), 316–322.

Andrade-Cetto, A.; Heinrich, M. Mexican Plants with Hypoglycaemic Effect Used in the Treatment of Diabetes. *J. Ethnopharmacol.* **2005,** *99* (3), 325–348.

Aneta, W.; Jan, O.; Magdalena, M.; Joanna, W. Phenolic Profile, Antioxidant and Antiproliferative Activity of Black and Red Currants (*Ribes* spp.) from Organic and Conventional Cultivation. *Int. J. Food Sci. Technol.* **2013,** *48* (4), 715–726.

Apostolou, A.; Stagos, D.; Galitsiou, E.; Spyrou, A.; Haroutounian, S.; Portesis, N.; et al. Assessment of Polyphenolic Content, Antioxidant Activity, Protection against ROS-Induced DNA Damage and Anticancer Activity of *Vitis vinifera* Stem Extracts. *Food Chem. Toxicol.* **2013,** *61,* 60–68.

Ayala-Zavala, J.; Vega-Vega, V.; Rosas-Domínguez, C.; Palafox-Carlos, H.; Villa-Rodriguez, J.; Siddiqui, M. W.; et al. Agro-industrial Potential of Exotic Fruit Byproducts as a Source of Food Additives. *Food Res. Int.* **2011,** *44* (7), 1866–1874.

Bagchi, D. *Nutraceutical and Functional Food Regulations in the United States and around the World*; Academic Press: Cambridge, MA, 2008.

Borrás-Linares, I.; Stojanović, Z.; Quirantes-Piné, R.; Arráez-Román, D.; Švarc-Gajić, J.; Fernández-Gutiérrez, A.; Segura-Carretero, A. *Rosmarinus officinalis* Leaves as a Natural Source of Bioactive Compounds. *Int. J. Mol. Sci.* **2014**, *15* (11), 20585–20606.

Brekhman, I. I. S. K. *Man and Biologically Active Substances: The Effect of Drugs, Diet and Pollution on Health*: Elsevier: Amsterdam, 2013.

Carr, A. G.; Mammucari, R.; Foster, N. A Review of Subcritical Water as a Solvent and Its Utilisation for the Processing of Hydrophobic Organic Compounds. *Chem. Eng. J.* **2011**, *172* (1), 1–17.

Carvalho, A. F. A.; de Oliva Neto, P.; Da Silva, D. F.; Pastore, G. M. Xylo-oligosaccharides from Lignocellulosic Materials: Chemical Structure, Health Benefits and Production by Chemical and Enzymatic Hydrolysis. *Food Res. Int.* **2013**, *51* (1), 75–85.

Cavalcanti, R. N.; Veggi, P. C.; Meireles, M. A. A. Supercritical Fluid Extraction with a Modifier of Antioxidant Compounds from Jabuticaba (*Myrciaria cauliflora*) Byproducts: Economic Viability. *Proc. Food Sci.* **2011**, *1*, 1672–1678.

Cerda, A.; Martínez, M. E.; Soto, C.; Poirrier, P.; Perez-Correa, J. R.; Vergara-Salinas, J. R.; Zúñiga, M. E. The Enhancement of Antioxidant Compounds Extracted from *Thymus vulgaris* Using Enzymes and the Effect of Extracting Solvent. *Food Chem.* **2013**, *139* (1), 138–143.

Contreras-Calderón, J.; Calderón-Jaimes, L.; Guerra-Hernández, E.; García-Villanova, B. Antioxidant Capacity, Phenolic Content and Vitamin C in Pulp, Peel and Seed from 24 Exotic Fruits from Colombia. *Food Res. Int.* **2011**, *44* (7), 2047–2053.

Corbin, C.; Fidel, T.; Leclerc, E. A.; Barakzoy, E.; Sagot, N.; Falguiéres, A.; et al. Development and Validation of an Efficient Ultrasound Assisted Extraction of Phenolic Compounds from Flax (*Linum usitatissimum* L.) Seeds. *Ultrason. Sonochem.* **2015**, *26*, 176–185.

Chandran, P.; Balaji, S. Phytochemical Investigation and Pharmocological Studies of the Flowers of *Pithecellobium dulce*. *Ethnobot. Leafl.* **2008**, *12*, 245–253.

Chang, R. C.-C.; Ho, Y.-S.; Yu, M.-S.; So, K.-F. *Medicinal and Nutraceutical Uses of Wolfberry in Preventing Neurodegeneration in Alzheimer's Disease*; 2010; pp 169–185.

Chaturvedi, S.; Sharma, P.; Garg, V. K.; Bansal, M. Role of Nutraceuticals in Health Promotion. *Health* **2011**, *4*, 5.

Cheigh, C.-I.; Yoo, S.-Y.; Ko, M.-J.; Chang, P.-S.; Chung, M.-S. Extraction Characteristics of Subcritical Water Depending on the Number of Hydroxyl Group in Flavonols. *Food Chem.* **2015**, *168*, 21–26.

Chen, J.; Song, Y.; Zhang, L. Lycopene/Tomato Consumption and the Risk of Prostate Cancer: A Systematic Review and Meta-Analysis of Prospective Studies. *J. Nutr. Sci. Vitaminol.* **2013**, *59* (3), 213–223.

Chiva-Blanch, G.; Visioli, F. Polyphenols and Health: Moving beyond Antioxidants. *J. Berry Res.* **2012**, *2* (2), 63–71.

Chu, Y.-F.; Sun, J.; Wu, X.; Liu, R. H. Antioxidant and Antiproliferative Activities of Common Vegetables. *J. Agric. Food Chem.* **2002**, *50* (23), 6910–6916.

Chun, O. K.; Kim, D. O.; Smith, N.; Schroeder, D.; Han, J. T.; Lee, C. Y. Daily Consumption of Phenolics and Total Antioxidant Capacity from Fruit and Vegetables in the American Diet. *J. Sci. Food Agric.* **2005**, *85* (10), 1715–1724.

Daliri, E. B.-M.; Lee, B. H. Current Trends and Future Perspectives on Functional Foods and Nutraceuticals. *Beneficial Microorganisms in Food and Nutraceuticals*; Springer: Berlin, 2015; pp 221–244.

Dang, Y. Y.; Zhang, H.; Xiu, Z. L. Microwave-Assisted Aqueous Two-Phase Extraction of Phenolics from Grape (*Vitis vinifera*) Seed. *J. Chem. Technol. Biotechnol.* **2014,** *89* (10), 1576–1581.

Das, L.; Bhaumik, E.; Raychaudhuri, U.; Chakraborty, R. Role of Nutraceuticals in Human Health. *J. Food Sci. Technol.* **2012,** *49* (2), 173–183.

de Vega, C.; Herrera, C. M.; Dötterl, S. Floral Volatiles Play a Key Role in Specialized Ant Pollination. *Perspect. Pl. Ecol., Evol. Syst.* **2014,** *16* (1), 32–42.

Delzenne, N. M.; Neyrinck, A. M.; Cani, P. D. Gut Microbiota and Metabolic Disorders: How Prebiotic Can Work? *Br. J. Nutr.* **2013,** *109* (S2), S81–S85.

Devalaraja, S.; Jain, S.; Yadav, H. Exotic Fruits as Therapeutic Complements for Diabetes, Obesity and Metabolic Syndrome. *Food Res. Int.* **2011,** *44* (7), 1856–1865.

Dillard, C. J.; German, J. B. Phytochemicals: Nutraceuticals and Human Health. *J. Sci. Food Agric.* **2000,** *80* (12), 1744–1756.

Do, Q. D.; Angkawijaya, A. E.; Tran-Nguyen, P. L.; Huynh, L. H.; Soetaredjo, F. E.; Ismadji, S.; Ju, Y.-H. Effect of Extraction Solvent on Total Phenol Content, Total Flavonoid Content, and Antioxidant Activity of *Limnophila aromatica*. *J. Food Drug Anal.* **2014,** *22* (3), 296–302.

Dominguez-Perles, R.; Teixeira, A.; Rosa, E.; Barros, A. Assessment of (Poly)phenols in Grape (*Vitis vinifera* L.) Stems by Using Food/Pharma Industry Compatible Solvents and Response Surface Methodology. *Food Chem.* **2014,** *164,* 339–346.

Dorta, E.; González, M.; Lobo, M. G.; Sánchez-Moreno, C.; de Ancos, B. Screening of Phenolic Compounds in By-product Extracts from Mangoes (*Mangifera indica* L.) by HPLC-ESI-QTOF-MS and Multivariate Analysis for Use as a Food Ingredient. *Food Res. Int.* **2014,** *57,* 51–60.

Dorta, E.; Lobo, M. G.; Gonzalez, M. Reutilization of Mango Byproducts: Study of the Effect of Extraction Solvent and Temperature on their Antioxidant Properties. *J. Food Sci.* **2012,** *77* (1), C80-C88.

Duba, K. S.; Casazza, A. A.; Mohamed, H. B.; Perego, P.; Fiori, L. Extraction of Polyphenols from Grape Skins and Defatted Grape Seeds Using Subcritical Water: Experiments and Modeling. *Food Bioprod. Process.* **2015,** *94,* 29–38.

Dureja, H.; Kaushik, D.; Kumar, V. Developments in Nutraceuticals. *Indian J. Pharmacol.* **2003,** *35* (6), 363–372.

Elleuch, M.; Bedigian, D.; Roiseux, O.; Besbes, S.; Blecker, C.; Attia, H. Dietary Fibre and Fibre-Rich By-products of Food Processing: Characterisation, Technological Functionality and Commercial Applications: A Review. *Food Chem.* **2011,** *124* (2), 411–421.

Faller, A.; Fialho, E. Polyphenol Content and Antioxidant Capacity in Organic and Conventional Plant Foods. *J. Food Compos. Anal.* **2010,** *23* (6), 561–568.

Feng, F.; Li, M.; Ma, F.; Cheng, L. Effects of Location within the Tree Canopy on Carbohydrates, Organic Acids, Amino Acids and Phenolic Compounds in the Fruit Peel and Flesh from Three Apple (*Malus × domestica*) Cultivars. *Hortic. Res.* **2014a,** *1,* 14019.

Feng, H.-L.; Tian, L.; Chai, W.-M.; Chen, X.-X.; Shi, Y.; Gao, Y.-S.; et al. Isolation and Purification of Condensed Tannins from Flamboyant Tree and their Antioxidant and Antityrosinase Activities. *Appl. Biochem. Biotechnol.* **2014b,** *173* (1), 179–192.

Fernandes, L.; Casal, S.; Cruz, R.; Pereira, J. A.; Ramalhosa, E. Seed Oils of Ten Traditional Portuguese Grape Varieties with Interesting Chemical and Antioxidant Properties. *Food Res. Int.* **2013,** *50* (1), 161–166.

Figuerola, F.; Hurtado, M. A. L.; Estévez, A. M.; Chiffelle, I.; Asenjo, F. Fibre Concentrates from Apple Pomace and Citrus Peel as Potential Fibre Sources for Food Enrichment. *Food Chem.* **2005,** *91* (3), 395–401.

Flórez, N.; Conde, E.; Domínguez, H. Microwave Assisted Water Extraction of Plant Compounds. *J. Chem. Technol. Biotechnol.* **2015,** *90* (4), 590–607.

Gawlik-Dziki, U.; Durak, A.; Jamioł, M.; Świeca, M. Interactions between Antiradical and Anti-Inflammatory Compounds from Coffee and Coconut Affected by Gastrointestinal Digestion—In Vitro Study. *LWT—Food Sci. Technol.* **2016,** *69,* 506–514.

Gengatharan, A.; Dykes, G. A.; Choo, W. S. Betalains: Natural Plant Pigments with Potential Application in Functional Foods. *LWT—Food Sci. Technol.* **2015,** *64* (2), 645–649.

Ghosh, A.; Sarkar, S.; Mandal, A. K.; Das, N. Neuroprotective Role of Nanoencapsulated Quercetin in Combating Ischemia–Reperfusion Induced Neuronal Damage in Young and Aged Rats. *PLoS ONE* **2013,** *8* (4), e57735.

González-Centeno, M.; Rosselló, C.; Simal, S.; Garau, M.; López, F.; Femenia, A. Physicochemical Properties of Cell Wall Materials Obtained from Ten Grape Varieties and their Byproducts: Grape Pomaces and Stems. *LWT—Food Sci. Technol.* **2010,** *43* (10), 1580–1586.

González-Montelongo, R.; Lobo, M. G.; González, M. Antioxidant Activity in Banana Peel Extracts: Testing Extraction Conditions and Related Bioactive Compounds. *Food Chem.* **2010,** *119* (3), 1030–1039.

Gupta, A.; Naraniwal, M.; Kothari, V. Modern Extraction Methods for Preparation of Bioactive Plant Extracts. *Int. J. Appl. Nat. Sci.* **2012,** *1* (1), 8–26.

Habib, H. M.; Kamal, H.; Ibrahim, W. H.; Al Dhaheri, A. S. Carotenoids, Fat Soluble Vitamins and Fatty Acid Profiles of 18 Varieties of Date Seed Oil. *Ind. Crops Prod.* **2013,** *42,* 567–572.

Hamerski, L.; Somner, G. V.; Tamaio, N. *Paullinia cupana* Kunth (Sapindaceae): A Review of Its Ethnopharmacology, Phytochemistry and Pharmacology. *J. Med. Plants Res.* **2013,** *7,* 2221–2229.

Hasler, C. M. Health Claims in the United States: An Aid to the Public or a Source of Confusion? *J. Nutr.* **2008,** *138* (6), 1216S–1220S.

Herrero, M.; Castro-Puyana, M.; Rocamora-Reverte, L.; Ferragut, J. A.; Cifuentes, A.; Ibáñez, E. Formation and Relevance of 5-Hydroxymethylfurfural in Bioactive Subcritical Water Extracts from Olive Leaves. *Food Res. Int.* **2012,** *47* (1), 31–37.

Hervert-Hernández, D.; García, O. P.; Rosado, J. L.; Goñi, I. The Contribution of Fruits and Vegetables to Dietary Intake of Polyphenols and Antioxidant Capacity in a Mexican Rural Diet: Importance of Fruit and Vegetable Variety. *Food Res. Int.* **2011,** *44* (5), 1182–1189.

Hijazi, A.; Bandar, H.; Rammal, H.; Hachem, A.; Saad, Z.; Badran, B. Techniques for the Extraction of Bioactive Compounds from Lebanese *Urtica dioica. Am. J. Phytomed. Clin. Therap.* **2013,** *1* (6), 507–513.

Hossain, M.; Barry-Ryan, C.; Martin-Diana, A. B.; Brunton, N. Optimisation of Accelerated Solvent Extraction of Antioxidant Compounds from Rosemary (*Rosmarinus officinalis* L.), Marjoram (*Origanum majorana* L.) and Oregano (*Origanum vulgare* L.) Using Response Surface Methodology. *Food Chem.* **2011,** *126* (1), 339–346.

Hsiao, P.-C.; Liaw, C.-C.; Hwang, S.-Y.; Cheng, H.-L.; Zhang, L.-J.; Shen, C.-C.; et al. Antiproliferative and Hypoglycemic Cucurbitane-Type Glycosides from the Fruits of *Momordica charantia. J. Agric. Food Chem.* **2013,** *61* (12), 2979–2986.

Huang, Y.; Chen, L.; Feng, L.; Guo, F.; Li, Y. Characterization of Total Phenolic Constituents from the Stems of *Spatholobus suberectus* Using LC-DAD-MSn and their Inhibitory Effect on Human Neutrophil Elastase Activity. *Molecules* **2013,** *18* (7), 7549–7556.

İnanç, A. L. Chlorophyll: Structural Properties, Health Benefits and Its Occurrence in Virgin Olive Oils. *Akad. Gıda* **2011,** *9* (2), 26–32.

Jaswir, I.; Noviendri, D.; Hasrini, R. F.; Octavianti, F. Carotenoids: Sources, Medicinal Properties and their Application in Food and Nutraceutical Industry. *J. Med. Plants Res.* **2011,** *5* (33), 7119–7131.

Johnson, L.; Strich, H.; Taylor, A.; Timmermann, B.; Malone, D.; Teufel-Shone, N.; et al. Use of Herbal Remedies by Diabetic Hispanic Women in the Southwestern United States. *Phytother. Res.* **2006,** *20* (4), 250–255.

Jung, C. H.; Cho, I.; Ahn, J.; Jeon, T. I.; Ha, T. Y. Quercetin Reduces High-Fat Diet-Induced Fat Accumulation in the Liver by Regulating Lipid Metabolism Genes. *Phytother. Res.* **2013,** *27* (1), 139–143.

Karim, A. A.; Azlan, A. Fruit Pod Extracts as a Source of Nutraceuticals and Pharmaceuticals. *Molecules* **2012,** *17* (10), 11931–11946.

Katalinic, V.; Mozina, S. S.; Generalic, I.; Skroza, D.; Ljubenkov, I.; Klancnik, A. Phenolic Profile, Antioxidant Capacity, and Antimicrobial Activity of Leaf Extracts from Six *Vitis vinifera* L. Varieties. *Int. J. Food Prop.* **2013,** *16* (1), 45–60.

Kazan, A.; Koyu, H.; Turu, I. C.; Yesil-Celiktas, O. Supercritical Fluid Extraction of *Prunus persica* Leaves and Utilization Possibilities as a Source of Phenolic Compounds. *J. Supercrit. Fluids* **2014,** *92,* 55–59.

Keefover-Ring, K.; Trowbridge, A.; Mason, C. J.; Raffa, K. F. Rapid Induction of Multiple Terpenoid Groups by Ponderosa Pine in Response to Bark Beetle-Associated Fungi. *J. Chem. Ecol.* **2016,** *42* (1), 1–12.

Khoddami, A.; Wilkes, M. A.; Roberts, T. H. Techniques for Analysis of Plant Phenolic Compounds. *Molecules* **2013,** *18* (2), 2328–2375.

Kim, S.-A.; Kim, S. H.; Kim, I. S.; Lee, D.; Dong, M.-S.; Na, C.-S.; et al. Simultaneous Determination of Bioactive Phenolic Compounds in the Stem Extract of *Rhus verniciflua* Stokes by High Performance Liquid Chromatography. *Food Chem.* **2013,** *141* (4), 3813–3819.

Krawitzky, M.; Arias, E.; Peiro, J.; Negueruela, A.; Val, J.; Oria, R. Determination of Color, Antioxidant Activity, and Phenolic Profile of Different Fruit Tissue of Spanish 'Verde Doncella'Apple Cultivar. *Int. J. Food Prop.* **2014,** *17* (10), 2298–2311.

Kubola, J.; Siriamornpun, S. Phenolic Contents and Antioxidant Activities of Bitter Gourd (*Momordica charantia* L.) Leaf, Stem and Fruit Fraction Extracts In Vitro. *Food Chem.* **2008,** *110* (4), 881–890.

Kumar, D.; Bhat, Z. A. Apigenin 7-Glucoside from Stachys tibetica Vatke and its Anxiolytic Effect in Rats. *Phytomedicine* **2014,** *21* (7), 1010–1014.

Liaotrakoon, W.; De Clercq, N.; Van Hoed, V.; Dewettinck, K. Dragon Fruit (*Hylocereus* spp.) Seed Oils: Their Characterization and Stability under Storage Conditions. *J. Am. Oil Chem. Soc.* **2013,** *90* (2), 207–215.

Liu, J.; Lu, J.-F.; Kan, J.; Wen, X.-Y.; Jin, C.-H. Synthesis, Characterization and In Vitro Anti-diabetic Activity of Catechin Grafted Inulin. *Int. J. Biol. Macromol.* **2014,** *64,* 76–83.

Liu, S.-C.; Lin, J.-T.; Wang, C.-K.; Chen, H.-Y.; Yang, D.-J. Antioxidant Properties of Various Solvent Extracts from Lychee (*Litchi chinenesis* Sonn.) Flowers. *Food Chem.* **2009,** *114* (2), 577–581.

Lockwood, G. B. The quality of Commercially Available Nutraceutical Supplements and Food Sources. *J. Pharm. Pharmacol.* **2011,** *63* (1), 3–10.

Lotfi, L.; Kalbasi-Ashtari, A.; Hamedi, M.; Ghorbani, F. Effects of Enzymatic Extraction on Anthocyanins Yield of Saffron Tepals (*Crocus sativus*) along with its Color Properties and Structural Stability. *J. Food Drug Anal.* **2015,** *23* (2), 210–218.

Low, D. Y.; D'Arcy, B.; Gidley, M. J. Mastication Effects on Carotenoid Bioaccessibility from Mango Fruit Tissue. *Food Res. Int.* **2015,** *67,* 238–246.

Ma, L.-Y.; Zhou, Q.-L.; Yang, X.-W. New SIRT1 Activator from Alkaline Hydrolysate of Total Saponins in the Stems–Leaves of *Panax ginseng. Bioorg. Med. Chem. Lett.* **2015,** *25* (22), 5321–5325.

Macfarlane, G. T.; Macfarlane, S. Manipulating the Indigenous Microbiota in Humans: Prebiotics, Probiotics, and Synbiotics. *The Human Microbiota: How Microbial Communities Affect Health and Disease*; 2013; pp 315–338.

Malaguti, M.; Dinelli, G.; Leoncini, E.; Bregola, V.; Bosi, S.; Cicero, A. F.; Hrelia, S. Bioactive Peptides in Cereals and Legumes: Agronomical, Biochemical and Clinical Aspects. *Int. J. Mol. Sci.* **2014,** *15* (11), 21120–21135.

Mallikarjun Gouda, K. G.; Udaya Sankar, K.; Sarada, R.; Ravishankar, G. Supercritical CO_2 Extraction of Functional Compounds from Spirulina and their Biological Activity. *J. Food Sci. Technol.* **2015,** *52* (6), 3627–3633.

Manchali, S.; Murthy, K. N. C.; Patil, B. S. Crucial Facts about Health Benefits of Popular Cruciferous Vegetables. *J. Funct. Foods* **2012,** *4* (1), 94–106.

Martínez-Las Heras, R.; Quifer-Rada, P.; Andrés, A.; Lamuela-Raventós, R. Polyphenolic Profile of Persimmon Leaves by High Resolution Mass Spectrometry (LC-ESI-LTQ-Orbitrap-MS). *J. Funct. Foods* **2016,** *23*, 370–377.

Matarese, F.; Cuzzola, A.; Scalabrelli, G.; D'Onofrio, C. Expression of Terpene Synthase Genes Associated with the Formation of Volatiles in Different Organs of *Vitis vinifera. Phytochemistry* **2014,** *105*, 12–24.

McCarty, M. F.; DiNicolantonio, J. J.; O'Keefe, J. H. Capsaicin May Have Important Potential for Promoting Vascular and Metabolic Health. *Open Heart* **2015,** *2* (1), e000262.

Meneses, M. A.; Caputo, G.; Scognamiglio, M.; Reverchon, E.; Adami, R. Antioxidant Phenolic Compounds Recovery from *Mangifera indica* L. By-products by Supercritical Antisolvent Extraction. *J. Food Eng.* **2015,** *163*, 45–53.

Messaoudi, R.; Abbeddou, S.; Mansouri, A.; Calokerinos, A. C.; Kefalas, P. Phenolic Profile and Antioxidant Activity of Date-Pits of Seven Algerian Date Palm Fruit Varieties. *Int. J. Food Prop.* **2013,** *16* (5), 1037–1047.

Middha, S. K.; Usha, T.; Pande, V. HPLC Evaluation of Phenolic Profile, Nutritive Content, and Antioxidant Capacity of Extracts Obtained from *Punica granatum* Fruit Peel. *Adv. Pharmacol. Sci.* **2013,** *2013*, Article ID 296236, 6 pages.

Mildner-Szkudlarz, S.; Bajerska, J.; Zawirska-Wojtasiak, R.; Górecka, D. White Grape Pomace as a Source of Dietary Fibre and Polyphenols and its Effect on Physical and Nutraceutical Characteristics of Wheat Biscuits. *J. Sci. Food Agric.* **2013,** *93* (2), 389–395.

Miron, T.; Herrero, M.; Ibáñez, E. Enrichment of Antioxidant Compounds from Lemon Balm (*Melissa officinalis*) by Pressurized Liquid Extraction and Enzyme-Assisted Extraction. *J. Chromatogr. A* **2013,** *1288*, 1–9.

Mlcek, J.; Rop, O. Fresh Edible Flowers of Ornamental Plants—A New Source of Nutraceutical Foods. *Trends Food Sci. Technol.* **2011,** *22* (10), 561–569.

Monroy, R.; Colín, H. El guamúchil Pithecellobium dulce (Roxb.) Benth, un ejemplo de uso múltiple. *Madera Bosq.* **2004,** *10* (1), 35–53.

Morel, I.; Lescoat, G.; Cogrel, P.; Sergent, O.; Pasdeloup, N.; Brissot, P.; et al. Antioxidant and Iron-Chelating Activities of the Flavonoids Catechin, Quercetin and Diosmetin on Iron-Loaded Rat Hepatocyte Cultures. *Biochem. Pharmacol.* **1993,** *45* (1), 13–19.

Moure, A.; Cruz, J. M.; Franco, D.; Domínguez, J. M.; Sineiro, J.; Domínguez, H.; et al. Natural Antioxidants from Residual Sources. *Food Chem.* **2001,** *72* (2), 145–171.

Muhammad, K.; Zahari, N. I. M.; Gannasin, S. P.; Adzahan, N. M.; Bakar, J. High Methoxyl Pectin from Dragon Fruit (*Hylocereus polyrhizus*) Peel. *Food Hydrocolloids* **2014**, *42*, 289–297.

Müller, L.; Caris-veyrat, C.; Lowe, G.; Böhm, V. Lycopene and Its Antioxidant Role in the Prevention of Cardiovascular Diseases—A Critical Review. *Crit. Rev. Food Sci. Nutr.* **2015**, *56* (11), 1868–1879.

Naczk, M.; Shahidi, F. Extraction and Analysis of Phenolics in Food. *J. Chromatogr. A* **2004**, *1054* (1), 95–111.

Nagmoti, D. M.; Kothavade, P. S.; Bulani, V. D.; Gawali, N. B.; Juvekar, A. R. Antidiabetic and Antihyperlipidemic Activity of *Pithecellobium dulce* (Roxb.) Benth Seeds Extract in Streptozotocin-Induced Diabetic Rats. *Eur. J. Integr. Med.* **2015**, *7* (3), 263–273.

Nawirska, A.; Kwaśniewska, M. Dietary Fibre Fractions from Fruit and Vegetable Processing Waste. *Food Chem.* **2005**, *91* (2), 221–225.

Nayak, B.; Dahmoune, F.; Moussi, K.; Remini, H.; Dairi, S.; Aoun, O.; Khodir, M. Comparison of Microwave, Ultrasound and Accelerated-Assisted Solvent Extraction for Recovery of Polyphenols from *Citrus sinensis* Peels. *Food Chem.* **2015**, *187*, 507–516.

Nyam, K. L.; Tan, C. P.; Karim, R.; Lai, O. M.; Long, K.; Man, Y. B. C. Extraction of Tocopherol-Enriched Oils from Kalahari Melon and Roselle Seeds by Supercritical Fluid Extraction (SFE-CO_2). *Food Chem.* **2010**, *119* (3), 1278–1283.

Olivas-Aguirre, F.; Wall-Medrano, A.; Gonzalez-Aguilar, G. A.; López-Díaz, J.; Alvarez-Parrilla, E.; De la Rosa, L. A.; Ramos-Jimenez, A. Hydrolyzable Tannins: Biochemistry, Nutritional & Analytical Aspects and Health Effects. *Nutr. Hosp.* **2014**, *31* (1), 55–66.

Otero-Pareja, M. J.; Casas, L.; Fernández-Ponce, M. T.; Mantell, C.; Ossa, E. J. Green Extraction of Antioxidants from Different Varieties of Red Grape Pomace. *Molecules* **2015**, *20* (6), 9686–9702.

Paes, J.; Dotta, R.; Barbero, G. F.; Martínez, J. Extraction of Phenolic Compounds and Anthocyanins from Blueberry (*Vaccinium myrtillus* L.) Residues Using Supercritical CO_2 and Pressurized Liquids. *J. Supercrit. Fluids* **2014**, *95*, 8–16.

Palafox-Carlos, H.; Yahia, E.; González-Aguilar, G. Identification and Quantification of Major Phenolic Compounds from Mango (*Mangifera indica*, cv. Ataulfo) fruit by HPLC–DAD–MS/MS-ESI and their individual contribution to the antioxidant activity during ripening. *Food Chem.* **2012**, *135* (1), 105–111.

Patel, S. *Emerging Bioresources with Nutraceutical and Pharmaceutical Prospects*: Springer, 2015.

Pitchai, D.; Roy, A.; Ignatius, C. In Vitro Evaluation of Anticancer Potentials of Lupeol Isolated from *Elephantopus scaber* L. on MCF-7 Cell Line. *J. Adv. Pharmaceut. Technol. Res.* **2014**, *5* (4), 179.

Prakash, D.; Gupta, C. Glucosinolates: The Phytochemicals of Nutraceutical Importance. *J. Complement Integr. Med.*, **2012**, *9*, Article 13.

Pujol, D.; Liu, C.; Gominho, J.; Olivella, M.; Fiol, N.; Villaescusa, I.; Pereira, H. The Chemical Composition of Exhausted Coffee Waste. *Ind. Crops Prod.* **2013**, *50*, 423–429.

Puri, M.; Sharma, D.; Barrow, C. J. Enzyme-Assisted Extraction of Bioactives from Plants. *Trends Biotechnol.* **2012**, *30* (1), 37–44.

Regulations, N. I. L. E. Ministerial Ordinance No. 41, July 1991. *Amendment to Ministerial Ordinance*; 1996, 33.

Reyes, L. F.; Villarreal, J. E.; Cisneros-Zevallos, L. The Increase in Antioxidant Capacity after Wounding Depends on the Type of Fruit or Vegetable Tissue. *Food Chem.* **2007**, *101* (3), 1254–1262.

Rodrigues, S.; Fernandes, F. A.; de Brito, E. S.; Sousa, A. D.; Narain, N. Ultrasound Extraction of Phenolics and Anthocyanins from Jabuticaba Peel. *Ind. Crops Prod.* **2015**, *69*, 400–407.

Rombaut, N.; Tixier, A. S.; Bily, A.; Chemat, F. Green Extraction Processes of Natural Products as Tools for Biorefinery. *Biofuels, Bioprod. Biorefining* **2014**, *8* (4), 530–544.

Romero-García, J. M.; Lama-Muñoz, A.; Rodríguez-Gutiérrez, G.; Moya, M.; Ruiz, E.; Fernández-Bolaños, J.; Castro, E. Obtaining Sugars and Natural Antioxidants from Olive Leaves by Steam-Explosion. *Food Chem.* **2016**, *210*, 457–465.

Ruiz-Montañez, G.; Ragazzo-Sánchez, J.; Calderón-Santoyo, M.; Velazquez-De La Cruz, G.; de León, J. R.; Navarro-Ocaña, A. Evaluation of Extraction Methods for Preparative Scale Obtention of Mangiferin and Lupeol from Mango Peels (*Mangifera indica* L.). *Food Chem.* **2014**, *159*, 267–272.

Saini, R. K.; Nile, S. H.; Park, S. W. Carotenoids from Fruits and Vegetables: Chemistry, Analysis, Occurrence, Bioavailability and Biological Activities. *Food Res. Int.* **2015**, *76*, 735–750.

Samet, I.; Villareal, M. O.; Motojima, H.; Han, J.; Sayadi, S.; Isoda, H. Olive Leaf Components Apigenin 7-Glucoside and Luteolin 7-Glucoside Direct Human Hematopoietic Stem Cell Differentiation towards Erythroid Lineage. *Differentiation* **2015**, *89* (5), 146–155.

Santos, D. T.; Veggi, P. C.; Meireles, M. A. A. Extraction of antioxidant compounds from Jabuticaba (*Myrciaria cauliflora*) Skins: Yield, Composition and Economical Evaluation. *J. Food Eng.* **2010**, *101* (1), 23–31.

Santos, J.; Herrero, M.; Mendiola, J.; Oliva-Teles, M.; Ibáñez, E.; Delerue-Matos, C.; Oliveira, M. Fresh-cut Aromatic Herbs: Nutritional Quality Stability during Shelf-Life. *LWT—Food Sci. Technol.* **2014**, *59* (1), 101–107.

Saura-Calixto, F.; Serrano, J.; Goñi, I. Intake and Bioaccessibility of Total Polyphenols in a Whole Diet. *Food Chem.* **2007**, *101* (2), 492–501.

Scalbert, A.; Johnson, I. T.; Saltmarsh, M. Polyphenols: Antioxidants and Beyond. *Am. J. Clin. Nutr.* **2005**, *81* (1), 215S–217S.

Schimpl, F. C.; Kiyota, E.; Mayer, J. L. S.; de Carvalho Gonçalves, J. F.; da Silva, J. F.; Mazzafera, P. Molecular and Biochemical Characterization of Caffeine Synthase and Purine Alkaloid Concentration in Guarana Fruit. *Phytochemistry* **2014**, *105*, 25–36.

Semwal, R. B.; Semwal, D. K.; Combrinck, S.; Viljoen, A. Butein: From Ancient Traditional Remedy to Modern Nutraceutical. *Phytochem. Lett.* **2015**, *11*, 188–201.

Sen, C. K.; Khanna, S.; Roy, S. Tocotrienols: Vitamin E beyond Tocopherols. *Life Sci.* **2006**, *78* (18), 2088–2098.

Sen, C. K.; Rink, C.; Khanna, S. Palm Oil-Derived Natural Vitamin E α-Tocotrienol in Brain Health and Disease. *J. Am. Coll. Nutr.* **2010**, *29* (sup3), 314S–323S.

Sen, S.; Asokkumar, K.; Umamaheswari, M.; Sivashanmugam, A.; Subhadradevi, V. Antiulcerogenic Effect of Gallic Acid in Rats and its Effect on Oxidant and Antioxidant Parameters in Stomach Tissue. *Indian J. Pharm. Sci.* **2013**, *75* (2), 149–155.

Setyaningsih, W.; Saputro, I.; Palma, M.; Barroso, C. Optimisation and Validation of the Microwave-Assisted Extraction of Phenolic Compounds from Rice Grains. *Food Chem.* **2015**, *169*, 141–149.

Shahidi, F. Nutraceuticals and Functional Foods: Whole versus Processed Foods. *Trends Food Sci. Technol.* **2009**, *20* (9), 376–387.

Shan, B.; Cai, Y. Z.; Sun, M.; Corke, H. Antioxidant Capacity of 26 Spice Extracts and Characterization of their Phenolic Constituents. *J. Agric. Food Chem.* **2005**, *53* (20), 7749–7759.

Sharma, S. K.; Bansal, S.; Mangal, M.; Dixit, A. K.; Gupta, R. K.; Mangal, A. Utilization of Food Processing By-products as Dietary, Functional and Novel Fibre: A Review. *Crit. Rev. Food Sci. Nutr.* (just-accepted), **2015**, *56* (10), 1647–1661.

Siddiqui, B. S.; Hasan, M.; Mairaj, F.; Mehmood, I.; Hafizur, R. M.; Hameed, A.; Shinwari, Z. K. Two New Compounds from the Aerial Parts of *Bergenia himalaica* Boriss and their Anti-Hyperglycemic Effect in Streptozotocin–Nicotinamide Induced Diabetic Rats. *J. Ethnopharmacol.* **2014**, *152* (3), 561–567.

Siro, I.; Kapolna, E.; Kapolna, B.; Lugasi, A. Functional Food. Product Development, Marketing and Consumer Acceptance—A Review. *Appetite* **2008**, *51* (3), 456–467.

Solana, M.; Boschiero, I.; Dall'Acqua, S.; Bertucco, A. A Comparison between Supercritical Fluid and Pressurized Liquid Extraction Methods for Obtaining Phenolic Compounds from *Asparagus officinalis* L. *J. Supercrit. Fluids* **2015**, *100*, 201–208.

Souto, A. A.; Campos, M. M.; Morrone, F. B.; da Silva, R. B. M.; de Sousa Maciel, I. *Composition Containing Resveratrol and/or Derivatives thereof and Plant Oil, Process for Producing Said Composition, Nutraceutical and/or Pharmaceutical Product, and Method for Enhancing the Potential of Resveratrol.* Google Patents, 2014.

Srivastava, S.; Sonkar, R.; Mishra, S. K.; Tiwari, A.; Balramnavar, V.; Mir, S.; et al. Antidyslipidemic and Antioxidant Effects of Novel Lupeol-Derived Chalcones. *Lipids.* **2013**, *48* (10), 1017–1027.

Swain, J.; Kumar Mishra, A. Location, Partitioning Behavior, and Interaction of Capsaicin with Lipid Bilayer Membrane: Study Using Its Intrinsic Fluorescence. *J. Phys. Chem. B* **2015**, *119* (36), 12086–12093.

Szakiel, A.; Niżyński, B.; Pączkowski, C. Triterpenoid Profile of Flower and Leaf Cuticular Waxes of Heather *Calluna vulgaris.* *Nat. Prod. Res.* **2013**, *27* (15), 1404–1407.

Tabart, J.; Kevers, C.; Sipel, A.; Pincemail, J.; Defraigne, J.-O.; Dommes, J. Optimisation of Extraction of Phenolics and Antioxidants from Black Currant Leaves and Buds and of Stability during Storage. *Food Chem.* **2007**, *105* (3), 1268–1275.

Thoo, Y. Y.; Ho, S. K.; Liang, J. Y.; Ho, C. W.; Tan, C. P. Effects of Binary Solvent Extraction System, Extraction Time and Extraction Temperature on Phenolic Antioxidants and Antioxidant Capacity from Mengkudu (*Morinda citrifolia*). *Food Chem.* **2010**, *120* (1), 290–295.

Tolba, M. F.; Azab, S. S.; Khalifa, A. E.; Abdel-Rahman, S. Z.; Abdel-Naim, A. B. Caffeic Acid Phenethyl Ester, a Promising Component of Propolis with a Plethora of Biological Activities: A Review on its Anti-inflammatory, Neuroprotective, Hepatoprotective, and Cardioprotective Effects. *IUBMB Life* **2013**, *65* (8), 699–709.

Torres-Moreno, H.; Velázquez, C.; Garibay-Escobar, A.; Curini, M.; Marcotullio, M.; Robles-Zepeda, R. Antiproliferative and Apoptosis Induction of Cucurbitacin-Type Triterpenes from *Ibervillea sonorae.* *Ind. Crops Prod.* **2015**, *77*, 895–900.

Tuncel, N. B.; Yılmaz, N. Optimizing the Extraction of Phenolics and Antioxidants from feijoa (*Feijoa sellowiana*, Myrtaceae). *J. Food Sci. Technol.* **2015**, *52* (1), 141–150.

Turkmen, N.; Sari, F.; Velioglu, Y. S. Effects of Extraction Solvents on Concentration and Antioxidant Activity of Black and Black Mate Tea Polyphenols Determined by Ferrous Tartrate and Folin–Ciocalteu Methods. *Food Chem.* **2006**, *99* (4), 835–841.

Uchôa-Thomaz, A. M. A.; Sousa, E. C.; Carioca, J. O. B.; Morais, S. M. D.; Lima, A. D.; Martins, C. G.; et al. Chemical Composition, Fatty Acid Profile and Bioactive Compounds of Guava Seeds (*Psidium guajava* L.). *Food Sci. Technol. (Campinas)* **2014**, *34* (3), 485–492.

Valls, J.; Pasamontes, N.; Pantaleón, A.; Vinaixa, S.; Vaqué, M.; Soler, A.; et al. Prospects of Functional Foods/Nutraceuticals and Markets. *Natural Products*; Springer: Berlin, 2013; pp 2491–2525.

Verhagen, H.; Vos, E.; Francl, S.; Heinonen, M.; van Loveren, H. Status of Nutrition and Health Claims in Europe. *Arch. Biochem. Biophys.* **2010,** *501* (1), 6–15.

Wang, L.; Weller, C. L. Recent Advances in Extraction of Nutraceuticals from Plants. *Trends Food Sci. Technol.* **2006,** *17* (6), 300–312. doi:http://dx.doi.org/10.1016/j.tifs.2005.12.004.

Wang, X.; Liu, R.; Zhang, W.; Zhang, X.; Liao, N.; Wang, Z.; et al. Oleanolic Acid Improves Hepatic Insulin Resistance via Antioxidant, Hypolipidemic and Anti-inflammatory Effects. *Mol. Cell. Endocrinol.* **2013,** *376* (1), 70–80.

Yeum, K.-J.; Russell, R. M. Carotenoid Bioavailability and Bioconversion. *Annu. Rev. Nutr.* **2002,** *22* (1), 483–504.

Yin, D.-D.; Yuan, R.-Y.; Wu, Q.; Li, S.-S.; Shao, S.; Xu, Y.-J.; et al. Assessment of Flavonoids and Volatile Compounds in Tea Infusions of Water Lily Flowers and their Antioxidant Activities. *Food Chem.* **2015,** *187,* 20–28.

Zawistowski, J.; Debasis, B. Regulation of Functional Foods in Selected Asian Countries in the Pacific Rim. In *Nutraceutical and Functional Food Regulations in the United States and Around the World*, first ed.; Bagchi, D.; Elsevier Inc.: USA, 2008; pp 365–401.

CHAPTER 5

HARVEST BY-PRODUCTS OF FRESH FRUITS AND VEGETABLES

A. E. QUIRÓS-SAUCEDA[1], G. R. VELDERRAIN-RODRIGUEZ[1],
J. ABRAHAM DOMÍNGUEZ-AVILA[1], H. PALAFOX-CARLOS[2],
J. F. AYALA-ZAVALA[1], and G. A. GONZALEZ-AGUILAR[1*]

[1]Centro de Investigación en Alimentación y Desarrollo, AC
(CIAD, AC), Carretera a la Victoria Km 0.6, La Victoria CP 83000,
Hermosillo, Sonora, Mexico

[2]Herbalife, Camino al Iteso No. 8900 Int. 1 A, Col. El Mante,
CP 45609, Tlaquepaque, Jalisco, Mexico

*Corresponding author. E-mail: gustavo@ciad.mx

CONTENTS

ABSTRACT

In many cases, the raw fresh fruits and vegetables are not consumed directly by humans, but first undergo processing to separate the desired value product from other constituents of the plant. The processing of plant foods into different products, or their consumption as such, results in the production of by-products which are usually wasted. However, it has been reported that these materials contain valuable substances such as pigments, sugars, organic acids, flavors, and bioactive compounds which possess different biological properties, such as antioxidants, enzymes, antimicrobial compounds, and fibers. In this sense, by-products of fresh fruits and vegetables have gained increasing interest since they can be used to isolate specific phytochemicals for application in nutraceutical supplements, dietary additives, new food, and pharmaceutical products, contributing to the recovery of agro-industrial process waste, with major industrial, economic, and environmental impact.

5.1 INTRODUCTION

Fruit and vegetable processing generates substantial quantities of waste/by-products that represent a major disposal problem for the industry concerned. For example, it is known to remove the juices from agricultural products by various methods, such as pressing. After removal of such juices, a considerable amount of by-products are produced, which is a problem since the plant material is usually prone to microbial spoilage, thus, limiting further exploitation (Oreopoulou and Tzia, 2007). For this reason, agricultural by-products have been commonly used only in animal feed or fertilizer, without economic benefits (Nikolic et al., 1986). By-products include remnants of fruits and vegetables, such as seeds, stems, peels, and flesh.

However, several authors have shown that plant foods by-products are also promising sources of compounds which may be used because of their favorable technological or nutritional properties (Fernández-López et al., 2004; Serna Cock and Torres León, 2015). They are composed principally by water, soluble sugars, dietary fiber, organic acids, amino acids and proteins, enzymes, minerals, oils, and lipids and also contain phenolic compounds and carotenoids (Ayala-Zavala et al., 2011). All these components are found in different amounts depending on the fraction of the agro-industrial by-product. Potential properties reported for these compounds include their uses as antioxidants, antimicrobials, flavors, thickeners, and nutraceuticals. In this sense, the extraction and use of these bioactive molecules are considered completely safe in comparison with synthetic compounds (Murthy and Naidu, 2012).

This chapter focuses primarily on describing the characterization of the plant foods derived by-products, as well as analyzes the potential uses as food additives and nutraceuticals.

5.2 FRESH FRUITS AND VEGETABLES BY-PRODUCTS PRODUCTION AND STATISTICS

The fruit and vegetable processing industry are among the most important activities that can be termed as agriculturally based (Dauthy, 1995). This activity is based on the transformation of fresh plant foods for the generation of preserved food products, such as wine, juices, jams, vegetable oil, potato starch, sugar production, among others (Sinha et al., 2012). However, from an environmental point of view, processing of fruits and vegetables produces large amount of solid waste, which they are commonly called by-products (Anal, 2013). Furthermore, large amount of fruits and vegetables by-products are also produced during the postharvest and packaging steps in the supply chain.

Once the crops have been harvested, they go through a series of steps to be packed and distributed raw or to be processed into other food products. During the supply chain, sorting is one of the primary reasons for by-products generation. Sorting is the removal of products based on quality or appearance criteria, including specifications for size, color, weight, blemish level, and Brix (a measure sugar content) (Gunders, 2012). As a result, this activity produces about 3% of potential by-products that could be processed and considered valuable products (FAO, 2011). However, most by-product production occurs during processing of fruits and vegetables, as well as by their consumption as such (Fig. 5.1). Trimming, peeling, cutting (size reduction), and pressing are processing steps aims to remove the parts of the fresh fruits and vegetables which are either not edible or difficult to digest (Dauthy, 1995). After removing the nonedible part, the edible part follows the food transformation process. The remaining waste produced (by-products) are peels, seeds, flesh, pomace, skin, and mill which were usually wasted, but nowadays, these by-products are converted to different high value-added compounds, particularly the fiber fraction (Ayala-Zavala et al., 2011).

The processing percentage of the harvested fruits and vegetables varies widely and depend on the type of the destined processing industry (fresh-cut, canning, juices, dried/dehydrating, etc.). For example, in the United States, more than half of total fruit and vegetable production goes into processing. Among fruits, over 70% of citrus production is processed, being oranges and grapes the top two fruit crops processed. Citrus fruits, primarily oranges, are processed mostly into juice, while grapes are processed primarily into juice, wine, and raisins. The quantity of grapes used for wine production alone makes up over one-third of all fruit processed, and raisins make up well over half of all dried fruit production (Lucier et al., 2006). Moreover, among vegetables, processing accounted for about 50% of total output during 2002–2004. Tomatoes (89% processed) and potatoes (68% of sales processed) are top two vegetables crops processed (USDA, 2012, 2014). Depending on the production process, 5–85% of fruits and vegetables by-products are produced (de Las Fuentes, 2002).

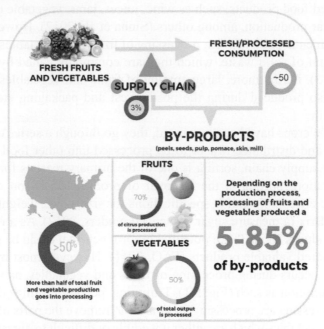

FIGURE 5.1   Harvest by-products of fresh fruits and vegetables.

5.3   CHARACTERIZATION OF THE DERIVED BY-PRODUCTS

Fruit and vegetables by-products are source of bioactive compounds that can be extracted and used as a food additives and/or to create higher value products such as dietary supplements and/or fortified foods (Ayala-Zavala et al., 2011). Therefore, the characterization of these agro-industrial by-products has been of utmost importance. Apple, grape, banana, mango, pineapple, citrus fruits, tomato, carrot, onion, olive, and potato are among the main generators of fruit and vegetables by-products, which are sources of pectin, dietary fiber, starch, oil, phenolic compounds (phenolic acids, flavonoids), carotenoids, anthocyanin pigments, and enzymes (Table 5.1).

In general, the solid residues (peels, seeds) generated from the processing of fruits and vegetables contain a remarkable dietary fiber fraction that has a great physiological and technological potential. Dietary fiber consists of remnants of the plant cells resistant to hydrolysis by the alimentary enzymes of man, and it is constituted by hemicelluloses, cellulose, lignin, oligosaccharides, pectins, gums, and waxes (Trowell, 1976). This bioactive compound has been used as ingredient in the food industry to improve the viscosity, texture, sensory characteristics, and shelf life of mainly breakfast cereal and bakery products such as integral breads and cookies (Elleuch et al., 2011; O'Shea et al., 2012). Moreover, production of pectin is considered the most reasonable way of utilizing apple pomace; also, fiber-pectins may easily be recovered from citrus fruits by-products like orange and lime peels (Schieber et al., 2001; Okino Delgado and Fleuri, 2016). Pectins are natural hydrocolloids that are principal structural elements of plant cell walls, and is also a component of soluble fibers which is well-documented that plays a significant role in the prevention of several illnesses (Trowell, 1976). Pectins are widely used as a gelling agents, stabilizers, and emulsifiers in the food industry as dairy products, desserts, and soft drinks. However, its applications extend to the cosmetic and pharmaceutical industries (May, 1990; Rubio-Senent et al., 2015).

Moreover, huge amounts of phenolic compounds with antioxidant activity have been identified in several fruits and vegetables by-products, mainly peels and seeds (Ayala-Zavala et al., 2011; Ignat et al., 2011). Phenolic compounds comprise a wide and diverse group of molecules classified as secondary metabolites in plants and their regular consumption has been associated with a reduced risk of a number of chronic diseases (Kris-Etherton et al., 2002). The protective biological properties of phenolic compounds on human health are based on their antioxidant, antiinflammatory, antiproliferative, antimicrobial, antimutagenic, antioangiogenic and neuroprotective actions, among others (Han et al., 2007). Many of these activities have been attributed to the potent properties of free-radical scavenging (e.g., ROS/RNS) and metal-ion chelating of phenolic compounds (Lü et al., 2010). In this sense, phenolic compounds present in harvest agricultural by-products can be used for the development of functional foods to promote the state of human well-being and as food additives to preserve flavor, taste, and appearance.

TABLE 5.1 Fruits and Vegetables By-products.

Fruit/ vegetable	Production process	By-products (%)	Type of by-products	Bioactive compounds	References
Citrus	Juice, canning, marmalade	50	Seeds, peels, flesh	Pectin, D-limonene, ethanol, cold-pressed oils, flavonoids, limonoids	Marin et al. (2007), Bousbia et al. (2009)
Grapes	Red and white wine	20–30	Pomace	Dietary fiber, pectins, ethanol, tartrates, citric acid, seed oil, anthocyanins, catechins, phenolic compounds	Mildner-Szkudlarz et al. (2013), Kabir et al. (2015)
Apple	Juice		Pomace	Pectin, catechins, hydroxycinnamates	Bhushan et al. (2008), Wang et al. (2014)
Mango	Juice, canning	35–60	Peels, seeds	Phenolic compounds (gallic and ellagic acids, gallates), phospholipids, pectin, rhamnose, antioxidant dietary fiber	Ajila and Rao (2013), de Lourdes García-Magaña et al. (2013)
Pineapple	Juice	40–50	Peels, core	Sucrose, starch, hemicellulose, bromelain, antioxidant dietary fiber	Larrauri et al. (1997b), Chaurasiya and Hebbar (2013), Pardo et al. (2014)
Banana	Peeling, cooking	30	Peels	Protein, ethanol, α-amylase, hemicellulases, cellulases, carotenoids	Paul and Sumathy (2013), Gu et al. (2014)
Guava	Juice	10–15	Peels, seeds, flesh	Low methoxylated pectin, oil (fatty acids), antioxidant dietary fiber	Prasad and Azeemoddin (1994), Jiménez-Escrig et al. (2001)
Passion fruit	Juice	75	Rind, seeds	Pectin, linoleic acid	Malacrida and Jorge (2012), Liew et al. (2014)
Mandarin	Peeled	16	Peels	Phenolic compounds, limonioid glucosides, pectin	Ayala-Zavala et al. (2010), Coll-Almela et al. (2015)
Cactus pear	Peeled	20–45	Spines, epidermis, glochids	Antioxidant dietary fiber, fiber (pectin, hemicellulose, cellulose), phenolic compounds, carotenoids	Bensadón et al. (2010)

TABLE 5.1 *(Continued)*

Fruit/ vegetable	Production process	By-products (%)	Type of by-products	Bioactive compounds	References
Coffee	Soluble coffee production	50	Flesh and husk	Fatty acids, cellulose, hemicellulose, phenolic compounds	Pujol et al. (2013)
Tomato	Juice	3–7	Skin, seeds, pomace	Lycopene, β-carotene, vitamin B12	Kumcuoglu et al. (2014), Stajčić et al. (2015)
Carrot	Juice	30–40	Pomace	β-Carotene, ethanol, phenolic compounds	Mustafa et al. (2012)
Olive	Oil extraction	50–100	Vegetation water, skin, seeds	Hydroxytyrosol derivatives, pectic oligosaccharides	Herrero et al. (2011), Lama-Muñoz et al. (2012)
Red beet	Juice	15–30	Pomace, peel	Phenolic compounds (p-coumaric and ferulic acids), betacyanins, betaxanthins	Schieber et al. (2001)
Potato	French fries, chips, puree	15–40	Peel	Phenolic acids (chlorogenic, gallic, protocatechuic and caffeic acids), α-amylase	Jadhav et al. (2013)
Onion	Peeled	15	Peel	Flavonoids (mainly quercetin), phenolic compounds, frutooligosaccharides and sulfur compounds	Roldán et al. (2008)

Antioxidant dietary fiber is a fiber concentrate containing significant amounts of natural antioxidants (mainly phenolic compounds) associated with nondigestible compounds (Saura-Calixto, 1998). This material combines the physiological properties of both dietary fiber and phenolic compounds and promises to be a potential food ingredient useful in enhancing the bioactive and technological properties of different products (Quirós-Sauceda et al., 2014). Spines, epidermis, and glochids from cactus pear, carrots peels, mango peel, and pineapple shells are some of the reported by-products that are source of antioxidant dietary fiber (Larrauri et al., 1997a, 1997b; Chantaro et al., 2008; Bensadón et al., 2010).

On the other hand, carotenoids are red, orange, and yellow lipid-soluble pigments found embedded in the membranes of chloroplasts and chromoplasts of plant cells, occur mainly on the surface of the plant tissues such as external pericarp and peels and play diverse roles in photobiology, photochemistry, and medicine (Edge et al., 1997; Ayala-Zavala et al., 2011). In human beings, carotenoids can serve several important biological activities as antioxidants, thus, are authorized as food ingredients to produce health products; furthermore, the extraction and use of carotenoids as colorant food additives is a common practice. Tomato-processing wastes (peel and seeds) contains a high concentration of carotenoids, mainly lycopene (70–80%) (Vági et al., 2007). It has been reported that carotenoids are mainly found in tomato peel than in other tomato products; in addition, tomato peel and seed show better antioxidant activities than that of tomato flesh (Toor and Savage, 2005). Tomato seeds contain 18–27% oil that shows physicochemical characteristics similar to any conventional oil (Al-Wandawi et al., 1985). Moreover, mango by-products, especially seeds and peels, are considered to be cheap sources of valuable food nutraceutical ingredients such as carotenoids. Ajila et al. (2007) reported that carotenoid content in mango peels is 4–8 times higher in ripe mango peels than in raw mango peels.

Anthocyanins, that form part of the group of flavonoids, comprise the largest group of water-soluble pigments that are widely distributed in the high plants, roots, caudexes, and leaves as well as flowers and fruits (Fan et al., 2008). They have a long history as part of the human diet and are associated to positive health effects, so humans use them for therapeutic purpose. Furthermore, anthocyanins are commonly used as natural antioxidants or colorants (Bridle and Timberlake, 1997). Natural valuable bioactive compounds, including anthocyanins, can be extracted from by-products (peel, flesh, and seed) of grape, pineapple, papaya, cashew apple, surinam cherry, among others (Corrales et al., 2008; da Silva et al., 2014). It is well known that by-products obtained after vinification of different grape varieties contains appreciable amounts of anthocyanins for potential development for nutraceutical purposes (Ky et al., 2014).

Agro-industrial wastes have been used as solid supports for the production of enzymes, using the solid-state fermentation technique. Examples include the production of cellulases, xylanases, ligninases, pectinases, lacases, and lipases enzymes (Sánchez et al., 2015). The fruit and vegetables by-products contains high levels of essential nutrients for growth microorganisms, which produce different enzymes that have numerous applications in industrial processes for food, drug, textile, and dye use (Das and Singh, 2004; Hasan et al., 2006). Some examples are mentioned below. Reddy et al. (2003) reported a lignolytic and cellulolytic enzymes formation during degradation of banana waste by *Pleurotus ostreatus* and *Pleurotus sajor-caju*. Also, grape pomace, the main waste in the wine industry, has been shown to be sole nutrient source to produce hydrolytic enzymes (cellulases, xylanases, and pectinases) using *Aspergillus awamori*. Mixture of olive mill wastes with winery wastes favors the production of endocellulases, endoxylanases, and feruloyl esterases with *Aspergillus niger*, *Aspergillus ibericus*, and *Aspergillus uvarum* (Salgado et al., 2014). Besides this process, some fruit by-products contain natural enzymes. Pineapple by-products, such as stems, are rich source of bromelain proteolytic enzyme, which is probably the most valuable and the most studied bioactive compound from pineapple fruit (Devakate et al., 2009). Bromelain is an aqueous extract that contains a complex mixture of proteases and nonproteases components, as well as several other substances in smaller quantities. The proteolytic enzymes are sulfhydryl proteases, since a free sulfhydryl group of a cysteine side chains is required for function. These enzymes perform an important role in proteolytic modulation of the cellular matrix in numerous physiological processes (Soares et al., 2012).

5.4 CONSIDERATION FOR PRACTICAL USES OF THE DERIVED BY-PRODUCTS

From economic and environmental points of view, adding value to harvest by-products of fruit and vegetables is always desirable. Such residues are usually used for animal feed; however, others several potential uses can be considered due the antioxidant, antimicrobial, sensorial, thickener, and health properties they present (Fig. 5.2) (Ayala-Zavala et al., 2010; O'Shea et al., 2012; Okino Delgado and Fleuri, 2016). One of the major potential uses can be as food additives, to make high-value products economically attractive (Fig. 5.3).

Peel
Dietary fiber, phenolic compounds, anthocyanins pigments, carotenoids

Flesh
Phenolic compounds, enzymes, dietary fiber, carotenoids

Seed
Phenolic compounds, oil

Steam
Phenolic compounds, enzymes, anthocyanins

FIGURE 5.2 Bioactive compounds in by-products of fresh fruits and vegetables.

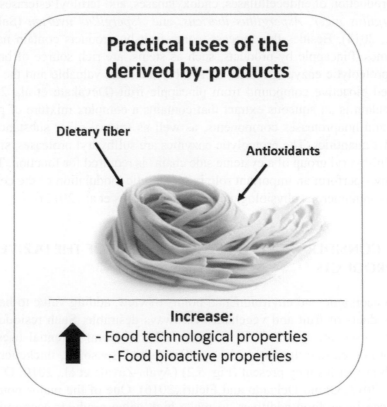

Practical uses of the derived by-products

Dietary fiber

Antioxidants

Increase:
- Food technological properties
- Food bioactive properties

FIGURE 5.3 Practical uses of the derived by-products of fresh fruits and vegetables.

5.4.1 FOOD ADDITIVES

Harvest fresh fruits and vegetables by-products have been used to protect a number of edible foodstuffs against bacterial/fungal growth or to increase their antioxidant capacity, both of which are related to their phenolic compounds content (Ayala-Zavala et al., 2008). Other uses are possible, such as thickening agents, which is mostly dependent on their dietary fiber content, with items of high pectin or cellulose content being appropriate options. In most cases, sensory parameters will be modified and can be perceived by the consumer, possibly affecting desirability.

5.4.1.1 ANTIMICROBIAL USES

Fish oil is valued because of its important essential ω-3 fatty acid content; however, these fatty acids are easily oxidized and can be spoiled due to bacterial growth. Chardonnay grape and black raspberry seed flours were used in the preservation of fish oil and compared to tocopherols. Both treatments showed total inhibition of *Escherichia coli* but null effect against *Lysteria monocytogenes* at 35°C, *L. monocytogenes* was only inhibited when stored at 4°C; tocopherols inhibited *E. coli* but not *L. monocytogenes*. The oxidative stability index (OSI) measured by the Rancimat method showed higher values for the Chardonnay grape seed flour (GSF) treatment, indicating extended shelf life under accelerated oxidation conditions. Appearance of the oil was affected by both seed extracts by increasing the light and yellowness (as determined by the Hunter *L*, *a*, *b* scale), but no sensory analysis was performed (Luther et al., 2007). Considering the null effect of tocopherols against one of the bacteria, and the higher OSI of Chardonnay seed flour, it can be thus considered a superior option than tocopherols.

Carambola (*Averrhoa carambola*, also known as star fruit) fruit pomace has shown antibacterial properties against *Bacillus* spp. when added to wheat bread at 5% concentration, such results are comparable to those of bread treated with preservatives. Addition of carambola pomace increased fiber content, a favorable parameter, while preserving sensory quality (Sudha et al., 2016). Carambola contains caramboxin (a phenylalanine-like neurotoxin) and high quantities of oxalic acid (Garcia-Cairasco et al., 2013), which limits its widespread consumption due to potential neurologic and/or renal toxicity; it is also a CYP450 inhibitor, which may induce drug interactions (Mallhi et al., 2015). Although the use of its by-products at low doses decreases the likelihood of potentially negative side effects, caution should nevertheless be exercised.

The residues from several fruits [orange (*Citrus sinensis*), passion fruit (*Passiflora edulis*), and watermelon (*Citrullus lanatus*)] and vegetables [lettuce (*Lactuca sativa*), courgette (*Cucurbita pepo*), carrot (*Daucus carota*), spinach (*Spinacea oleracea*), mint (*Mentha* sp.), taro (*Colocasia esculenta*), cucumber (*Cucumis sativus*), and rocket (*Eruca sativa*)] obtained after preparing juices were processed into flour and incorporated into biscuits and cereal bars. Microbiologic analysis showed no bacterial or mycotic contamination for up to 90 days in most of the varieties prepared, while those that did show evidence of contamination, a high humidity content was considered as the significant variable. These results are within local legal regulations and would allow for commercialization. Consumer acceptance was not hindered by the plant food residue, with an average evaluation of 6/9, and consumers expressing likelihood of purchase (Ferreira et al., 2015).

Wheat bread prepared with essential oils extracted from citrus peels has shown a significant decrease in bacterial and mold growth. Sensory parameters were affected by the oil treatment, most notably odor; however, this may be circumvented by spraying the wrapping materials with the oils instead of incorporating them into the flour, which also decreased microorganism growth (Salim-ur-Rehman et al., 2007). An interesting point is raised here, which is that it may not always be necessary to add agricultural by-products to the food itself, but that preservative effect may be still be achieved by spraying/coating the materials in contact with the food, this may be particularly useful when other parameters are overwhelmingly decreased. In addition, Cruz-Valenzuela et al. (2013) evaluated the antimicrobial and antioxidant potential of orange peel and seed extracts as additives to preserve quality of fresh-cut orange. Treated fruit showed the highest values of total phenolic content, flavonoids, and antioxidant activity and the lowest mesophilic bacteria and total molds and yeast compared to controls. This work demonstrates the potential of seed of by-products rich in antioxidant and antimicrobial compounds that can be used to preserve quality of fresh-cut oranges.

5.4.1.2 ANTIOXIDANT USES

The wine industry discards grape by-products that remain after the grapes are processed into wine; GSF prepared from such waste was incorporated into frankfurters (0%, 0.5%, 1%, 2%, 3%, 4%, or 5%) to enhance antioxidant capacity. Oxidation of the meat was minimized with the addition of GSF, which is likely due to its high anthocyanins concentration. Some parameters such as color were negatively impacted, while others such as total protein and fiber were improved; acceptability was reduced with the added GSF due to changes in their appearance (Ozvural and Vural, 2011). A similar study was performed using grape seed extract (GSE) as an additive to a precooked, frozen, stored meat model prepared from beef and pork meat to compare GSE against common antioxidant ingredients ascorbic acid and propyl gallate. After 4 months of storage at −18°C, rancid odors and flavors of GSE-treated meal were significantly lower than the control and the ascorbic acid treatments, and similar to those treated with propyl gallate. Beef flavor was also preserved by the GSE treatment. Products of oxidation (as determined by the thiobarbituric acid-reactive substances assay) remained stable or were decreased by the GSE, similarly to propyl gallate, thereby demonstrating adequate antioxidant activity on meat products that is comparable to commonly used antioxidants (Kulkarni et al., 2011).

Cauliflower residues have been incorporated into a ready-to-eat extruded snack in substitution of wheat flour (5–20%). A statistically significant increase in antioxidant capacity by virtue of the phenolic compounds was evident, however, in-vitro protein digestibility and fiber content was concomitantly reduced. Acceptability remained up to 10% cauliflower addition (Stojceska et al., 2008). Brewer's spent grain has also been incorporated into extruded snacks (0–30%), resulting in an increase of total antioxidant capacity of the product, and contrary to cauliflower, it also increases fiber and protein content (Stojceska et al., 2008).

Extraction of antioxidant compounds from mango kernels (of phenolic and phospholipid nature) has allowed them to be incorporated into buffalo ghee, a clarified butter of Indian origin. Addition of either the phenolic or phospholipid molecules prolonged ghee's stability, but a synergistic effect was documented when both kinds of compounds were simultaneously added (Puravankara et al., 2000). Ghee produced from buffalo milk is more prone to oxidation as compared to that made from cow milk; it is therefore an item whose shelf life can increase in parallel with its antioxidant capacity.

Watermelon rinds or sharlyn melon peels were used as wheat flour (2.5%, 5.0%, or 7.5%) or fat (5%, 10%, or 15%) substituents in cake preparation to improve antioxidant capacity. An optimal value of 5% wheat flour or fat substitution was determined for watermelon rind, which had higher antioxidant capacity values as compared to sharlyn melon. The increase in antioxidant capacity was reflected in an extended shelf life of up to 21 days (Al-Sayed and Ahmed, 2013). In addition to antioxidant compounds,

watermelon rind contains high citrulline concentrations, which is metabo-
lized to nitric oxide, a potent vasodilator that can reduce blood pressure.

Vega-Vega et al. (2013) reported that the addition of ethanolic extract of
"Haden" mango seeds to fresh-cut mango flesh increased the total phenolic
(7.4 times) and flavonoids (3.1 times) concentration. In addition, the treated
fruit presented highest antioxidant capacity and a microbial reduction of
80% of mesophilic plate count and 97% of total molds. These results demonstrate the potential of phenolic compounds derived from mango seed as antimicrobials and antioxidants.

5.4.1.3 THICKENER USES

Milled cauliflower curds and stems have been utilized as a source of fiber
with thickening properties in different food models (beef burgers, béchamel
sauce, tomato sauce, carrot pâté, and others). Incorporation of cauliflower
yields a firmer texture in beef burgers and increased viscosity in the béchamel
and tomato sauces; however, the effect in béchamel sauce was perceived only
when cauliflower was cooked together with the other ingredients, otherwise
a decrease in viscosity was evident. It was determined that cauliflower can
be used as a partial substitute for common thickening agents xanthan gum
(bacterial origin) and carrageenan (algal origin) for tomato sauce and carrot
pâté, respectively, although residual taste as still perceivable in both items
(Femenia et al., 1997).

Lemon and orange peels have been used as a source of pectin for jams
prepared at low, household scale. Recovered pectin was 15% and 21% for
orange and lemon peels, respectively, which yielded final products that were
significantly different from commercially available jams, specifically, a
lower gel strength was reported. Although they showed adequate acceptability scores, their values were lower than commercial products (Sulieman
et al., 2013).

Tomato flesh is produced as waste after tomatoes are processed into
ketchup and other products and is normally discarded. Tomato flesh powder
has been used as a thickening agent (up to 10% w/w) in commercial ketchup
preparations because of its dietary fiber content. A high water absorbing
capacity was documented (6.6 mL water/g of powder), which as related to an
apparent increase in viscosity; increase in L values (luminosity) and change
in a/b ratio was apparent but still of within acceptable range (Farahnaky et
al., 2008). Similarly, dried tomato peels have also been used as thickening
agents in sauces, reporting adequate rheological and sensory characteristics

(Donegà et al., 2015). The use of tomato waste as thickening agent in tomato products has obvious advantages, for example, no foreign smells or tastes will be added, while a higher incorporation of tomato into the final product implies decreased refuse.

5.4.1.4   SENSORIAL IMPACT

Aslam et al. (2014) improved the protein, fiber, and total phenolic content of biscuits by supplementing them with mango peel and kernel powder. The final product remained acceptable to consumers with up to 10% mango peel and 5% mango kernel, and in fact, a pleasant mango flavor was reported at said values. When exceeding the 10% and 5% threshold for peel and kernel, respectively, bitter taste and darker coloration reduced desirability. A similar study used mango peel and kernel powder to fortify cookies and found that 20% powder as the optimum value that increased total phenolic content and fiber and still preserved sensory characteristics, while higher percentages imparted an undesired bitter taste (Bandyopadhyay et al., 2014).

Replacing water with the aqueous extracts of the by-products of some vegetables (chicory, cabbage, celery, fennel, olive leaf, or grape marc) has been reported as an alternative to increase the phenolic content of bread. Although the main goal is achieved, the majority of sensory parameters are also altered, for example, darker (chicory and grape marc) or lighter (cabbage and celery) crumb tonalities were detected, grape marc produced a sweeter taste, salty (cabbage and celery), bitter (olive leaf), or astringent (chicory) tastes were also reported (Baiano et al., 2015).

Ground chicken breast and thigh was refrigerated (4°C) or cooked and then refrigerated with or without 0.1% GSE for up to 14 days; they were then frozen (−18°C) for another 14 days. The patties were cooked or reheated after storage, which was followed by sensory analysis; undesired smells and flavors were decreased by the grape seed treatment in breast meat; the effect on thigh meat was minimal, failing at reducing undesired smells and flavors while also affecting color toward darker and redder meat tonalities (Brannan, 2009). These results demonstrate that desired effects can be achieved (preservation in this case) in some types of meats but not others, even when the meat is of the same origin, yet certain undesired alterations will still be evident.

Raw or cooked lemon albedo (also called pith) was added to beef bologna sausages (up to 10%); results showed an increase in fiber and protein content and a decrease in fat content and residual nitrates, all changes which can be considered favorable or healthier. Change in fat content was easily perceived by the evaluators and was linked to a reduced juiciness perception; odor and taste were unaffected, hardness was increased, lightness and redness was significantly altered and resulted in a lighter product. Acceptability was adequate up to 5% uncooked and 7.5% cooked albedo (Fernandez-Gines et al., 2004).

Evidence shows that inclusion of fruits and vegetables by-products can significantly alter sensory perception both positively and negatively. The impact on baked products such as bread or cookies may by positive or neutral, since they are produced from plant sources, the incorporation of another plant item is easier to mask, or its impact may be minor (Salim-ur-Rehman et al., 2007; Ferreira et al., 2015; Sudha et al., 2016). On the contrary, the sensory parameters of animal products enriched with agricultural by-products such as meat, will almost certainly be more noticeable and may result in decreased acceptance by the consumer (Ozvural and Vural, 2011). Explicit disclosure that a product is enriched with added-value ingredients may compensate for altered characteristics, for example, the nitrate reduction reported for sausages which can be transformed into nitrosamines that favor tumorigenesis (Fernandez-Gines et al., 2004). Similarly, low-fat meat products can be perceived as healthier, which has motivated the food industry to search for fat substituents, some of which are of fruit and vegetables origin (Resurreccion, 2004), while reduced-sugar yogurts can be supplemented with inulin or grains that increase fiber content (Hoppert et al., 2013). Consumer acceptance should be taken into account when supplementing any item, in particular when doing so with elements labeled as "by-products," which can evoke unappealing mental images, thus, turning the consumer away from the product.

5.4.2 NUTRACEUTICAL USES

Current health trends and people's concerns about their own dietary habits are the principal reasons for the growth of functional foods market and the development of nutraceutical products. Kalra (2003) proposed as definition of functional foods, those foods in which scientific intelligence is applied when being cooked or prepared with or without knowledge of how or why it is being used. On the other hand, this author also proposed as a definition to nutraceutical product, "any functional food that aids in the prevention or treatment of any disease or human disorder." However, the term "nutraceutical" was first coined by Stephen DeFelice, founder and chairman of the Foundation for Innovation in Medicine located in Cranford, New Jersey (Dureja et al., 2003). Nevertheless, despite the fact that the definition proposed by Kalra in 2003 is based on the Stephen DeFelice's definition, it is unclear and makes it hard to distinguish between functional foods and regular processed foods.

However, Stephen DeFelice defined in an interview on October 2011 that nutraceuticals are even foods or part of foods, such as a dietary supplement, that has medical or health benefit. This definition is consistent with the one in the OED, as a foodstuff, food additive, or dietary supplement that has beneficial physiological effects but is not essential to the diet. In that sense, nutraceuticals have been more successful as dietary supplements, as they have an advantage over medicine because they avoid side effects, as they are usually naturally dietary components. Likewise, the definition of dietary supplement is consistent with the definition of nutraceutical. As the Dietary Supplement, Health and Education Act defined, dietary supplements are products intended to supplement the diet that bears or contains one or more ingredients such as vitamins, herbal or botanical, concentrate or extract, amino acids, dietary fiber, or even a combination of these ingredients.

In that sense, from the point of view of this last definition above and as it has been mentioned previously in this chapter, the harvest by-products of fruits and vegetables represents a rich source of more than one of the ingredients that usually comprise dietary supplements, and therefore it makes those by-products an excellent source for nutraceutical products development. By-product's nutraceuticals are usually as Health Canada established nutraceuticals definition "prepared from foods, but sold in the form of pills or powders, or in other medicinal forms not usually associated with foods" (Keservani et al., 2010). Hence, functional ingredients used in nutraceutical products are usually extracted, isolated, and concentrated to sell them as dietary supplements in either a form of powders or pills. As follows, the extraction and use of fruits and vegetables by-products to obtain fibers and antioxidants for nutraceutical products development will be discussed.

5.4.2.1 DIETARY FIBERS

The group of compounds denominated as dietary fibers includes a mixture of plant carbohydrate polymers, both oligosaccharides and polysaccharides, for example, cellulose, hemicellulose, pectic substances, gums, resistant starch, inulin, among other noncarbohydrates compounds associated (lignin, waxes, polyphenols, and resistant protein) (Elleuch et al., 2011). The use of fibers as food additives has an impact on the food technological and physiological functionality in functional foods and nutraceutical products. According to the American Dietetic Association, the recommended intake for adults range from 25 to 30 g/day or 13 g per 1000 kcal, with an insoluble/soluble ratio of 3:1. Hence, fiber rich nutraceutical products are consumed when the individuals desires to prevent or treat physiological diseases and disorders such as high blood cholesterol levels, reduce glycemic and insulinemic responses, enhance intestinal function, among other chronic disorders such as cardiovascular diseases (Borderías et al., 2005). Among the most promising fruits by-products are the peels of citrus, apple, banana, and passion fruit, because of their content of insoluble and soluble dietary fiber (do Espírito Santo et al., 2012).

Despite there is a great variety of fruit and vegetable raw materials from where dietary fiber could be extracted and transformed in a nutraceutical product, some sources may have certain advantages over others considering the final nutraceutical products characteristics. The main characteristics of a commercialized dietary fiber nutraceutical product according to Larrauri (1999) are total dietary fiber higher than 50%, moisture lower than 9%, low content of lipids, a low caloric value (lower than 8.36 kJ/g), and neutral flavor and taste. For example, citrus waste products has an advantage over others by-products source of dietary fiber, such as cereals, and this is that citrus by-products has a higher proportion of soluble dietary fiber. Likewise, the estimated residue amount after citrus industrial processing is around 15×10^6 t, being almost half of a whole citrus fruit discarded as waste (Marín et al., 2007). Lario et al. (2004) reported that lemon by-products are source of high dietary fiber powder, and this powder has good functional and microbial quality. On the other hand, Chau and Huang (2004) reported that raw passion fruit seeds are a rich source of insoluble dietary fiber, around 64 g per 100 g.

Dietary fiber powders are usually obtained from the dried and grinded plant tissues, either harvest by-products, pomace, or seeds. For example, Ovando-Martinez et al. (2009) used dried unripe banana powder as food additive to pasta (spaghetti) formulations. These authors found that formulations containing 45% of banana flour had 26.18 ± 0.26 and 3.37 ± 0.22 of insoluble and soluble indigestible fractions, respectively. Hence, pasta products containing banana flour exhibit a low rate of carbohydrate enzymatic hydrolysis and could be used as an option of low glycemic index foods. Likewise, other authors such as Aparicio-Saguilán et al. (2007) isolated banana resistant starch to incorporate it in cookies formulations. They found that the sensorial acceptation of cookies was higher when wheat flour and banana resistant starch were in 15:85 proportions, respectively. These authors reported that using this proportion of banana-resistant starch powder, they achieve cookies with higher dietary fiber content compared with control cookies increasing from $13.4 \pm 1.50\%$ to $20.5 \pm 1.04\%$ of insoluble dietary fiber, but decreasing from $3.4 \pm 0.19\%$ to $2.92 \pm 1.04\%$ of soluble dietary fiber.

Similar studies have used dietary fiber powders as food additives in other food products, such as meats. For example, Sáyago-Ayerdi et al. (2009) found that adding 2% of grape antioxidant dietary fiber powder to raw and cooked chicken hamburgers was sensorially accepted, while it increases the dietary fiber content in those food products and also inhibits oxidation, increasing its lipid stability and shelf life. On the other hand, the use of dietary fiber in snacks formulation is also showing promising results. As Zamora-Gasga et al. (2014) reported, the combination of native agave fructans, as soluble dietary fiber source, and ground agave fiber, as source of insoluble dietary fiber, reduce the sugar content in granola bars to 10 g per 100 g preserving consumer preference. Likewise, Blancas-Benitez et al. (2015) evaluated the addition of dried "Ataulfo" mango concentrate and its dried by-products and found that it incorporation (75:25, respectively) to starch-molded mango snacks the dietary fiber and phenolic compounds content reducing its sugar content without affecting physical–chemical parameters.

Mildner-Szkudlarz et al. (2013) added powdered white grape pomace as a food additive to wheat biscuits, to increase its total dietary fiber content. These authors found that the sensorially accepted powdered concentration was up to 10%, which means that per 1 kg of wheat biscuits, the food products would have 64.86 g of total dietary fiber. In a similar study performed by Mustafa et al. (2016), bamboo powder was used as food additive in cookies in up to only 6% (1.48% of total dietary fiber) according to sensorial acceptance values. Furthermore, Acosta-Estrada et al. (2014) used 9% of nejayote powder from white dent maize as food additive in pan-bread formulations, resulting in an increase up to 54% of dietary fiber without any changes in the overall baking performance and bread quality. However, even when the previously mentioned examples of dietary fiber are bakery products but it could also be successfully added to beverages, confections, dairy, frozen dairy, meat, pasta, and soups (Quirós-Sauceda et al., 2014). Nevertheless, the use of dietary fiber powders as food additives may reduce it's shelf life according to the type of food where it is contained, being that a great advantage of the intake as a dietary supplement powder.

Scientific reports suggest that dietary fiber supplementation has effects on weight loss for individuals on weight-reducing diets. As it is described by Anderson et al. (2009) in trials where dietary fiber was provided in the form of tablets and given three times daily (4.5–20 g/day), the weight loss achieved was greater than the achieved with placebo. Moreover, Hozumi et al. (1995) evaluated the long-term dietary fiber supplementation in rats fed a high cholesterol diet and found that it may help retard or prevent the atheromatous formation found in cholesterol-fed diabetic rats. However, the results obtained from dietary fiber supplementation may vary from one population to another, as well as the age and metabolism of the individual taking that supplementation. For example, Streppel et al. (2005) found that fiber supplementation at an average dose of 11.5 g/day changed systolic blood pressure by −1.13 mmHg and diastolic blood pressure by −1.26 mmHg. Likewise, these authors found that the reductions in blood pressure tended to be larger in individuals above 40 years old and in hypertensive populations rather than in younger and in normotensive ones.

Hence, to achieve most of the health-related benefits from dietary fiber nutraceuticals, it is essential to consider if the intake of this is as a food additive or dietary supplement. Likewise, it should be noticed that the results may vary from one individual to other; therefore, it is suggested that medical check-ups should be performed before and after supplementation to evaluate the real effects and assure that the ingested dose is correct for the desired effect.

5.4.2.2   ANTIOXIDANTS

Human metabolism involves a series of changes and reactions, where the oxidant by-products of normal metabolism cause damage to biomolecules, such as DNA, proteins, and lipids. The damage to these molecules appears to be the main cause of aging and degenerative diseases of aging such as cancer, cardiovascular disease, cataracts, immune system decline, and brain dysfunction (Frei, 2012). Aerobic organisms, humans are victims of oxygen because of the formation of oxygen free radicals and other reactive oxygen species (ROS) initiates a chain reaction that results in damage to biomolecules. In that sense, a theory called "Free Radical Theory of Aging," also known as oxidative damage theory, has emerged. This theory suggests that free radicals and other ROS, formed unavoidably in human metabolism and arising due to the action of various exogenous factors, damage biomolecules, causing age-related diseases and aging (Sadowska-Bartosz and Bartosz, 2014). Therefore, several studies are evaluating the regular intake of antioxidants and how this consumption protects biomolecules and prevent age-related diseases.

Among the most common antioxidants used as nutraceuticals are carotenoids, vitamin E (tocopherols and tocotrie-nols), ascorbic acid, lipoic acids, and polyphenols. Despite the chemical differences among these compounds, they are all phytochemicals with antioxidant capacity naturally present in foods (Prakash and Sharma, 2014). Carotenoids, for example, constitute a ubiquitous group of isoprenoid pigments; they are very efficient physical quenchers of singlet oxygen and scavengers of other ROS. Some of these compounds also can be transformed into vitamin A, only α, β, ε-carotene possess vitamin A activity, being β-carotene the most active. On the other hand, vitamin C is a leading natural antioxidant that can scavenge ROS and has anticarcinogenic effects. This vitamin has been exhaustively studied due to its strong antioxidant activity, among other health-related benefits, such as deposition of type IV collagen in the basement membrane, stimulating endothelial proliferation and inhibiting apoptosis, scavenging radical species and sparing endothelial cell-derived nitric oxide to help modulate blood flow (May and Harrison, 2013). However, due to their strong antioxidant activity, along with their abundance in foods and large diversity of chemical structures, phenolic compounds are one of the main compounds of interest for using in nutraceutical products.

However, so far a large part of studies investigating the effectiveness of antioxidant supplementation therapy in humans shows contrasting results, usually because of the often-limited statistic power of the studies, patient's background, and the bioavailability of the antioxidants. Even when isolated fruit and vegetable antioxidants, such as flavonoids, have shown anti-inflammatory, antiviral, and anticancer activity, the effect depends on the patient's background and the bioavailability of the type of flavonoid ingested (Marin et al., 2002). Shui and Leong (2006) proposed the use of the waste generated after star fruit juice processing. These authors compared the residue of 10 star fruits with a bottle of pycnogenol containing 40 pills. Pycnogenol is an extract from maritime pine bark and it comprises 80–85 wt% of proanthocynidins, which are phenolic compounds. This anthocyanidins are polymers of flavonoids, and some of them are related to diverse health benefits. For example, (−)-epigallocatechin gallate found in green tea is an anthocyanidin that have shown to protect mammalian cells from free-radical-mediated oxidative stress and to cause apoptosis of cancer cells.

Another example is the supplementation with carotenoids and antioxidant vitamins, such as vitamin C. Duthie et al. (1996) supplemented 50–59-year-old men for 20 weeks with vitamin C (100 mg/day), vitamin E (280 mg/day), and β-carotene (25 mg/day) and found that the oxidative damage in the limphocytes DNA decreased. Likewise, Santanam et al. (2013)

supplemented 59 women with vitamin E (1200 IU) and vitamin C (1000 mg), and they found that the administration of these antioxidants reduces chronic pelvic pain in women with endometriosis and inflammatory markers in the peritoneal fluid. On the other hand, Michailidis et al. (2013) supplemented men with a thiol-based antioxidant, to enhance the activity of glutathione, which is a potent endogenous antioxidant in human body. In that study, they performed a double-blind, crossover design, 10 men received placebo or N-acetylcysteine (20 mg/kg/day) after muscle-damaging exercise. However, although thiol-based antioxidant supplementation enhances the glutathione availability in skeletal muscle, it disrupts the skeletal muscle inflammatory response and repair capability.

5.6   CONCLUDING REMARKS

The increasing technologies and research around the world are setting the harvest by-products as an economic opportunity rather than an ecological contamination risk. Several harvest by-products, and fruit and vegetable wastes generated by food industry have been characterized and some of them have been identified as sources of certain group or specific compounds with health related properties. In the growing market of functional foods, and as the trend in human diet involves the use of natural compounds, by-product phytochemicals represents an economical opportunity to farmers and food-processing plants. The first step to exploit these by-products is to characterize the raw material, to identify the phytochemical embedded and the yield of extraction from it. Once the by-product is properly characterized and studied, it can be decided which food product may be enriched with the extracted phytochemicals.

A practical use for by-product phytochemicals is its use as nutraceutical products. As nutraceuticals, by-product phytochemicals can be used as food additives in diverse food products, such as bakery, beverages, dairy, frozen dairy, meat, pasta, among others. However, as nutraceuticals are not necessarily associated to food products, phytochemicals may be extracted, isolated, and concentrated as powders or tablets used as dietary supplements. Hence, the most common phytochemicals extracted from harvest by-products used as dietary supplements are dietary fiber and antioxidants. However, the current information about the efficacy of these dietary supplements not always show desirable effects, indicating that there's more research work needed to do to obtain more desirable effects. As the efficacy of nutraceuticals may vary from one individual to another, more studies should be carried to propose alternatives that enhance phytochemical bioavalability and increase the chances to achieve desirable effect in health.

KEYWORDS

- by-products
- phytochemicals
- fruits
- vegetables
- food additives

REFERENCES

Acosta-Estrada, B. A.; Lazo-Vélez, M. A.; Nava-Valdez, Y.; Gutiérrez-Uribe, J. A.; Serna-Saldívar, S. O. Improvement of Dietary Fiber, Ferulic Acid and Calcium Contents in Pan Bread Enriched with Nejayote Food Additive from White Maize (Zea mays). J. Cer. Sci. 2014, 60 (1), 264–269.

Ajila, C.; Bhat, S.; Rao, U. P. Valuable Components of Raw and Ripe Peels from Two Indian Mango Varieties. Food Chem. 2007, 102 (4), 1006–1011.

Ajila, C.; Rao, U. P. Mango Peel Dietary Fibre: Composition and Associated Bound Phenolics. J. Funct. Foods 2013, 5 (1), 444–450.

Al-Sayed, H. M. A.; Ahmed, A. R. Utilization of Watermelon Rinds and Sharlyn Melon Peels as a Natural Source of Dietary Fiber and Antioxidants in Cake. Ann. Agric. Sci. 2013, 58 (1), 83–95.

Al-Wandawi, H.; Abdul-Rahman, M.; Al-Shaikhly, K. Tomato Processing Wastes as Essential Raw Materials Source. J. Agric. Food Chem. 1985, 33 (5), 804–807.

Anal, A. K. Food Processing By-products. Handbook of Plant Food Phytochemicals; John Wiley & Sons Ltd.: Ames, IA, 2013; pp 180–197.

Anderson, J. W.; Baird, P.; Davis, R. H.; Ferreri, S.; Knudtson, M.; Koraym, A.; Waters, V.; Williams, C. L. Health Benefits of Dietary Fiber. Nutr. Rev. 2009, 67 (4), 188–205.

Aparicio-Saguilán, A.; Sayago-Ayerdi, S. G.; Vargas-Torres, A.; Tovar, J.; Ascencio-Otero, T. E.; Bello-Pérez, L. A. Slowly Digestible Cookies Prepared from Resistant Starch-Rich Lintnerized Banana Starch. J. Food Compos. Anal. 2007, 20 (3), 175–181.

Aslam, H. K. W.; Raheem, M. I. U.; Rabia Ramzan, A. S.; Shoaib, M.; Sakandar, H. A. Utilization of Mango Waste Material (Peel, Kernel) to Enhance Dietary Fiber Content and Antioxidant Properties of Biscuit. J. Glob. Innov. Agric. Soc. Sci. 2014, 2 (2), 76–81.

Ayala-Zavala, J.; Vega-Vega, V.; Rosas-Domínguez, C.; Palafox-Carlos, H.; Villa-Rodriguez, J.; Siddiqui, M. W.; Dávila-Aviña, J.; González-Aguilar, G. Agro-industrial Potential of Exotic Fruit Byproducts as a Source of Food Additives. Food Res. Int. 2011, 44 (7), 1866–1874.

Ayala-Zavala, J. F.; del Toro-Sánchez, L.; Alvarez-Parrilla, E.; Soto-Valdez, H.; Martín-Belloso, O.; Ruiz-Cruz, S.; González-Aguilar, G. Natural Antimicrobial Agents Incorporated in Active Packaging to Preserve the Quality of Fresh Fruits and Vegetables. *Stewart Postharv. Rev.* **2008,** *4* (3), 1–9.

Ayala-Zavala, J.; Rosas-Domínguez, C.; Vega-Vega, V.; González-Aguilar, G. Antioxidant Enrichment and Antimicrobial Protection of Fresh-Cut Fruits Using their Own Byproducts: Looking for Integral Exploitation. *J. Food Sci.* **2010,** *75* (8), R175–R181.

Baiano, A.; Viggiani, I.; Terracone, C.; Romaniello, R.; Del Nobile, M. A. Physical and Sensory Properties of Bread Enriched with Phenolic Aqueous Extracts from Vegetable Wastes. *Czech. J. Food Sci.* **2015,** *33* (3), 247–253.

Bandyopadhyay, K.; Chakraborty, C.; Bhattacharyya, S. Fortification of Mango Peel and Kernel Powder in Cookies Formulation. *J. Acad. Ind. Res.* **2014,** *2* (12), 661–664.

Bensadón, S.; Hervert-Hernández, D.; Sáyago-Ayerdi, S. G.; Goñi, I. By-products of *Opuntia ficus-indica* as a Source of Antioxidant Dietary Fiber. *Plant Foods Hum. Nutr.* **2010,** *65* (3), 210–216.

Bhushan, S.; Kalia, K.; Sharma, M.; Singh, B.; Ahuja, P. S. Processing of Apple Pomace for Bioactive Molecules. *Crit. Rev. Biotechnol.* **2008,** *28* (4), 285–296.

Blancas-Benitez, F. J.; de Jesús Avena-Bustillos, R.; Montalvo-González, E.; Sáyago-Ayerdi, S. G.; McHugh, T. H. Addition of Dried 'Ataulfo' Mango (*Mangifera indica* L.) By-products as a Source of Dietary Fiber and Polyphenols in Starch Molded Mango Snacks. *J. Food Sci. Technol.* **2015,** *52* (11), 7393–7400.

Borderías, A. J.; Sánchez-Alonso, I.; Pérez-Mateos, M. New Applications of Fibres in Foods: Addition to Fishery Products. *Trends Food Sci. Technol.* **2005,** *16* (10), 458–465.

Bousbia, N.; Vian, M. A.; Ferhat, M. A.; Meklati, B. Y.; Chemat, F. A New Process for Extraction of Essential Oil from Citrus Peels: Microwave Hydrodiffusion and Gravity. *J. Food Eng.* **2009,** *90* (3), 409–413.

Brannan, R. G. Effect of Grape Seed Extract on Descriptive Sensory Analysis of Ground Chicken during Refrigerated Storage. *Meat Sci.* **2009,** *81* (4), 589–595.

Bridle, P.; Timberlake, C. Anthocyanins as Natural Food Colours—Selected Aspects. *Food Chem.* **1997,** *58* (1), 103–109.

Coll-Almela, L.; Saura-López, D.; Laencina-Sánchez, J.; Schols, H. A.; Voragen, A. G.; Ros-García, J. M. Characterisation of Cell-Wall Polysaccharides from Mandarin Segment Membranes. *Food Chem.* **2015,** *175*, 36–42.

Corrales, M.; Toepfl, S.; Butz, P.; Knorr, D.; Tauscher, B. Extraction of Anthocyanins from Grape By-products Assisted by Ultrasonics, High Hydrostatic Pressure or Pulsed Electric Fields: A Comparison. *Innov. Food Sci. Emerg. Technol.* **2008,** *9* (1), 85–91.

Cruz-Valenzuela, M. R.; Carrazco-Lugo, D. K.; Vega-Vega, V.; Gonzalez-Aguilar, G. A.; Ayala-Zavala, J. F. Fresh-cut Orange Treated with its Own Seed By-products Presented Higher Antioxidant Capacity and Lower Microbial Growth. *Int. J. Postharv. Technol. Innov.* **2013,** *3* (1), 13–27.

Chantaro, P.; Devahastin, S.; Chiewchan, N. Production of Antioxidant High Dietary Fiber Powder from Carrot Peels. *LWT—Food Sci. Technol.* **2008,** *41* (10), 1987–1994.

Chau, C.; Huang, Y. Characterization of Passion Fruit Seed Fibres—A Potential Fibre Source. *Food Chem.* **2004,** *85* (2), 189–194.

Chaurasiya, R. S.; Hebbar, H. U. Extraction of Bromelain from Pineapple Core and Purification by RME and Precipitation Methods. *Sep. Purif. Technol.* **2013,** *111*, 90–97.

da Silva, L. M. R.; de Figueiredo, E. A. T.; Ricardo, N. M. P. S.; Vieira, I. G. P.; de Figueiredo, R. W.; Brasil, I. M.; Gomes, C. L. Quantification of Bioactive Compounds in Pulps and By-Products of Tropical Fruits from Brazil. *Food Chem.* **2014**, *143*, 398–404.

Das, H.; Singh, S. K. Useful Byproducts from Cellulosic Wastes of Agriculture and Food Industry—A Critical Appraisal. *Crit. Rev. Food Sci. Nutr.* **2004**, *44* (2), 77–89.

Dauthy, M. E. *Fruit and Vegetable Processing*; Food and Agriculture Organization of the United Nations: Rome, 1995.

de Las Fuentes, L. *AWARENET: Agro-Food Wastes Minimisation and Reduction Network*. 1. International Conference on Waste Management and the Environment, 2002.

de Lourdes García-Magaña, M.; García, H. S.; Bello-Pérez, L. A.; Sáyago-Ayerdi, S. G.; de Oca, M. M.-M. Functional Properties and Dietary Fiber Characterization of Mango Processing By-products (*Mangifera indica* L., cv Ataulfo and Tommy Atkins). *Plant Foods Hum. Nutr.* **2013**, *68* (3), 254–258.

Devakate, R.; Patil, V.; Waje, S.; Thorat, B. Purification and Drying of Bromelain. *Sep. Purif. Technol.* **2009**, *64* (3), 259–264.

do Espírito Santo, A. P.; Cartolano, N. S.; Silva, T. F.; Soares, F. A.; Gioielli, L. A.; Perego, P.; Converti, A.; Oliveira, M. N. Fibers from Fruit By-products Enhance Probiotic Viability and Fatty Acid Profile and Increase CLA Content in Yoghurts. *Int. J. Food Microbiol.* **2012**, *154* (3), 135–144.

Donegà, V.; Marchetti, M. G.; Pedrini, P.; Costa, S.; Tamburini, E. Valorization of Tomato Dried Peels Powder as Thickening Agent in Tomato Purees. *J. Food Process. Technol.* **2015**, *6* (11), 1–7.

Dureja, H.; Kaushik, D.; Kumar, V. Developments in Nutraceuticals. *Indian J. Pharmacol.* **2003**, *35* (6), 363–372.

Duthie, S. J.; Ma, A.; Ross, M. A.; Collins, A. R. Antioxidant Supplementation Decreases Oxidative DNA Damage in Human Lymphocytes. *Cancer Res.* **1996**, *56* (6), 1291–1295.

Edge, R.; McGarvey, D.; Truscott, T. The Carotenoids as Anti-oxidants—A Review. *J. Photochem. Photobiol. B: Biol.* **1997**, *41* (3), 189–200.

Elleuch, M.; Bedigian, D.; Roiseux, O.; Besbes, S.; Blecker, C.; Attia, H. Dietary fibre and fibre-rich by-products of food processing: Characterisation, Technological Functionality and Commercial Applications: A Review. *Food Chem.* **2011**, *124* (2), 411–421.

Fan, G.; Han, Y.; Gu, Z.; Chen, D. Optimizing Conditions for Anthocyanins Extraction from Purple Sweet Potato Using Response Surface Methodology (RSM). *LWT—Food Sci. Technol.* **2008**, *41* (1), 155–160.

FAO. *Global Food Losses and Food Waste*, 2011, www.fao.org/ag/ags/ags-division/publications/publication/en/?dyna_fef%5Buid%5D=74045.

Farahnaky, A.; Abbasi, A.; Jamalian, J.; Mesbahi, G. The Use of Tomato Pulp Powder as a Thickening Agent in the Formulation of Tomato Ketchup. *J. Text. Stud.* **2008**, *39* (2), 169–182.

Femenia, A.; Lefebvre, A. C.; Thebaudin, J. Y.; Robertson, J. A.; Bourgeois, C. M. Physical and Sensory Properties of Model Foods Supplemented with Cauliflower Fiber. *J. Food Sci.* **1997**, *62* (4), 635–639.

Fernandez-Gines, J. M.; Fernandez-Lopez, J.; Sayas-Barbera, E.; Sendra, E.; Perez-Alvarez, J. A. Lemon Albedo as a New Source of Dietary Fiber: Application to Bologna Sausages. *Meat Sci.* **2004**, *67* (1), 7–13.

Fernández-López, J.; Fernández-Ginés, J.; Aleson-Carbonell, L.; Sendra, E.; Sayas-Barberá, E.; Pérez-Alvarez, J. Application of Functional Citrus By-products to Meat Products. *Trends Food Sci. Technol.* **2004**, *15* (3), 176–185.

Ferreira, M. S. L.; Santos, M. C. P.; Moro, T. M. A.; Basto, G. J.; Andrade, R. M. S.; Goncalves, E. C. B. A. Formulation and Characterization of Functional Foods Based on Fruit and Vegetable Residue Flour. *J. Food Sci. Technol.—Mysore* **2015,** *52* (2), 822–830.

Frei, B. *Natural Antioxidants in Human Health and Disease,* Elsevier Science, 2012.

Garcia-Cairasco, N.; Moyses-Neto, M.; Del Vecchio, F.; Oliveira, J. A. C.; dos Santos, F. L.; Castro, O. W.; Arisi, G. M.; Dantas, M.; Carolino, R. O. G.; Coutinho-Netto, J.; Dagostin, A. L. A.; Rodrigues, M. C. A.; Leao, R. M.; Quintiliano, S. A. P.; Silva, L. F.; Gobbo-Neto, L.; Lopes, N. P. Elucidating the Neurotoxicity of the Star Fruit. *Angew. Chem.—Int. Ed.* **2013,** *52* (49), 13067–13070.

Gu, S.; Zhu, K.; Luo, C.; Jiang, Y.; Xu, Y. Microwaves-Assisted Extraction of Polyphenols from Banana Peel. *Med. Plant* **2014,** *5* (1), 21.

Gunders, D. Wasted: How America is Losing Up to 40 Percent of Its Food from Farm to Fork to Landfill. *Natural Resources Defense Council Issue Paper. August.* This Report Was Made Possible through the Generous Support of The California Endowment, 2012.

Han, X.; Shen, T.; Lou, H. Dietary Polyphenols and their Biological Significance. *Int. J. Mol. Sci.* **2007,** *8* (9), 950–988.

Hasan, F.; Shah, A. A.; Hameed, A. Industrial Applications of Microbial Lipases. *Enzyme Microb. Technol.* **2006,** *39* (2), 235–251.

Herrero, M.; Temirzoda, T. N.; Segura-Carretero, A.; Quirantes, R.; Plaza, M.; Ibañez, E. New Possibilities for the Valorization of Olive Oil By-products. *J. Chromatogr. A* **2011,** *1218* (42), 7511–7520.

Hoppert, K.; Zahn, S.; Janecke, L.; Mai, R.; Hoffmann, S.; Rohm, H. Consumer Acceptance of Regular and Reduced-Sugar Yogurt Enriched with Different Types of Dietary Fiber. *Int. Dairy J.* **2013,** *28* (1), 1–7.

Hozumi, T.; Yoshida, M.; Ishida, Y.; Mimoto, H.; Sawa, J.; Doi, K.; Kazumi, T. Long-Term Effects of Dietary Fiber Supplementation on Serum Glucose and Lipoprotein Levels in Diabetic Rats Fed a High Cholesterol Diet. *Endocr. J.* **1995,** *42* (2), 187–192.

Ignat, I.; Volf, I.; Popa, V. I. A Critical Review of Methods for Characterisation of Polyphenolic Compounds in Fruits and Vegetables. *Food Chem.* **2011,** *126* (4), 1821–1835.

Jadhav, S. A.; Kataria, P. K.; Bhise, K. K.; Chougule, S. A. Amylase Production from Potato and Banana Peel Waste. *Int. J. Curr. Microbiol. App. Sci.* **2013,** *2* (11), 410–414.

Jiménez-Escrig, A.; Rincón, M.; Pulido, R.; Saura-Calixto, F. Guava Fruit (*Psidium guajava* L.) as a New Source of Antioxidant Dietary Fiber. *J. Agric. Food Chem.* **2001,** *49* (11), 5489–5493.

Kabir, F.; Sultana, M. S.; Kurnianta, H. Antimicrobial Activities of Grape (*Vitis vinifera* L.) Pomace Polyphenols as a Source of Naturally Occurring Bioactive Components. *Afr. J. Biotechnol.* **2015,** *14* (26), 2157–2161.

Kalra, E. K. Nutraceutical—Definition and Introduction. *AAPS Pharmsci.* **2003,** *5* (3), 27–28.

Keservani, R. K.; Kesharwani, R. K.; Vyas, N.; Jain, S.; Raghuvanshi, R.; Sharma, A. K. Nutraceutical and Functional Food as Future Food: A Review. *Pharm. Lett.* **2010,** *2* (1), 106–116.

Kris-Etherton, P. M.; Hecker, K. D.; Bonanome, A.; Coval, S. M.; Binkoski, A. E.; Hilpert, K. F.; Griel, A. E.; Etherton, T. D. Bioactive Compounds in Foods: Their Role in the Prevention of Cardiovascular Disease and Cancer. *Am. J. Med.* **2002,** *113* (9), 71–88.

Kulkarni, S.; DeSantos, F. A.; Kattamuri, S.; Rossi, S. J.; Brewer, M. S. Effect of Grape seed Extract on Oxidative, Color and Sensory Stability of a Pre-cooked, Frozen, Re-heated Beef Sausage Model System. *Meat Sci.* **2011,** *88* (1), 139–144.

Kumcuoglu, S.; Yilmaz, T.; Tavman, S. Ultrasound Assisted Extraction of Lycopene from Tomato Processing Wastes. *J. Food Sci. Technol.* **2014,** *51* (12), 4102–4107.

Ky, I.; Lorrain, B.; Kolbas, N.; Crozier, A.; Teissedre, P.-L. Wine By-products: Phenolic Characterization and Antioxidant Activity Evaluation of Grapes and Grape Pomaces from Six Different French Grape Varieties. *Molecules* **2014,** *19* (1), 482–506.

Lama-Muñoz, A.; Rodríguez-Gutiérrez, G.; Rubio-Senent, F.; Fernández-Bolaños, J. Production, Characterization and Isolation of Neutral and Pectic Oligosaccharides with Low Molecular Weights from Olive By-products Thermally Treated. *Food Hydrocolloids* **2012,** *28* (1), 92–104.

Lario, Y.; Sendra, E.; Garcí, J.; Fuentes, C.; Sayas-Barberá, E.; Fernández-López, J.; Pérez-Alvarez, J. Preparation of High Dietary Fiber Powder from Lemon Juice By-products. *Innov. Food Sci. Emerg. Technol.* **2004,** *5* (1), 113–117.

Larrauri, J. New Approaches in the Preparation of High Dietary Fibre Powders from Fruit By-products. *Trends Food Sci. Technol.* **1999,** *10* (1), 3–8.

Larrauri, J. A.; Rupérez, P.; Calixto, F. S. Pineapple Shell as a Source of Dietary Fiber with Associated Polyphenols. *J. Agric. Food Chem.* **1997a,** *45* (10), 4028–4031.

Larrauri, J. A.; Rupérez, P.; Saura-Calixto, F. Mango Peel Fibres with Antioxidant Activity. *Zeitsch. Lebensmitteluntersuch. Forsch. A* **1997b,** *205* (1), 39–42.

Liew, S. Q.; Chin, N. L.; Yusof, Y. A. Extraction and Characterization of Pectin from Passion Fruit Peels. *Agric. Agric. Sci. Proc.* **2014,** *2,* 231–236.

Lü, J. M.; Lin, P. H.; Yao, Q.; Chen, C. Chemical and Molecular Mechanisms of Antioxidants: Experimental Approaches and Model Systems. *J. Cell. Mol. Med.* **2010,** *14* (4), 840–860.

Lucier, G.; Pollack, S.; Ali, M.; Perez, A. *Fruit and Vegetable Backgrounder*, US Department of Agriculture, Economic Research Service, 2006.

Luther, M.; Parry, J.; Moore, J.; Meng, J. H.; Zhang, Y. F.; Cheng, Z. H.; Yu, L. L. Inhibitory Effect of Chardonnay and Black Raspberry Seed Extracts on Lipid Oxidation in Fish Oil and their Radical Scavenging and Antimicrobial Properties. *Food Chem.* **2007,** *104* (3), 1065–1073.

Malacrida, C. R.; Jorge, N. Yellow Passion Fruit Seed Oil (*Passiflora edulis* f. *flavicarpa*), Physical and Chemical Characteristics. *Braz. Arch. Biol. Technol.* **2012,** *55* (1), 127–134.

Mallhi, T. H.; Sarriff, A.; Adnan, A. S.; Khan, Y. H.; Qadir, M. I.; Hamzah, A. A.; Khan, A. H. Effect of Fruit/Vegetable–Drug Interactions on CYP450, OATP and *p*-Glycoprotein: A Systematic Review. *Trop. J. Pharm. Res.* **2015,** *14* (10), 1927–1935.

Marin, F.; Frutos, M.; Pérez-Alvarez, J.; Martinez-Sánchez, F.; Del Rio, J. Flavonoids as Nutraceuticals: Structural Related Antioxidant Properties and their Role on Ascorbic Acid Preservation. *Stud. Nat. Prod. Chem.* **2002,** *26,* 741–778.

Marín, F. R.; Soler-Rivas, C.; Benavente-García, O.; Castillo, J.; Pérez-Alvarez, J. A. By-products from Different Citrus Processes as a Source of Customized Functional Fibres. *Food Chem.* **2007,** *100* (2), 736–741.

May, C. D. Industrial Pectins: Sources, Production and Applications. *Carbohydr. Polym.* **1990,** *12* (1), 79–99.

May, J. M.; Harrison, F. E. Role of Vitamin C in the Function of the Vascular Endothelium. *Antioxid. Redox Signal.* **2013,** *19* (17), 2068–2083.

Michailidis, Y.; Karagounis, L. G.; Terzis, G.; Jamurtas, A. Z.; Spengos, K.; Tsoukas, D.; Chatzinikolaou, A.; Mandalidis, D.; Stefanetti, R. J.; Papassotiriou, I. Thiol-Based Antioxidant Supplementation Alters Human Skeletal Muscle Signaling and Attenuates its Inflammatory Response and Recovery after Intense Eccentric Exercise. *Am. J. Clin. Nutr.* **2013,** *98* (1), 233–245.

Mildner-Szkudlarz, S.; Bajerska, J.; Zawirska-Wojtasiak, R.; Górecka, D. White Grape Pomace as a Source of Dietary Fibre and Polyphenols and its Effect on Physical and Nutraceutical Characteristics of Wheat Biscuits. *J. Sci. Food Agric.* **2013,** *93* (2), 389–395.

Murthy, P. S.; Naidu, M. M. Recovery of Phenolic Antioxidants and Functional Compounds from Coffee Industry By-products. *Food Bioprocess Technol.* **2012,** *5* (3), 897–903.

Mustafa, A.; Trevino, L. M.; Turner, C. Pressurized Hot Ethanol Extraction of Carotenoids from Carrot By-products. *Molecules* **2012,** *17* (2), 1809–1818.

Mustafa, U.; Naeem, N.; Masood, S.; Farooq, Z. Effect of Bamboo Powder Supplementation on Physicochemical and Organoleptic Characteristics of Fortified Cookies. *LWT—Food Sci. Technol.* **2016,** *4* (1), 7–13.

Nikolic, A.; Cuperlovic, M.; Milijic, Z.; Djordjevic, D.; Krsmanovic, J. The Potential Nutritive Value for Ruminants of Some Fibrous Residues from the Food Processing Industry. *Acta Vet. (Belgrade)* **1986,** *36*, 13–22.

O'Shea, N.; Arendt, E. K.; Gallagher, E. Dietary Fibre and Phytochemical Characteristics of Fruit and Vegetable By-products and their Recent Applications as Novel Ingredients in Food Products. *Innov. Food Sci. Emerg. Technol.* **2012,** *16*, 1–10.

Okino Delgado, C. H.; Fleuri, L. F. Orange and Mango By-products: Agro-industrial Waste as Source of Bioactive Compounds and Botanical versus Commercial Description—A Review. *Food Rev. Int.* **2016,** *32* (1), 1–14.

Oreopoulou, V.; Tzia, C. Utilization of Plant By-products for the Recovery of Proteins, Dietary Fibers, Antioxidants, and Colorants. *Utilization of By-products and Treatment of Waste in the Food Industry*, Springer: Berlin, 2007; pp 209–232.

Ovando-Martinez, M.; Sáyago-Ayerdi, S.; Agama-Acevedo, E.; Goñi, I.; Bello-Pérez, L. A. Unripe Banana Flour as an Ingredient to Increase the Undigestible Carbohydrates of Pasta. *Food Chem.* **2009,** *113* (1), 121–126.

Ozvural, E. B.; Vural, H. Grape Seed Flour is a Viable Ingredient to Improve the Nutritional Profile and Reduce Lipid Oxidation of Frankfurters. *Meat Sci.* **2011,** *88* (1), 179–183.

Pardo, M. E. S.; Cassellis, M. E. R.; Escobedo, R. M.; García, E. J. Chemical Characterisation of the Industrial Residues of the Pineapple (*Ananas comosus*). *J. Agric. Chem. Environ.* **2014,** *3* (02), 53.

Paul, M. S.; Sumathy, V. J. H. Production of Amylase from Banana Peels with *Bacillus subtilis* Using Solid State Fermentation. *Int. J. Curr. Microbiol. App. Sci.* **2013,** *2* (10), 195–206.

Prakash, D.; Sharma, G. *Phytochemicals of Nutraceutical Importance*, CABI: Wallingford, 2014.

Prasad, N.; Azeemoddin, G. Characteristics and Composition of Guava (*Psidium guajava* L.) Seed and Oil. *J. Am. Oil Chem. Soc.* **1994,** *71* (4), 457–458.

Pujol, D.; Liu, C.; Gominho, J.; Olivella, M.; Fiol, N.; Villaescusa, I.; Pereira, H. The Chemical Composition of Exhausted Coffee Waste. *Ind. Crops Prod.* **2013,** *50*, 423–429.

Puravankara, D.; Boghra, V.; Sharma, R. S. Effect of Antioxidant Principles Isolated from Mango (*Mangifera indica* L.) Seed Kernels on Oxidative Stability of Buffalo Ghee (Butter-Fat). *J. Sci. Food Agric.* **2000,** *80* (4), 522–526.

Quirós-Sauceda, A.; Palafox-Carlos, H.; Sáyago-Ayerdi, S.; Ayala-Zavala, J.; Bello-Perez, L. A.; Álvarez-Parrilla, E.; De La Rosa, L.; González-Córdova, A.; González-Aguilar, G. Dietary Fiber and Phenolic Compounds as Functional Ingredients: Interaction and Possible Effect after Ingestion. *Food Funct.* **2014,** *5* (6), 1063–1072.

Reddy, G.; Babu, P. R.; Komaraiah, P.; Roy, K.; Kothari, I. Utilization of Banana Waste for the Production of Lignolytic and Cellulolytic Enzymes by Solid Substrate Fermentation

Using Two Pleurotus Species (*P. ostreatus* and *P. sajor-caju*). *Process Biochem.* **2003**, *38* (10), 1457–1462.

Resurreccion, A. V. A. Sensory Aspects of Consumer Choices for Meat and Meat Products. *Meat Sci.* **2004**, *66* (1), 11–20.

Roldán, E.; Sánchez-Moreno, C.; de Ancos, B.; Cano, M. P. Characterisation of Onion (*Allium cepa* L.) By-products as Food Ingredients with Antioxidant and Antibrowning Properties. *Food Chem.* **2008**, *108* (3), 907–916.

Rubio-Senent, F.; Rodríguez-Gutiérrez, G.; Lama-Muñoz, A.; Fernández-Bolaños, J. Pectin Extracted from Thermally Treated Olive Oil By-products: Characterization, Physico-Chemical Properties, In Vitro Bile Acid and Glucose Binding. *Food Hydrocolloids* **2015**, *43*, 311–321.

Sadowska-Bartosz, I.; Bartosz G. Effect of Antioxidants Supplementation on Aging and Longevity. *BioMed Res. Int.* **2014**, *2014*, 17 pages.

Salgado, J. M.; Abrunhosa, L.; Venâncio, A.; Domínguez, J. M.; Belo, I. Screening of Winery and Olive Mill Wastes for Lignocellulolytic Enzyme Production from *Aspergillus* Species by Solid-State Fermentation. *Biomass Convers. Bioref.* **2014**, *4* (3), 201–209.

Salim-ur-Rehman; Hussain, S.; Nawaz, H.; Ahmad, M. M.; Murtaza, M. A.; Rizvi1, A. J. Inhibitory Effect of Citrus Peel Essential Oils on the Microbial Growth of Bread. *Pak. J. Nutr.* **2007**, *6* (6), 558–561.

Sánchez, S. R.; Sánchez, I. G.; Arévalo-Villena, M.; Pérez, A. B. Production and Immobilization of Enzymes by Solid-State Fermentation of Agroindustrial Waste. *Bioprocess Biosyst. Eng.* **2015**, *38* (3), 587–593.

Santanam, N.; Kavtaradze, N.; Murphy, A.; Dominguez, C.; Parthasarathy, S Antioxidant Supplementation Reduces Endometriosis-Related Pelvic Pain in Humans. *Transl. Res.* **2013**, *161* (3), 189–195.

Saura-Calixto, F. Antioxidant Dietary Fiber Product: A New Concept and A Potential Food Ingredient. *J. Agric. Food Chem.* **1998**, *46* (10), 4303–4306.

Sáyago-Ayerdi, S.; Brenes, A.; Goñi, I. Effect of Grape Antioxidant Dietary Fiber on the Lipid Oxidation of Raw and Cooked Chicken Hamburgers. *LWT—Food Sci. Technol.* **2009**, *42* (5), 971–976.

Schieber, A.; Stintzing, F.; Carle, R. By-products of Plant Food Processing as a Source of Functional Compounds—Recent Developments. *Trends Food Sci. Technol.* **2001**, *12* (11), 401–413.

Serna Cock, L.; Torres León, C. Agro-industrial Potential of Peels of Mango (*Mangifera indica*) Keitt and Tommy Atkins. *Acta Agron.* **2015**, *64* (2), 110–115.

Shui, G.; Leong, L. P. Residue from Star Fruit as Valuable Source for Functional Food Ingredients and Antioxidant Nutraceuticals. *Food Chem.* **2006**, *97* (2), 277–284.

Sinha, N.; Sidhu, J.; Barta, J.; Wu, J.; Cano, P. *Fruits and Fruit Processing*; John Wiley & Sons: Ames, IA, 2012.

Soares, P. A.; Vaz, A. F.; Correia, M. T.; Pessoa Jr, A.; Carneiro-da-Cunha, M. G. Purification of Bromelain from Pineapple Wastes by Ethanol Precipitation. *Sep. Purif. Technol.* **2012**, *98*, 389–395.

Stajčić, S.; Ćetković, G.; Čanadanović-Brunet, J.; Djilas, S.; Mandić, A.; Četojević-Simin, D. Tomato Waste: Carotenoids Content, Antioxidant and Cell Growth Activities. *Food Chem.* **2015**, *172*, 225–232.

Stojceska, V.; Ainsworth, P.; Plunkett, A.; İbanoğlu, E.; İbanoğlu, Ş. Cauliflower By-products as a New Source of Dietary Fibre, Antioxidants and Proteins in Cereal Based Ready-to-Eat Expanded Snacks. *J. Food Eng.* **2008**, *87* (4), 554–563.

Streppel, M. T.; Arends, L. R.; van't Veer, P.; Grobbee, D. E.; Geleijnse, J. M. Dietary Fiber and Blood Pressure: A Meta-Analysis of Randomized Placebo-Controlled Trials. *Arch. Intern. Med.* **2005,** *165* (2), 150–156.

Sudha, M. L.; Viswanath, P.; Siddappa, V.; Rajarathnam, S.; Shashirekha, M. N. Control of Rope Spore Forming Bacteria Using Carambola (*Averrhoa carambola*) Fruit Pomace Powder in Wheat Bread Preparation. *Quality Assur. Saf. Crops Foods.* **2016,** *8* (4), 555–564.

Sulieman, A. M. E.; Khodari, K. M. Y.; Salih, Z. A. Extraction of Pectin from Lemon and Orange Fruits Peels and its Utilization in Jam Making. *Int. J. Food Sci. Nutr. Eng.* **2013,** *3* (5), 81–84.

Toor, R. K.; Savage, G. P. Antioxidant Activity in Different Fractions of Tomatoes. *Food Res. Int.* **2005,** *38* (5), 487–494.

Trowell, H. Definition of Dietary Fiber and Hypotheses that It Is a Protective Factor in Certain Diseases. *Am. J. Clin. Nutr.* **1976,** *29* (4), 417–427.

USDA. *Tomatoes*; USDA, 2012.

USDA. *Potatoes*; USDA, 2014.

Vági, E.; Simándi, B.; Vásárhelyiné, K.; Daood, H.; Kéry, Á.; Doleschall, F.; Nagy, B. Supercritical Carbon Dioxide Extraction of Carotenoids, Tocopherols and Sitosterols from Industrial Tomato By-products. *J. Supercrit. Fluids* **2007,** *40* (2), 218–226.

Vega-Vega, V.; Silva-Espinoza, B. A.; Cruz-Valenzuela, M. R.; Bernal-Mercado, A. T.; González-Aguilar, G. A.; Vargas-Arispuro, I.; Corrales-Maldonado, C. G.; Ayala-Zavala, J. F. Antioxidant Enrichment and Antimicrobial Protection of Fresh-Cut Mango Applying Bioactive Extracts from their Seeds By-products. *Food Nutr. Sci.* **2013,** *4* (8A), 197.

Wang, X.; Chen, Q.; Lü, X. Pectin Extracted from Apple Pomace and Citrus Peel by Subcritical Water. *Food Hydrocolloids* **2014,** *38*, 129–137.

Zamora-Gasga, V. M.; Bello-Pérez, L. A.; Ortíz-Basurto, R. I.; Tovar, J.; Sáyago-Ayerdi, S. G. Granola Bars Prepared with *Agave tequilana* Ingredients: Chemical Composition and In Vitro Starch Hydrolysis. *LWT—Food Sci. Technol.* **2014,** *56* (2), 309–314.

CHAPTER 6

WINERY AND GRAPE JUICE EXTRACTION BY-PRODUCTS

F. J. VÁZQUEZ-ARMENTA, A. T. BERNAL-MERCADO,
R. PACHECO-ORDAZ, G. A. GONZALEZ-AGUILAR, and
J. F. AYALA-ZAVALA*

Centro de Investigacion en Alimentacion y Desarrollo, A.C. (CIAD, AC), Carretera a la Victoria Km 0.6, La Victoria, Hermosillo, Sonora 83000, Mexico

*Corresponding author. E-mail: jayala@ciad.mx

CONTENTS

ABSTRACT

Grapes belong to the world's largest fruit crops and is one of the most diffuse fruits in the world both as fresh fruit (table grape) and processed in wine and grape juice. Industrial processing of grapes generates large quantities of solid residues that consist of a mixture of grape skins and seed namely grape pomace or grape marc, and stems or stalks. These by-products, which are commonly discarded, are an important source of dietary fiber and phenolic compounds, with antimicrobial and antioxidant properties. For this reason, scientific efforts have focused on the characterization of the chemical components of waste grape by-products to obtain high value-added ingredients for their application in the food industry and at the same time, to have economic benefits and beneficial impact on the environment. So, in this chapter, we present an overview of the characterization of bioactive compounds from grape by-products and its uses as natural food additives with antimicrobial and antioxidant properties. In addition, we discuss the potential of phenolic compounds from grape by-products to be used for development of nutraceutical products to improve human health.

6.1   TYPES OF PROCESSING AND STATISTICS IN PRODUCTION

Grape (*Vitis vinifera* L.) is a true berry, classed in a group of several seeded fleshy fruits. The berries are organized into a cluster and each berry is attached to the rachis by a small pedicel containing the vessels which supply the berry with water and nutrients (Ribéreau-Gayon et al., 2006). Grapes belong to the world's largest fruit crops with a global production of around 77.18×10^6 t in 2013 (FAOSTAT, 2015). China, the United States, Italy, Spain, and France are the main producers around the world providing about 48% of the global production (FAOSTAT, 2015). Grape is one of the most diffuse fruit in the world both as fresh fruit (table grape) and processed in wine and grape juice.

Grape juice and winemaking have different processes, among them the fermentation process, which leads to alcohol generation, as the main one. For red wine production (Fig. 6.1), grapes after harvesting are destemming and crushing for grape juice extraction, followed by maceration in presence of grape skins and seeds (Bautista-Ortín et al., 2005). The main purpose of maceration is the extraction of color compounds (anthocyanins and phenolic substances) from the solid components of the grape (Bautista-Ortín et al., 2005). Next, a primary fermentation (alcoholic fermentation) is carried out by yeast cells (1–2 weeks in the presence of grape skins), and their ability to absorb phenolic compounds may result in a higher or lower phenolic content in the wine and in its by-product. The secondary fermentation (malolactic fermentation) is conducted by lactic acid bacteria, in the absence of grape skins, which are removed by pressing the must. Thus, grape skins and seeds are discarded as by-product that is namely "grape pomace" or "grape marc." Finally, after adequate aging, red wine are clarified, blended, and filtrated for bottling.

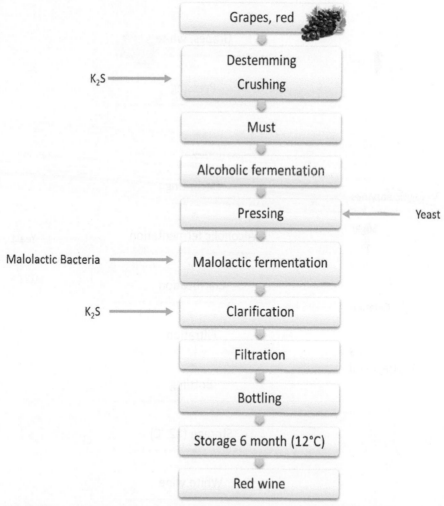

FIGURE 6.1 Flow diagram describing the elaboration of red wine from grapes.

On the other side, although red wines are obtained by the alcoholic fermentation of musts in the presence of solid parts of the berry (skins and seeds), white wines are exclusively produced by the fermentation of grape juice (Ribéreau-Gayon et al., 2006). The main steps processing of white wine are crushing, must, alcoholic fermentation, pressing, clarification, filtration, bottling, and storage (Fig. 6.2). Initially, the grapes are crushed and the stems are removed, leaving liquid must that flows. In white wine, all the grape skins are separated from the "must" by filters or centrifuges before the must undergoes fermentation. During the fermentation process,

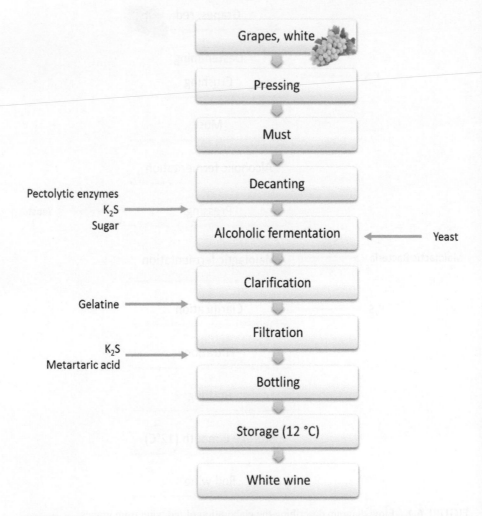

FIGURE 6.2 Flow diagram describing the elaboration of white wine from grapes.

wild yeast are used to turn the sugar in the must into alcohol. After crushing and fermentation, wine needs to be stored, filtered, and properly aged.

In the other hand, the methods for the grape-juice preparation are very different to the varieties used and local traditions. In general, the main steps in juice preparations are destemming, crushing, depectination, pressing, pasteurization, setting, and filtration, as is shown in Figure 6.3 (Bates et al., 2001). First, bunches of grapes are separated from the stem. This step yield

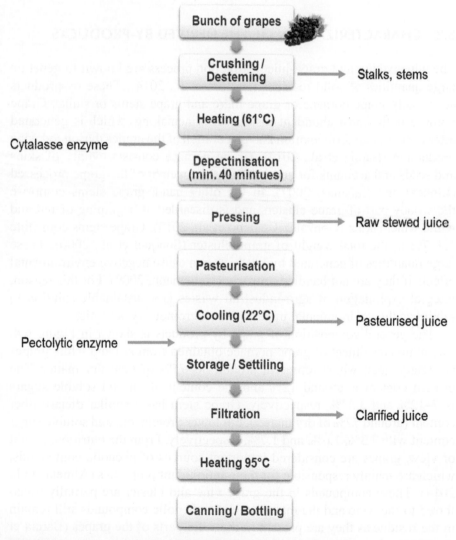

FIGURE 6.3 Process describing the preparation of juice from grapes.

high amounts of by-products, such as stalks and stems. Then, the crushed fruits are heated to 60–63°C, at this moment pectolytic enzyme are added to break down the pectin to make the grape pulp ready for pressing. This part of the process helps to extract color from the skins into the juice. Later, the juice is pasteurized with a flash-heated at 80–85°C and then rapidly cooled. The processing continues once the argols have settled and then the sediment can be filtered. At the end, the juice is heated at 95°C and ready to bottled or canned (Bates et al., 2001).

6.2   CHARACTERIZATION OF THE DERIVED BY-PRODUCTS

The winemaking and grape-juice extraction process are known to generate large quantities of solid residues (Brahim et al., 2014). These by-products are namely grape pomace or grape marc and grape stems or stalks. Grape pomace is the most abundant residue of winemaking, which is generated after concomitant fermentation and maceration of the grapes during red wine production (Barcia et al., 2014). Grape pomace consists mainly of skins and seeds and accounts for about 20% of the weight of the grape processed (Llobera and Cañellas, 2007). In the other hand, grape stems comprise the woody part of grape clusters and is discarded at beginning of red and white winemaking (González-Centeno et al., 2013). Grape stems constitute 1.4–7% of the total weight of grape cluster (Souquet et al., 2000). These large quantities of generated by-products can cause negative environmental effects if they are not handled properly (Nellemann, 2009). For this reason, integral exploitation of agro-industrial wastes is a sustainable solution to minimize the environmental impact of the agroindustry activities.

The general composition of grape by-products is shown in Figure 6.4. The main constituent of grape pomace obtained from red and white grapes is dietary fiber, which represent around 50–75% of total dry matter. The protein content is around 12% and the content of oil and soluble sugars is 7–12% and 3.27%, respectively, Grape stem has a similar dietary fiber content (around 77% of dry matter), but lower protein, oil, and soluble sugar content with 7.3%, 1.6%, and 1.7%, respectively. From the nutritional point of view, grapes are considered important sources of phenolic compounds, which are mainly responsible for their antioxidant properties (Almela et al., 2014). These compounds in the grape vine and cluster are partially transferred to the wine and the majority of the phenolic compounds still remain in the residue as they are present in the solid parts of the grapes (Barcia et al., 2014).

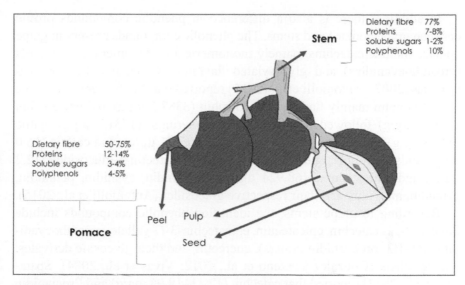

FIGURE 6.4   Proximal composition of grape by-products.

The determination of total phenolic content (TPC) is the starting point for the characterization of grape by-products because it gives useful information about the relative composition of the sample. There are several spectrophotometric methods based on different principles for the quantification of phenolics in plant samples which are used to determine diverse structural groups (Fontana et al., 2013). Generally, grape pomace has 4–5% of phenolic compounds, whereas in grape stems, the phenolic content is twofold higher than grape pomace. For example, Llobera and Cañellas (2007) reported that TPC of Manto Negro grape stem (11.6 g GAE/100 g dw) is four times higher than that for the pomace sample (2.63 g GAE/100 g dw). Similarly, TPC and total proanthocyanidins content of stems from 10 different cultivars of *V. vinifera* (Cabernet Sauvignon, Callet, Chardonnay, Macabeu, Manto Negro, Merlot, Parellada, Premsal blanc, Syrah, and Tempranillo) ranged from 4.7 to 11.52 g GAE/100 g dw and from 0.079 to 0.202 g of tannins/100 g dw, respectively (González-Centeno et al., 2012). In contrast, the TPC of grape pomaces from Chardonnay, Macabeu, Parellada and Premsal Blanc grape varieties ranged from 3.09 to 4.65 g GAE/100 g dw, and the total proanthocyanidin content from 0.05 to 0.092 g of tannins/100 g dw (González-Centeno et al., 2013). These differences could be attributed to the fact that vine stems are directly discarded during the winemaking process; thus, their original phenolic composition is preserved almost intact (Brahim et al., 2014).

In addition, there is also a difference in phenolic compounds profile between grape pomace and stems. The phenolic compounds present in grape pomace include catechins, namely monomeric and oligomeric flavan-3-ols (proanthocyanidins) and glycosylated flavonols. Catechins (Llobera and Cañellas, 2007). Antoniolli et al. (2015) reported that Malbec grape-pomace extract contain mainly flavanols as catechin (3387.5 µg/g) and epicatechin (1763.4 µg/g) followed by phenolic acids; syringic (1731.7 µg/g), gallic (252.8 µg/g), p-coumaric (64.6 µg/g), ferulic (24.1 µg/g), and caffeic (16.0 µg/g) acids, flavonols; quercetin (557.3 µg/g), quercetin-3-glucoside (112.2 µg/g), and anthocyanins (68,924 µg/g); delphinidin, cyanidin, petunidin, peonidin, malvidin, and their respective glucosides (Antoniolli et al., 2015).

Regarding to grape stems, the identified phenolic compounds include flavonoids, as catechin, epicatechin, epicatechin-3-O-gallate, proanthocyanidins (B1–B3, procyanidin dimers), quercetin, and their glycoside derivates, among others (Gonzalez-Centeno et al., 2012; Vivas et al., 2004). Spatafora et al. (2013) reported that catechin (12.11–19.95 mg/g) and Procianidin B1 (6.19–13.73 mg/g) are the most abundant phenolic compounds in stems from Sicillian cultivars (Nerello Mascalese and Nero d'Avola). Barros et al. (2014) describes the phenolic content of grape stems from red and white cultivars grown in northern Portugal, being quercetin-3-O-glucuronide, malvidin derivatives, and epicatechin as the main metabolites, representing from 54% to 75% of the TPC in all cultivars analyzed, whereas gallic acid has been considered as the main hydroxybenzoic acid in grape stems, and hydroxycinnamic acids (p-coumaric, ferulic, caffeic, and their derivates) are present only in trace amounts (Anastasiadi et al., 2012).

Since the production of secondary metabolites by vine plants depends not only on genetic characteristics but also on growth environmental conditions, the content of the polyphenols in grapes may vary between different cultivars, being influenced by locations, harvest time, and the growth environment. Also, the phenolic composition of grape by-products from juice processing and from wine making may be significantly different depending on the type of wine produced. During red wine vintage, the crushed grapes are macerated with juice, and the majority of soluble phenols dissolve in the wine, leaving complex polyphenols in the pomace. Anthocyanin content of grape pomace varies with wine vinification method and contact time. The longer contact time, the lower amount (Yu and Ahmedna, 2013). So, the accurate characterization of phenolic compounds of grape processing by-products could be useful to determine its potential application in food industry.

6.3   CONSIDERATION FOR PRACTICAL USES OF THE DERIVED BY-PRODUCTS

Grape processing industry generates large amounts of waste; however, only a fraction of this waste is exploited for other purposes and the majority are discarded (Muñoz-González et al., 2014). This causes great economic problems for the industry and also, if they are not properly managed can eventually cause serious environmental pollution problems. Therefore, it is imperative to search alternative uses for wastes derived from grapes processing.

Grape waste, consisting of peel, seed, stems, and unused pulp, are important source of phenolic compounds which exhibit various bioactive properties such as antioxidants and antimicrobial (Barcia et al., 2014). This makes them an excellent option for using them in the production of food additives, drugs, disinfectants, among others. Added to this, it is the perception of consumers about the health risks due to the use of synthetic compounds in these kind of products (Ayala-Zavala et al., 2011). Thus, the economics of processing crops could be improved by developing higher value use for their by-products (Ayala-Zavala et al., 2011).

In the latest years, scientific works carried out on the characterization of the chemical components of waste grape by-products has allowed looking for different applications in trying to obtain high value-added ingredients for their utilization in food industry (Muñoz-González et al., 2014). In this context, the integral exploitation of grapes represents a successful opportunity to obtain nutrients and phenolic compounds, and at the same time, to have economic benefits and beneficial impact on the environment.

6.4   WINERY AND GRAPE JUICE EXTRACTION BY-PRODUCTS AS SOURCE OF ANTIMICROBIAL AND ANTIOXIDANT AGENTS

One of the major issues in the food industry is the prevention of contamination due to the presence of pathogenic bacteria during food production, due to its direct consequences to public health (Newell et al., 2010). Only in the United States from 2000 to 2008, pathogenic microorganism caused 9.4 million episodes of foodborne illness, 55,961 hospitalizations, and 1351 deaths. Leading causes of death were nontyphoidal *Salmonella* spp. (28%) followed by *Listeria monocytogenes* (19%) (Scallan et al., 2011). Additionally, the Center for Science in the Public Interest maintains a record of foodborne illness outbreaks which stressed that *Salmonella* spp., *Escherichia coli*, and *L. monocytogenes* are the main pathogenic bacteria associated with these outbreaks.

In addition, the presence of deteriorative bacteria can compromise the organoleptic attributes of fresh food and foodstuffs (Ayala-Zavala et al., 2011). Microbial spoilage is caused by the proliferation of microorganisms, which is characterized by the production of pectinolytic enzymes and metabolites, resulting in visual and textural defects, as well as the generation of off-odors, severely limiting the safety and shelf-life of foods, together with the risk of pathogenic microbial growth (Regaert et al., 2009). Chemical synthetic additives can reduce food decay, but consumption of more food that has been formulated with chemical preservatives has increased consumer concern and created a demand for more natural and minimally processed food. As a result, there has been a great interest in naturally produced antimicrobial agents (Cleveland et al., 2001).

In this context, grape by-products could be an important source of natural extracts with antimicrobial activity against pathogenic bacteria (Table 6.1). Supercritical extracts obtained from Merlot pomace at 300 bar/50°C showed antimicrobial activity with a minimum inhibitory concentration (MIC) of 625 μg/mL against *Staphylococcus aureus*, and 1000 μg/mL against *E. coli* and *Pseudomonas aeruginosa* (Oliveira et al., 2013). The main components from the extracts were gallic acid, *p*-OH-benzoic acid, vanillic acid, and epicatechin (Oliveira et al., 2013). In another study, *S. aureus* was more sensitive to ethanolic and methanolic extracts of Petit Verdot and Pinot Noir grape marcs (MICs = 3.13–6.25 mg/mL) than *L. monocytogenes* (MICs = 12.5–>25 mg/mL) (Martin et al., 2012). The total phenolic compounds of methanolic and ethanolic extracts ranged from 161.9 to 229.2 mg GAE/g dw, epicatechin being the most abundant phenolic compound, whereas gallic acid, caffeic acid, and syringic acid were found in trace amounts (Martin et al., 2012). Also, grape pomace extracts inhibited glucosyltransferases B and C (70–85% inhibition), responsible of synthesis of extracellular polysaccharides in *Streptococcus mutans*, at concentrations as low as 62.5 μg/mL. Furthermore, grape pomace extracts inhibited the glycolytic pH-drop by *S. mutans* cells without affecting the bacterial viability, an effect that can be attributed to partial inhibition of F-ATPase activity (30–65% inhibition at 125 μg/mL) (Thimothe et al., 2007).

Katalinić et al. (2010) determined the phenolic composition of grape skin extracts from seven red grape and seven white grape varieties grown in Croatia and evaluated its antimicrobial and antioxidant properties. Total phenolic, flavonoids, and catechins content in red grape varieties averaged 875 mg GAE/kg, 686 mg GAE/kg, and 474 mg CE/kg, respectively, whereas white grape varieties presents a TPC = 1851 mg GAE/kg, total flavonoids = 1332 mg GAE/kg, total catechins = 775 mg CE/kg and total anthocyanins content = 763 mg of malvidin-3-glucoside equivalents. In addition, individual phenolic compounds (+)-catechin, (−)-epicatechin, epicatechin gallate, procyanidin B1 and procyanidin B2, quercetin glucoside, and resveratrol was variety dependent. Grape-skin extracts showed antibacterial activity against *S. aureus*, *Bacillus cereus*, *E. coli* O157:H7, *S. infantis*, and *Campylobacter coli* with MICs in the range of 0.014–0.59 mg GAE/mL, with lower MICs of white cultivars, especially against *Campylobacter* and *Salmonella* (Katalinić et al., 2010).

TABLE 6.1   Antibacterial Properties of Extracts from Grape By-products.

By-product extract	Tested organism	MIC	References
Seed	*E. coli*	25 mg/mL	Cheng et al. (2012)
	S. aureus	0.78 mg/mL	
	S. aureus	0.23–44 GAE/mL	Katalinić et al. (2010)
	B. cereus	0.12–0.34 GAE/mL	
	E. coli	0.15–0.44 GAE/mL	
	S. infantis	0.15–0.44 GAE/mL	
	E. coli	0.04–0.25 GAE/mL	
Stem	*L. monocytogenes*	3.45 mg/mL	Anastasiadi et al. (2008)
	L. monocytogenes	18 mg/mL	Vázquez-Armenta et al. (2017)
	S. aureus	16 mg/mL	
	E. coli O157:H7	18 mg/mL	
	S. enterica ser. Typhimurium	16 mg/mL	
Pomace	*S. aureus*	625 µg/mL	Oliveira et al. (2013)
	E. coli	1000 µg/mL	
	P. aeruginosa	1000 µg/mL	
	E. coli	25 mg/mL	Cheng et al. (2012)
	S. aureus	0.78 mg/mL	
	L. monocytogenes	12.5 mg/mL	Martin et al. (2012)

MIC, minimum inhibitory concentration is the lowest concentration of an antimicrobial compound that will inhibit the visible growth of a microorganism after overnight incubation.

With regard to grape stems, a study carried out by Vázquez-Armenta et al. (2017) showed that ethanolic extract of table grape var. Red Globe possess antibacterial activity against *L. monocytogenes*, *S. aureus*, *S. enterica* ser.

Typhimurium and *E. coli* O157:H7 at MIC range of 16–18 mg/mL. Additionally, it was shown that this extract affects the different phases of bacterial growth such lag time and growth rate. The antibacterial properties was attributed to phenolic content (TPC = 37.25 mg GA/g; TFC = 98.07 mg QE/g) and its main constituents: rutin (212.57 µg/g), gallic acid (184.10 µg/g), chlorogenic acid (173.26 µg/g), caffeic acid (50.59 µg/g), catechin (34.31 µg/g) and ferulic acid (8.19 µg/g). Also, it has been reported that methanolic extract of red grape var. Mandilaria stems has antimicrobial activity against *L. monocytogenes* with a MIC of 0.345% (w/v). The antilisterial effect was related to TPC of extract (536.8 mg GAE/g) and its main constituents: (+)-catechin (85.81 mg/g), procyanidin B3 (31.55 mg/g), ε-viniferin (31.42 mg/g), and transresveratrol (17.56 mg/g) (Anastasiadi et al. 2008).

The antimicrobial properties of extracts from grape by-products also has been evaluated in diverse food systems (Table 6.2). Vázquez-Armenta et al. (2017) showed that the application of stem extract from grapes var. Red Globe, rich in phenolic compounds, was effective to reduce the populations

TABLE 6.2 Antimicrobial Properties of Extracts from Grape By-products Applied on Different Food Systems.

By-product extract	Food matrix	Concentration used	Antimicrobial properties	References
Stem	Lettuce and spinach	25 mg/mL	Reduce the populations of *L. monocytogenes*, *S. aureus*, *E. coli*, and *Salmonella* in lettuce (0.859–1.884 log reduction) and spinach (0.843–2.605 log reduction)	Vázquez-Armenta et al. (2017)
Seed	Cucumber and lettuce	1 mg/mL	Reduced the counts of *Salmonella* spp. and *L. monocytogenes* 2.5–2.8 log CFU/g	Xu et al. (2007)
Seed	Tomatoes	1.25 mg/mL	Reduced the counts of *L. monocytogenes* ≈ 2 log CFU/g	Bisha et al. (2010)
Pomace	Beef patties	100 mg/g	Decrease the total counts of foodborne pathogens (*Enterobacteriaceae*, coliform bacteria, and *S. aureus*) and food spoilage microorganisms (yeast and molds, *Micrococcaceae*, aerobic mesophilic, psychrotrophic, lipolytic, and proteolytic bacteria)	Sagdic et al. (2011)

of *L. monocytogenes*, *S. aureus*, *E. coli*, and *Salmonella* in lettuce (0.859–1.884 log reduction) and spinach (0.843–2.605 log reduction). Xu et al. (2007) showed that the application of grape seed extract at 1 mg/mL in fresh cucumber and lettuce reduce the counts of *Salmonella* spp. and *L. monocytogenes* approximately 2.5–2.8 log CFU/g. Similarly, *L. monocytogenes* was reduced by ≈2 log CFU/g on tomatoes surfaces exposed to commercial grape-seed extract solution (0.125%) (Bisha et al., 2010). These results indicated that grape by-product extracts are effective to reduce the pathogenic load in fresh vegetables, a commodity group of highest concern from a microbiological safety perspective (FAO/WHO, 2008), because they are often grown in the open field and vulnerable to contamination from soil, sewage, water used for irrigation, and contact with (feces of) wildlife (FAO/WHO, 2008).

In addition, it has been reported that grape pomace extracts incorporated into beef patties at 10% decrease the total counts of foodborne pathogens (*Enterobacteriaceae*, coliform bacteria, and *S. aureus*) and food spoilage microorganisms (yeast and molds, *Micrococcaceae*, aerobic mesophilic, psychrotrophic, lipolytic, and proteolytic bacteria) after 48 h of storage at 4°C (Sagdic et al., 2011), whereas grape seed extracts incorporated at 1% into pea starch films reduced the growth of *Brochothrix thermosphacta* (1.3 log CFU/mL reductions) in pork loins (Corrales et al., 2009). The antimicrobial effect could be attributed to migration of phenolic compounds of grape seed extracts inside the meat (Corrales et al., 2009). These studies suggest that grape by-product extracts could be useful as antimicrobial agents to prevent the deterioration of meat products as well as to inhibit some foodborne pathogens.

The antimicrobial properties of grape by-product extracts are attributed to phenolic content. The antimicrobial mode of action of phenolic compounds is suggested to act on multiple cellular targets. Gallic and ferulic acids, for example, led to irreversible changes in membrane permeability in *E. coli*, *P. aeruginosa*, *S. aureus*, and *L. monocytogenes*, through hydrophobicity changes, decrease of negative surface charge, and occurrence of local rupture or pore formation in the cell membranes with consequent leakage of essential intracellular constituents (Borges et al., 2013). Also, gallic acid effectively permeabilized the bacterial membrane of *Salmonella* spp., by chelation of divalent cations from the outer membrane (Nohynek et al., 2006).

In the other hand, isoquercitrin and rutin inhibit the catalytic activity of topoisomerase IV, an essential enzyme for cell survival in *E. coli* (Bernard et al., 1997), whereas the antimicrobial activity of quercetin is attributed to inhibition of DNA gyrase, interacting with DNA or with ATP-binding site of the enzyme (Plaper et al., 2003). Quercetin and apigenin also function as reversible inhibitors that are competitive with the substrate ATP of D-alanine:D-alanine ligase, an essential enzyme that catalyzes the ligation of D-Ala–D-Ala in the assembly of peptidoglycan precursors, whereas they are noncompetitive with the other substrate D-Ala (Wu et al., 2008).

In addition to the antimicrobial properties, it has been reported that grape by-products and its constituents possess bioactive properties as antioxidant and radical scavenging. Free radicals/reactive oxygen species (ROS) have been a threat in food systems by decreasing the self-stability. Free radicals are chemical species having one or more unpaired electrons due to which they are highly unstable and cause damage to other molecules. ROS include superoxide radicals, hydroxyl radicals, singlet oxygen, and hydrogen peroxide which are often generated as by-products of biological reactions or initiated by exposure of lipids to light, heat, ionizing radiation, metal ions, or metalloprotein catalysts (Prakash et al., 2015).

The presence of ROS in food products could let to oxidation process that brings undesirable changes related to food quality, organoleptic characteristics, safety, and nutritional value. Food undergone extensive oxidation has major defects and no consumer acceptability. Oxidation in foods is manifested mainly through discoloration and generation of off-flavors, while alteration of principal components, such as vitamins, lipids, and proteins, is not always apparent (Prakash et al., 2015). Lipid oxidation generates a series of chemical reactions that can alter physicochemical parameters, sensorial attributes (odor, color, flavor), and shelf-life in meat and meat products (Garrido et al., 2011), whereas a consequence of protein oxidation, essential amino acids may be lost, and protein digestibility may decrease. These protein alterations are detrimental to the overall quality of fresh meat and meat products. For this reason, the shielding of food matrices against oxidative deterioration is of undisputed technological importance (Prakash et al., 2015).

To avoid oxidative reaction in foodstuff, various techniques have been applied with such applications of sodium and potassium metabisulfite (Sgroppo et al., 2010). Sulfite is commonly used in meat industry due to its ability to delay the microbial spoilage and discoloration resulting from myoglobin oxidation. Although sulfites possess exceptional good technological properties, their exposure could led allergic and respiratory reactions, especially for sensitive individuals (Garcia-Lomillo et al., 2016). In response to recent demand for natural products, the meat and poultry industry is actively seeking natural solutions to minimize oxidative rancidity and increase shelf-life of this type of products (Karre et al., 2013).

Recent investigation has focused toward identification of novel anti-oxidants from natural sources. Antioxidants from fruit by-products may be grouped in accordance to their mode of action, that is, as acidulants, reducing and/or chelating agents, enzyme inhibitors, and free-radical scavengers. Data reported in literature highlight that extracts obtained from grape by-products are important source of phenolic compounds with free-radical scavenger activity. Garrido et al. (2011) showed that pomace extract from grape var. Monastrell has a total anthocyanins content of 816.06 ± 17.76 mg/L with an antioxidant activity of 63.36 ± 4.55 mmol TE/L. Also, Pinot Noir grape pomace has a TPC of 67.74 ± 6.91 mg GAE/g and antiradical scavenger capacity of 37.46 ± 1.86 ascorbic acid equivalents/g and 91.78 ± 4.58 mg TE/g (Tseng and Zhao, 2013), whereas Malbec grape pomace extract with TPC = 165.7 ± 30.2 mg GAE/g have an oxygen radical absorbance capacity of 2756 μmol TE/g (Antoniolli et al., 2015).

de Sá et al. (2014) reported the antioxidant activity of Fernão Pires grape stems extracts (EC_{50} = 0.052–0.090 mg/mM of DPPH·) is correlated (r = 0.89) to TPC (17.0–19.0 mg GAE/g dw) and total phenol index (20.0–40.0 mg (+)-catechin equivalents/g dw). Similarly, the TPC and total proanthocy-anidins of the stems of 10 grape varieties (Cabernet Sauvignon, Callet, Chardonnay, Macabeu, Manto Negro, Merlot, Parellada, Premsal blanc, Syrah, and Tempranillo) have been reported to range from 47.04 to 115.25 mg GAE/g and from 79.1 to 202.3 mg tannins/g with antioxidant capacities of 99.7–253.2, 145.4–378.6, 65.4–170.1, and 101.9–282.1 mg TE/g, measured by ABTS, CUPRAC, FRAP, and ORAC assays, respectively (Gonzalez-Centeno et al., 2012).

The antioxidant mode of action of flavonoids and phenolic acids is through scavenging free radicals directly by hydrogen atom donation. The free radical P–O· may react with a second radical, acquiring a stable quinone structure. However, the antioxidant activity depends on the arrangement of functional groups on its core structure. Both the configuration and total number of hydroxyl groups substantially influence the mechanism of the antioxidant activity (Procházková et al., 2011).

Numerous studies concluded that grape by-product extracts are effective antioxidant in raw and cooked meat products (Table 6.3). Grape pomace extract (obtained by methanolic extraction + high–low instantaneous pressure) added at 0.06% in pork burgers inhibited lipid oxidation (7 times lower than control) and maintain color parameters (a^* values, related to redness color), during 6 days of storage at 4°C (Garrido et al., 2011). The color stability of the a^* value during storage could be related with the antioxidant effect of grape pomace extract, since secondary products generated in lipid oxidation processes (e.g., aldehydes) strongly promote myoglobin deterioration (Garrido et al., 2011).

TABLE 6.3 Antioxidant Activity of Extracts from Grape By-products Applied on Different Food Systems.

By-product	Food matrix	Concentration used	Antioxidant properties	References
Pomace	Pork burgers	0.6 mg/g	Reduced 7 times the lipid oxidation compared to control and maintained color parameters during 6 days at 4°C	Garrido et al. (2011)
Skin	Beef patties	20 mg/g	Protected against protein radical formation and myosin cross-linking	Garcia-Lomillo et al. (2016)
Dietary fiber from pomace	Raw and cooked chicken breast hamburger	20 mg/g	Increased 2 and 2.5-fold the radical scavenging capacity of the treated product and inhibit 36.6% lipid oxidation	Sáyago-Ayerdi et al. (2009)

In the same manner, milled red grape skins with a total phenol content of 9.9 mg GAE/g protected against protein radical formation and myosin cross-linking by the addition of 2.0% (w/w) in beef patties stored for 15 days at 4°C in a high oxygen atmosphere (70% O_2 and 30% CO_2) (Garcia-Lomillo et al., 2016), whereas skins from white grape with total phenol of 4.0 mg GAE/g only protected against myosin cross-linking (Garcia-Lomillo et al., 2016). Addition of 2% of grape antioxidant dietary fiber to raw and cooked chicken breast hamburger increase 2.0 and 2.5-fold the radical scavenging capacity of the treated product, respectively, and inhibit lipid oxidation in 36.6% compared with control (Sáyago-Ayerdi et al., 2009).

Foodstuffs containing high amount of fat with oil-in-water emulsions can be readily oxidized during processing and storage, which led to the formation of undesirable volatile compounds (Tseng and Zhao, 2013). In this sense, grape by-product extracts has been demonstrated to retard lipid oxidation in high lipid content foods. For example, fortification of yogurt, Italian and Thousand Island salad dressings with grape pomace resulted in 35–65% reduction of peroxide values in all samples, also increased dietary fiber content (0.94–3.6%; w/w product), TPC (958–1340 mg GAE/kg product), and antioxidant activity (710–936 mg AAE/kg product) (Tseng and Zhao, 2013). Also, a study carried out by Spigno et al. (2013) showed that phenolic grape pomace extract incorporated as oil-in-water nanoemulsion improved the shelf-life of hazelnut paste by inhibiting its oxidation, by increasing extract dispersability in the paste and preserving the antioxidant activity. Since the prevention of oxidative deterioration high lipid content foods and meat products is essential if their quality and shelf-life are to be guaranteed, these studies demonstrated that grape by-products are very effective inhibitors of lipid oxidation and has potential to be used as a natural antioxidants in food industry.

6.5   NUTRACEUTICAL USES OF WINERY AND GRAPE JUICE EXTRACTION BY-PRODUCTS

The diet provides us with certain compounds that are supposed to be beneficial to our health. These compounds can be naturally taken from foods such as vegetables and fruits. Unfortunately, only a modest amount of these compounds are obtained from plant foods. To obtain greater health benefit, it is necessary to complement our diet with enriched foods or supplements such as nutraceuticals. Nutraceuticals are products that contain bioactive compounds isolated from a food matrix (mostly plant foods), which are ingested in capsules, pills, tablets, or liquid form to treat a specific disease or disorders (Kalra, 2003).

Grapes are a promising source for the development of nutraceutical products, due to its unique combination of phytochemicals (flavonoids, tannins, phenolic acids, resveratrol, and dietary fiber) (Teixeira et al., 2014). Most of these bioactive compounds are present in the fruit skin, seeds (Mendes et al., 2013). For this reason, by-products generated from wine and grape juice production represent an inexpensive source of bioactive compounds with nutraceutical purposes.

Grape pomace is a valuable by-product due to its great amount of phytochemical such anthocyanins, flavonols, resveratrol, and dietary fiber (Yu and Ahmedna, 2013). Fiber consists in all nondigestible carbohydrates and lignin that are intrinsic and intact in plants, and possess beneficial physiological effects in humans (Slavin, 2013). Most of the fiber in grape is accumulated in the skin, pulp and seed, which after the manufacture of grape juice and wine remains as pomace. The consumption of fiber always has been associated with health benefits. Nowadays, there is a growing interest for food plants with bioactive compounds associated with the fiber, because of their positives effects in human health. Grape pomace is great source of dietary fiber and phenolic compounds and it serves as a dietary supplement that combines the benefits of both phytochemical in the prevention of cardiovascular diseases and cancer (Zhu et al., 2015).

Dietary fiber has some biological and protective effects; it plays an important role in the maintaining the homeostasis in the gastrointestinal tract by modulating gastrointestinal transit time, fecal weight, fecal acidity, short-chain fatty acids productions and fecal bile acids (Touriño et al., 2009). Also, the consumption of dietary fiber affects the populations of the microbiota that inhabit the gastrointestinal tract, the modulation of the gut microbiota by dietary fiber is associated with health benefits (Slavin, 2013). A study revealed that the intake of grape antioxidant dietary fiber can modulate mucosal apoptosis by modulating the cellular redox environment (López-Oliva et al., 2010). Moreover, grape antioxidant fiber promotes the growth of *Lactobacillus reuteri* and *Lactobacillus acidophilus* in male Wistar rats after 4 weeks (Pozuelo et al., 2012). On the other hand, anticarcinogenic activity also has been reported; lyophilized red grape pomace rich in antioxidant dietary fiber shows a chemopreventive effect and potential antitumoral activity in a mouse model (Sánchez-Tena et al., 2013) However, more clinical trials are needed to confirm these beneficial effects.

Other benefit of dietary fiber from grape by-products is the antioxidant activity which has been used to control oxidative stress in obese rats (Fernández-Iglesias et al., 2014). Also it has been reported that a high consumption of a grape supplement, rich in dietary fiber, prevent cardiovascular disease by reducing the levels LDL, apolipoprotein-B, and serum lipids in plasma (Tomé-Carneiro et al., 2012). Also, grape antioxidant dietary fiber affects the expression of pro- and anti-apoptotic Bcl-2 proteins which attenuates the mitochondrial apoptotic pathway in the colonic mucosa (López-Oliva et al., 2013). However, the exact mechanism of how grape-antioxidant dietary fiber act on the different afflictions have not yet be clarified.

Numerous studies have demonstrate that grape phenolic compounds particularly flavonoids, have many health benefits such as antimutagenic and anticarcinogenic activity, antioxidant and anti-inflammatory activities and prevention and delay of cardiovascular diseases (Bak et al., 2013; Feregrino-Perez et al., 2012; Tomé-Carneiro et al., 2012; Yu and Ahmedna (2013). Some studies have also shown that taking grape seed extract reduce food intake in rats and energy intake in human; in the same way, increase lifespan and retard the onset of age-related markers (Shukitt-Hale et al., 2006). However, antioxidant properties are the most notable bioactivity of phenolic compounds from grape pomace. The antioxidative characteristics have been widely studied, including scavenging of free radicals, inhibition of lipid oxidation, reduction in hydroperoxide formation to name a few (Xia et al., 2010).

Plant phenolic compounds can enhance the endogenous anti-oxidative system, by improving the antioxidant balance, and effectively prevent oxidative damage. Flavonoids are the phenolic compounds most studied in this area due to its capacity to decrease lipid peroxidation and increased plasma total antioxidant capacity, they also attenuate stress-sensitive signaling pathways pro-oxidant enzymes and induct antioxidant enzymes including superoxide dismutase, catalase, and glutathione peroxidase (Lu et al., 2013). Evidence suggest that phenolic compounds beyond of the direct antioxidant capacities in scavenging of free radicals mainly act by direct interactions with important cellular receptors or key signaling pathways, which may result in modification of the redox status of the cell and may trigger a series of redox-dependent reactions (Scalbert et al., 2005). However, the uptake of phenolic compounds by the gastrointestinal tract is incomplete and their circulating levels in plasma low.

Nonabsorbed phenolic compounds reach the colon by directly passing through the small intestine (Manach et al., 2005). In the colon, phenolic compounds are fermented and deconjugated by bacterial enzymes and are able to inhibit growth of potentially pathogenic bacteria, suppress many virulence factors such as neutralization of bacterial toxins, inhibition of biofilm formation, reduction of adhesion, and also show synergism with antibiotics (Rodriguez Vaquero et al., 2010).

Phenolic compounds from grape by-products could act as prebiotic-type substances, meaning that they increase the amount of healthy bacteria in the gut. The exact mechanisms of how phenolic compounds promote bacterial growth has not been elucidated; however, it has been proposed that selected bacteria possess specific enzymes that can metabolize phenolic compounds to obtain energy as well as enhance consumptions of nutrients such as sugars (García-Ruiz et al., 2008). The main enzymes involved in this process are α-rhamnosidase, β-glucosidase, and β-glucuronidase. Other enzymes such as esterases, hydrogenases, dehydroxylase, demethylase, descarboxylase, and isomerase are in charge of breakdown the backbone structure of phenolic compounds (Selma et al., 2014). Additionally, phenolic compounds can enhance the adhesion ability of beneficial bacteria. Some of these compounds can mimic acetylated homserine lactones, phenolic compounds works as regulatory factors of biofilm formation involved in bacterial adhesion (Huber et al., 2003). Furthermore, some bacteria are able to metabolize hydroxicinamic acids by the reduction of the carbon–carbon double bound producing hydroxyphenylpropionic acids, which can be further descarboxylated to p-ethyphenols (Alberto et al., 2004). For these bioactive properties, phenolic compounds from grape by-products could be used for the development of nutraceutical products to improve human health.

6.6 CONCLUDING REMARKS

The negative environmental effects of improperly handled by-products generated during processing of grapes to obtain wine and grape juice can be minimized if an integral exploitation is achieved. Grape by-products as pomace and stalks are rich source of phenolic compounds and dietary fiber that could be used to obtain high valuable food ingredients, natural additives and nutraceuticals. Integral exploitation of grapes by-products represents a successful opportunity to have economic benefits for agro-industrial activity with beneficial impact on the environment.

KEYWORDS

- grape juice extraction
- maceration
- pomace
- alcoholic fermentation
- by-products

REFERENCES

Alberto, M. R.; Gómez-Cordovés, C.; Manca de Nadra, M. C. Metabolism of Gallic Acid and Catechin by *Lactobacillus hilgardii* from Wine. *J. Agric. Food Chem.* **2004,** *52*, 6465–6469.

Almela, C.; Espert, M.; Ortolá, M.; Castelló, M. Influence of Minimally Processed Grapes Washing with Lemon Essential Oil. *Int. Food Res. J.* **2014,** *21*, 1851–1857.

Anastasiadi, M.; Chorianopoulos, N. G.; Nychas, G.-J. E.; Haroutounian, S. A. Antilisterial Activities of Polyphenol-Rich Extracts of Grapes and Vinification Byproducts. *J. Agric. Food Chem.* **2008,** *57*, 457–463.

Anastasiadi, M.; Pratsinis, H.; Kletsas, D.; Skaltsounis, A.-L.; Haroutounian, S. A. Grape Stem Extracts: Polyphenolic Content and Assessment of their In Vitro Antioxidant Properties. *LWT—Food Sci. Technol.* **2012,** *48*, 316–322.

Antoniolli, A.; Fontana, A. R.; Piccoli, P.; Bottini, R. Characterization of Polyphenols and Evaluation of Antioxidant Capacity in Grape Pomace of the cv. Malbec. *Food Chem.* **2015,** *178*, 172–178.

Ayala-Zavala, J.; Vega-Vega, V.; Rosas-Domínguez, C.; Palafox-Carlos, H.; Villa-Rodriguez, J.; Siddiqui, M. W.; Dávila-Aviña, J.; González-Aguilar, G. Agro-industrial Potential of Exotic Fruit Byproducts as a Source of Food Additives. *Food Res. Int.* **2011**, *44*, 1866–1874.

Bak, M.-J.; Truong, V. L.; Kang, H.-S.; Jun, M.; Jeong, W.-S. Anti-inflammatory Effect of Procyanidins from Wild Grape (*Vitis amurensis*) Seeds in LPS-Induced RAW 264.7 Cells. *Oxid. Med. Cell. Longevity* **2013**, *2013*, 11 pp.

Barcia, M. T.; Pertuzatti, P. B.; Rodrigues, D.; Gómez-Alonso, S.; Hermosín-Gutiérrez, I.; Godoy, H. T. Occurrence of Low Molecular Weight Phenolics in *Vitis vinifera* Red Grape Cultivars and their Winemaking By-products from São Paulo (Brazil). *Food Res. Int.* **2014**, *62*, 500–513.

Barros, A.; Gironés-Vilaplana, A.; Teixeira, A.; Collado-González, J.; Moreno, D. A.; Gil-Izquierdo, A.; Rosa, E.; Domínguez-Perles, R. Evaluation of Grape (*Vitis vinifera* L.) Stems from Portuguese Varieties as a Resource of (Poly)phenolic Compounds: A Comparative Study. *Food Res. Int.* **2014**, *65*, 375–384.

Bates, R. P.; Morris, J. R.; Crandall, P. G. General Juice Manufacture Principles. In *Principles and Practices of Small- and Medium-scale Fruit Juice Processing*; FAO, Ed.; FAO Agricultural Services Bulletin, 2001; p. 37.

Bautista-Ortín, A. B.; Martínez-Cutillas A.; Ros-García, J. M.; López-Roca, J. M.; Gómez-Plaza, E. Improving Colour Extraction and Stability in Red Wines: the Use of Maceration Enzymes and Enological Tannins. Int. J. Food Sci. Tech. **2005**, *40,* 867–878.

Bernard, F.-X.; Sable, S.; Cameron, B.; Provost, J.; Desnottes, J.-F.; Crouzet, J.; Blanche, F. Glycosylated Flavones as Selective Inhibitors of Topoisomerase IV. *Antimicrob. Agents Chemother.* **1997**, *41*, 992–998.

Bisha, B.; Weinsetel, N.; Brehm-Stecher, B. F.; Mendonca, A. Antilisterial Effects of Gravinol-s Grape Seed Extract at Low Levels in Aqueous Media and Its Potential Application as a Produce Wash. *J. Food Protect.* **2010**, *73*, 266–273.

Borges, A.; Ferreira, C.; Saavedra, M. J.; Simoes, M. Antibacterial Activity and Mode of Action of Ferulic and Gallic Acids against Pathogenic Bacteria. *Microb. Drug Resist.* **2013**, *19*, 256–265.

Brahim, M.; Gambier, F.; Brosse, N. Optimization of Polyphenols Extraction from Grape Residues in Water Medium. *Ind. Crops Prod.* **2014**, *52*, 18–22.

Cheng, V. J.; Bekhit A. E. A.; McConnell, M.; Mros, S.; Zhao, J. Effect of Extraction Solvent, Waste Fraction and Grape Variety on the Antimicrobial and Antioxidant Activities of Extracts from Wine Residue from Cool Climate. *Food Chem.* **2012**, *134*, 474–482.

Cleveland, J.; Montville, T. J.; Nes, I. F.; Chikindas, M. L. Bacteriocins: Safe, Natural Antimicrobials for Food Preservation. *Int. J. Food Microbiol.* **2001**, *71*, 1–20.

Corrales, M.; Han, J. H.; Tauscher, B. Antimicrobial Properties of Grape Seed Extracts and their Effectiveness after Incorporation into Pea Starch Films. *Int. J. Food Sci. Technol.* **2009**, *44*, 425–433.

de Sá, M.; Justino, V.; Spranger, M. I.; Zhao, Y. Q.; Han, L.; Sun, B. S. Extraction Yields and Anti-oxidant Activity of Proanthocyanidins from Different Parts of Grape Pomace: Effect of Mechanical Treatments. *Phytochem. Anal.* **2014**, *25*, 134–140.

FAO/WHO. Microbiological Hazards in Fresh Fruits and Vegetables. *Microbiological Risk Assessment Series.* FAO/WHO, 2008.

FAOSTAT. *FAO Statistical Databases & Data-sets*, 2015.

Feregrino-Perez, A. A.; Torres-Pacheco, I.; Vazquez-Cruz, M. A. Functional Properties and Quality Characteristics of Bioactive Compounds in Berries: Biochemistry, Biotechnology, and Genomics. *Food Reach. Int.* **2012**, *54*, 1195–1207.

Fernández-Iglesias, A.; Pajuelo, D.; Quesada, H.; Díaz, S.; Bladé, C.; Arola, L.; Salvadó, M. J.; Mulero, M. Grape Seed Proanthocyanidin Extract Improves the Hepatic Glutathione Metabolism in Obese Zucker Rats. *Mol. Nutr. Food Res.* **2014,** *58,* 727–737.

Fontana, A. R.; Antoniolli, A.; Bottini, R. Grape Pomace as a Sustainable Source of Bioactive Compounds: Extraction, Characterization, and Biotechnological Applications of Phenolics. *J. Agric. Food Chem.* **2013,** *61,* 8987–9003.

Garcia-Lomillo, J.; González-SanJosé, M. L.; Skibsted, L. H.; Jongberg, S. Effect of Skin Wine Pomace and Sulfite on Protein Oxidation in Beef Patties during High Oxygen Atmosphere Storage. *Food Bioprocess Technol.* **2016,** *9,* 1–11.

García-Ruiz, A.; Bartolomé, B.; Martínez-Rodríguez, A. J.; Pueyo, E.; Martín-Álvarez, P. J.; Moreno-Arribas, M. Potential of Phenolic Compounds for Controlling Lactic Acid Bacteria Growth in Wine. Food Control **2008,** *19,* 835–841.

Garrido, M. D.; Auqui, M.; Martí, N.; Linares, M. B. Effect of Two Different Red Grape Pomace Extracts Obtained under Different Extraction Systems on Meat Quality of Pork Burgers. *LWT—Food Sci. Technol.* **2011,** *44,* 2238–2243.

González-Centeno, M. A. R.; Jourdes, M.; Femenia, A.; Simal, S.; Rosselló, C.; Teissedre, P.-L. Characterization of Polyphenols and Antioxidant Potential of White Grape Pomace Byproducts (*Vitis vinifera* L.). *J. Agric. Food Chem.* **2013,** *61,* 11579–11587.

Gonzalez-Centeno, M. R.; Jourdes, M.; Femenia, A.; Simal, S.; Rossello, C.; Teissedre, P. L. Proanthocyanidin Composition and Antioxidant Potential of the Stem Winemaking Byproducts from 10 Different Grape Varieties (*Vitis vinifera* L.). *J. Agric. Food Chem.* **2012,** *60,* 11850–11858.

Huber, B.; Eberl, L.; Feucht, W.; Polster, J. Influence of Polyphenols on Bacterial Biofilm Formation and Quorum-Sensing. *Zeitsch. Naturforsch. C* **2003,** *58,* 879–884.

Kalra, E. K. Nutraceutical—Definition and Introduction. *AAPS Pharmsci.* **2003,** *5,* 27–28.

Karre, L.; Lopez, K.; Getty, K. J. K. Natural Antioxidants in Meat and Poultry Products. *Meat Sci.* **2013,** *94,* 220–227.

Katalinić, V.; Možina, S. S.; Skroza, D.; Generalić, I.; Abramovič, H.; Miloš, M.; Ljubenkov, I.; Piskernik, S.; Pezo, I.; Terpinc, P.; Boban, M. Polyphenolic Profile, Antioxidant Properties and Antimicrobial Activity of Grape Skin Extracts of 14 *Vitis vinifera* Varieties Grown in Dalmatia (Croatia). *Food Chem.* **2010,** *119,* 715–723.

López-Oliva, M. E.; Agis-Torres, A.; Goni, I.; Muñoz-Martínez, E. Grape Antioxidant Dietary Fibre Reduced Apoptosis and Induced a Pro-reducing Shift in the Glutathione Redox State of the Rat Proximal Colonic Mucosa. *Br. J. Nutr.* **2010,** *103,* 1110–1117.

López-Oliva, M. E.; Pozuelo, M. J.; Rotger, R.; Muñoz-Martínez, E.; Goni, I. Grape Antioxidant Dietary Fibre Prevents Mitochondrial Apoptotic Pathways by Enhancing Bcl-2 and Bcl-x L Expression and Minimising Oxidative Stress in Rat Distal Colonic Mucosa. *Br. J. Nutr.* **2013,** *109,* 4–16.

Lu, M.-F.; Xiao, Z.-T.; Zhang, H.-Y. Where Do Health Benefits of Flavonoids Come From? Insights from Flavonoid Targets and their Evolutionary History. *Biochem. Biophys. Res. Commun.* **2013,** *434,* 701–704.

Llobera, A.; Cañellas, J. Dietary Fibre Content and Antioxidant Activity of Manto Negro Red Grape (*Vitis vinifera*): Pomace and Stem. *Food Chem.* **2007,** *101,* 659–666.

Manach, C.; Williamson, G.; Morand, C.; Scalbert, A.; Rémésy, C. Bioavailability and Bioefficacy of Polyphenols in Humans. I. Review of 97 Bioavailability Studies. *Am. J. Clin. Nutr.* **2005,** *81,* 230S–242S.

Martin, J. G. P.; Porto, E.; Corrêa, C. B.; De Alencar, S. M.; Da Gloria, E. M.; Cabral, I.; De Aquino, L. Antimicrobial Potential and Chemical Composition of Agro-industrial Wastes. *J. Nat. Prod.* **2012**, *5*, 27–36.

Mendes, J. A.; Prozil, S. O.; Evtuguin, D. V.; Lopes, L. P. C. Towards Comprehensive Utilization of Winemaking Residues: Characterization of Grape Skins from Red Grape Pomaces of Variety Touriga Nacional. *Ind. Crops Prod.* **2013**, *43*, 25–32.

Muñoz-González, C.; Rodríguez-Bencomo, J. J.; Martín-Álvarez, P. J.; Moreno-Arribas, M. V.; Pozo-Bayón, M. Á. Recovery of Aromatic Aglycones from Grape Pomace Winemaking By-products by Using Liquid–Liquid and Pressurized-Liquid Extraction. *Food Anal. Methods* **2014**, *7*, 47–57.

Nellemann, C. *The Environmental Food Crisis: The Environment's Role in Averting Future Food Crises: A UNEP Rapid Response Assessment.* UNEP/Earthprint, 2009.

Newell, D. G.; Koopmans, M.; Verhoef, L.; Duizer, E.; Aidara-Kane, A.; Sprong, H.; Opsteegh, M.; Langelaar, M.; Threfall, J.; Scheutz, F.; der Giessen, J. v.; Kruse, H. Food-Borne Diseases—The Challenges of 20 Years Ago Still Persist While New Ones Continue To Emerge. *Int. J. Food Microbiol.* **2010**, *139*, Supplement, S3–S15.

Nohynek, L. J.; Alakomi, H.-L.; Kähkönen, M. P.; Heinonen, M.; Helander, I. M.; Oksman-Caldentey, K.-M.; Puupponen-Pimiä, R. H. Berry Phenolics: Antimicrobial Properties and Mechanisms of Action against Severe Human Pathogens. *Nutr. Cancer* **2006**, *54*, 18–32.

Oliveira, D. A.; Salvador, A. A.; Smânia, A.; Smânia, E. F.; Maraschin, M.; Ferreira, S. R. Antimicrobial Activity and Composition Profile of Grape (*Vitis vinifera*) Pomace Extracts Obtained by Supercritical Fluids. *J. Biotechnol.* **2013**, *164*, 423–432.

Plaper, A.; Golob, M.; Hafner, I.; Oblak, M.; Šolmajer, T.; Jerala, R. Characterization of Quercetin Binding Site on DNA Gyrase. *Biochem. Biophys. Res. Commun.* **2003**, *306*, 530–536.

Pozuelo, M. J.; Agis-Torres, A.; Hervert-Hernández, D.; Elvira López-Oliva, M.; Muñoz-Martínez, E.; Rotger, R.; Goni, I. Grape Antioxidant Dietary Fiber Stimulates *Lactobacillus* Growth in Rat Cecum. *J. Food Sci.* **2012**, *77*, H59–H62.

Prakash, B.; Kedia, A.; Mishra, P. K.; Dubey, N. K. Plant Essential Oils as food Preservatives to Control Moulds, Mycotoxin Contamination and Oxidative Deterioration of Agri-food Commodities—Potentials and Challenges. *Food Control* **2015**, *47*, 381–391.

Procházková, D.; Boušová, I.; Wilhelmová, N. Antioxidant and Prooxidant Properties of Flavonoids. *Fitoterapia* **2011**, *82*, 513–523.

Regaert, P.; Jacxsens, L.; Vandekinderen, I.; Baert, L.; Devlieghere, F. Microbiological and Safety Aspects of Fresh-Cut Fruits and Vegetables. In *Advances in Fresh-Cut Fruits and Vegetables Processing*; Martín-Belloso, O., Fortuny, R. S., Eds.; Taylor and Francis Group: Milton Park, 2009.

Ribéreau-Gayon, P.; Dubourdieu, D.; A. Lonvaud, B. D. The Grape and Its Maduration. In *Handbook of Enology, the Microbiology of Wine and Vinifications*; Ribéreau-Gayon, P., Dubourdieu, D. A., Lonvaud, B. D., Eds.; John Wiley & Sons: Hoboken, NJ, 2006.

Rodriguez Vaquero, M. J.; Aredes Fernandez, P. A.; Manca de Nadra, M. C.; Strasser de Saad, A. M. Phenolic Compound Combinations on *Escherichia coli* Viability in a Meat System. *J. Agric. Food Chem.* **2010**, *58*, 6048–6052.

Sagdic, O.; Ozturk, I.; Yilmaz, M. T.; Yetim, H. Effect of Grape Pomace Extracts Obtained from Different Grape Varieties on Microbial Quality of Beef Patty. *J. Food Sci.* **2011**, *76*, M515–M521.

Sánchez-Tena, S.; Reyes-Zurita, F. J.; Díaz-Moralli, S.; Vinardell, M. P.; Reed, M.; García-García, F.; Dopazo, J.; Lupiáñez, J. A.; Günther, U.; Cascante, M. Maslinic Acid-Enriched

Diet Decreases Intestinal Tumorigenesis in Apc (Min/+) Mice Through Transcriptomic and Metabolomic Reprogramming. PLoS ONE **2013**, *8*, e59392.

Sáyago-Ayerdi, S. G.; Brenes, A.; Goñi, I. Effect of Grape Antioxidant Dietary Fiber on the Lipid Oxidation of Raw and Cooked Chicken Hamburgers. *LWT—Food Sci. Technol.* **2009**, *42*, 971–976.

Scalbert, A.; Johnson, I. T.; Saltmarsh, M. Polyphenols: Antioxidants and Beyond. *Am. J. Clin. Nutr.* **2005**, *81*, 215S–217S.

Scallan, E.; Hoekstra, R. M.; Angulo, F. J.; Tauxe, R. V.; Widdowson, M.-A.; Roy, S. L.; Jones, J. L.; Griffin, P. M. Foodborne Illness Acquired in the United States—Major Pathogens. *Emerg. Infect. Dis.* **2011**, *17*, 7–15.

Selma, M. V.; Beltrán, D.; García-Villalba, R.; Espín, J. C.; Tomás-Barberán, F. A. Description of Urolithin Production Capacity from Ellagic Acid of Two Human Intestinal *Gordonibacter* Species. *Food Funct.* **2014**, *5*, 1779–1784.

Sgroppo, S.; Vergara, L.; Tenev, M. Effects of Sodium Metabisulphite and Citric Acid on the Shelf Life of Fresh Cut Sweet Potatoes. *Spanish J. Agric. Res.* **2010**, *8*, 686–693.

Shukitt-Hale, B.; Carey, A.; Simon, L.; Mark, D. A.; Joseph, J. A. Effects of Concord Grape Juice on Cognitive and Motor Deficits in Aging. *Nutrition* **2006**, *22*, 295–302.

Slavin, J. Fiber and Prebiotics: Mechanisms and Health Benefits. Nutrients 2013, 5, 1417–1435.

Souquet, J.-M.; Labarbe, B.; Le Guernevé, C.; Cheynier, V.; Moutounet, M. Phenolic Composition of Grape Stems. *J. Agric. Food Chem.* **2000**, *48*, 1076–1080.

Spatafora, C.; Barbagallo, E.; Amico, V.; Tringali, C. Grape Stems from Sicilian *Vitis vinifera* Cultivars as a Source of Polyphenol-Enriched Fractions with Enhanced Antioxidant Activity. *LWT—Food Sci. Technol.* **2013**, *54*, 542–548.

Spigno, G.; Donsì, F.; Amendola, D.; Sessa, M.; Ferrari, G.; De Faveri, D. M. Nanoencapsulation Systems to Improve Solubility and Antioxidant Efficiency of a Grape Marc Extract into Hazelnut Paste. *J. Food Eng.* **2013**, *114*, 207–214.

Teixeira, A.; Baenas, N.; Dominguez-Perles, R.; Barros, A.; Rosa, E.; Moreno, D. A.; Garcia-Viguera, C. Natural Bioactive Compounds from Winery By-products as Health Promoters: A Review. *Int. J. Mol. Sci.* **2014**, *15*, 15638–15678.

Thimothe, J.; Bonsi, I. A.; Padilla-Zakour, O. I.; Koo, H. Chemical Characterization of Red Wine Grape (*Vitis vinifera* and *Vitis* Interspecific Hybrids) and Pomace Phenolic Extracts and their Biological Activity against *Streptococcus mutans*. *J. Agric. Food Chem.* **2007**, *55*, 10200–10207.

Tomé-Carneiro, J.; Gonzálvez, M.; Larrosa, M.; García-Almagro, F. J.; Avilés-Plaza, F.; Parra, S.; Yáñez-Gascón, M. J.; Ruiz-Ros, J. A.; García-Conesa, M. T.; Tomás-Barberán, F. A. Consumption of a Grape Extract Supplement Containing Resveratrol Decreases Oxidized LDL and ApoB in Patients Undergoing Primary Prevention of Cardiovascular Disease: A Triple-Blind, 6-Month Follow-Up, Placebo-Controlled, Randomized Trial. *Mol. Nutr. Food Res.* **2012**, *56*, 810–821.

Touriño, S.; Fuguet, E.; Vinardell, M. P.; Cascante, M.; Torres, J. L. Phenolic Metabolites of Grape Antioxidant Dietary Fiber in Rat Urine. *J. Agric. Food Chem.* **2009**, *57*, 11418–11426.

Tseng, A.; Zhao, Y. Wine Grape Pomace as Antioxidant Dietary Fibre for Enhancing Nutritional Value and Improving Storability of Yogurt and Salad Dressing. *Food Chem.* **2013**, *138*, 356–365.

Vázquez-Armenta, F. J.; Silva-Espinoza, B. A.; Cruz-Valenzuela, M. R.; González-Aguilar, G. A.; Nazzaro, F.; Fratianni, F.; Ayala-Zavala, J. F. Antibacterial and Antioxidant Properties of Grape Stem Extract Applied as Disinfectant in Fresh Leafy Vegetables. *J. Food Sci. Technol.* **2017**, *54*, 3192–3200.

Vivas, N.; Nonier, M.-F.; de Gaulejac, N. V.; Absalon, C.; Bertrand, A.; Mirabel, M. Differentiation of Proanthocyanidin Tannins from Seeds, Skins and Stems of Grapes (*Vitis vinifera*) and Heartwood of Quebracho (*Schinopsis balansae*) by Matrix-Assisted Laser Desorption/ Ionization Time-of-Flight Mass Spectrometry and Thioacidolysis/Liquid Chromatography/ Electrospray Ionization Mass Spectrometry. *Anal. Chim. Acta* **2004**, *513*, 247–256.

Wu, D.; Kong, Y.; Han, C.; Chen, J.; Hu, L.; Jiang, H.; Shen, X. D-Alanine:D-Alanine Ligase as a New Target for the Flavonoids Quercetin and Apigenin. *Int. J. Antimicrob. Agents* **2008**, *32*, 421–426.

Xia, E.-Q.; Deng, G.-F.; Guo, Y.-J.; Li, H.-B. Biological Activities of Polyphenols from Grapes. *Int. J. Mol. Sci.* **2010**, *11*, 622–646.

Xu, W.; Qu, W.; Huang, K.; Guo, F.; Yang, J.; Zhao, H.; Luo, Y. Antibacterial Effect of Grapefruit Seed Extract on Food-Borne Pathogens and its Application in the Preservation of Minimally Processed Vegetables. *Postharv. Biol. Technol.* **2007**, *45*, 126–133.

Yu, J.; Ahmedna, M. Functional Components of Grape Pomace: Their Composition, Biological Properties and Potential Applications. *Int. J. Food Sci. Technol.* **2013**, *48*, 221–237.

Zhu, F.; Du, B.; Zheng, L.; Li, J. Advance on the Bioactivity and Potential Applications of Dietary Fibre from Grape Pomace. *Food Chem.* **2015**, *186*, 207–212.

CHAPTER 7

BY-PRODUCTS FROM MINIMAL PROCESSING OF FRESH FRUITS AND VEGETABLES

M. R. TAPIA-RODRIGUEZ[1], M. G. GOÑI[2,3],
G. A. GONZÁLEZ-AGUILAR[1], and J. F. AYALA-ZAVALA[1*]

[1]Centro de Investigación en Alimentación y Desarrollo, A.C.
(CIAD, AC), Carretera a la Victoria Km 0.6, La Victoria, Hermosillo,
Sonora 83000, México

[2]Grupo de Investigación en Ingeniería en Alimentos, Facultad de
Ingeniería, Universidad Nacional de Mar del Plata, Buenos Aires,
Argentina

[3]Consejo Nacional de Investigaciones Científicas y Técnicas
(CONICET), CABA, Buenos Aires, Argentina

*Corresponding author. E-mail: jayala@ciad.mx

CONTENTS

ABSTRACT

As a consequence of awareness on the potential use of by-products from food processing, both from the environmental and from the economical point of view, several studies are focusing on finding possible technical applications. In the minimally processed industry, there is a double objective for those studies: reduce environmental impact of the waste obtained from the food processing and increasing revenue for the producers by giving added value to otherwise undeveloped resources. This way, instead of spending money on waste management, those residues could be processed to obtain important compounds with added value that may have several uses.

In later years, minimally processing of fruits and vegetables production has increased, thus a large amount of by-product is available but currently underdeveloped and all those unused resources are wasted like seeds, leaves, or steams. Moreover, many of those by-products obtained from fruits and vegetables minimal processing are important sources of bioactive compounds of added value for the pharmaceutical or food industry, especially those with antioxidant activity due to the growing interest in functional foods and nutraceutical products and their economic value. In the present chapter, several by-products are presented along with their uses and applications to describe the advantages of the total usage of fruits and vegetables. On other hand, by-products obtained from fruits and vegetables minimally processed are important sources of fiber and bioactive compounds with antioxidant capacity that could be used as food additives in functional foods or to produce nutraceuticals. Many of these compounds extracted have presented antimicrobial activity and have potential applications in active packaging, edible coatings or as additives in foods to increase shelf life. Polyphenols, fibers, vitamins, and small proteins are frequently found in the by-products of fruits and vegetables. Future research should focus on the extraction procedure and the purification of those bioactive compounds obtained in the by-products of minimally processed fruits and vegetables.

7.1 INTRODUCTION

In recent years, there has been an increase in the environmental awareness along with the realization of the economic and nutritional importance of by-products from food processing. Therefore, many studies were focused in finding those by-products and as a consequence reducing the environmental impact of food processing and increasing revenue for the producers by giving added value to otherwise undeveloped resources. Minimally processed fruits and vegetables are an important source of organic waste. This aspect of the industry could be considered as a positive or a negative aspect, depending on if it is being used to its maximal capacity of it is just discarded. Instead of spending money on waste management, such as transportation fees, storage, or chemical treatment, those residues could be processed to obtain important compounds with added value that may have several uses, such as functional ingredients or in pharmacology or cosmetology, among other potential uses.

Minimally processed fruits and vegetables are a growing market; therefore, a number of by-products available for further processing is also increasing. The economic value of those unused resources should compensate the amount of money invested in the processing and purification of the compounds obtained from them. Given the growing interest in functional foods and nutraceutical products and their economic value, residues of fruit and vegetable production, which are rich in organic compounds, are an important source of by-products (Balasundram et al., 2006). In the present chapter, several by-products are presented along with their uses and applications to describe the advantages of total usage of fruits and vegetables.

7.2   TYPES OF PROCESSING AND PRODUCTION

Demand for minimal processed fruits and vegetables have grown strongly in the developing countries over the past few years with a 14% of commercial production of vegetable products, after the most produced, canned (40%), and frozen fruit and vegetables (36%), due to their increasing industrialization and greater participation in world commerce (IBIS, 2015). The minimal processing by-products of fruits and vegetables are considered waste material and they are constituted by a significant part of peels, seeds, stems, leaves, etc. The increasing production of these by-products' source depends on the different regions and processing countries (Ayala-Zavala et al., 2011).

However, these by-products are mainly obtained by the minimal operations of processing: washing, peeling, and cutting. The increasing production of this organic waste represents a growing problem since the plant material is usually prone to microbial spoilage (Szebiotko, 1985). Therefore, agro-industrial waste is often utilized as feed or as fertilizer. However, demand for feed may be varying and dependent on agricultural yields. The problem of disposing by-products is further aggravated by legal restrictions. Thus, efficient, inexpensive, and environmentally sound utilization of these materials is becoming more important especially since profitability and jobs may suffer (Lowe and Buckmaster, 1995).

The reprocessing of those by-products offers some beneficial applications rather than merely discharging the waste materials. However, this exploitation needs cross-cutting technologies for further exploration of these agro-waste sources of bioactive compounds and their applications in food industry as a promising field. Fruits or vegetables processing result in various amounts of by-products depending on the raw material. In Table 7.1, some examples of by-products obtained from minimally processed fruits and vegetables are shown.

TABLE 7.1   By-products from Processed Fruits and Vegetables Suitable for Reprocessing.

Residue plant material	By-products	References
Apple	Peel rich in pectin	Schieber et al. (2003)
Peach	Kernel rich in protein and oil content	Oreopoulou and Tzia (2007)
Pear	Peel rich in pectin	Schieber et al. (2003)
Grape	Dietary fiber, grape seed oil, and stems rich in phenolic compounds	Larrauri et al. (1997)
Citric fruits	Peel rich in pectin and dietary fibers, seeds with oil content	Garau et al. (2007)
Mango	Peel rich in pectin and dietary fibers, seed with oil content	Sruamsiri and Silman (2009)
Carrot	Peel rich in carotenoids	Schieber et al. (2001)

7.3   CHARACTERIZATION OF THE DERIVED BY-PRODUCTS

The minimally processed fruits and vegetables are fresh products characterized as ready-to-eat products, making them very useful in the food industry for satisfying the demands of healthy foods to conserve or improve the human nutrition (Rico et al., 2007). These products comprise mixtures of fruits and vegetables, which can be consumed in salads or fruit cocktail. In Figure 7.1, it is shown that how in both cases, the by-products generated are around 50% (peel, seeds, stem, and leaves) in fruits and 20–30% (peel, seeds, and leaves) in vegetable products (Schieber et al., 2001).

The by-products of plant processing are an important source of fiber and other bioactive compounds; these residues can be used as food additives in the production of functional foods with benefits, such as antioxidant, antimicrobial potential, or to improve the organoleptic properties of some products. There are several studies that focus on the characterization of plant material by-products; the most applied method for by-product characterization is solvent extraction, used for the recovery of bioactive compounds like antioxidants and other food additives like colorants. This extraction uses several organic solvents and their mix choice depends on the raw material and the polarity of the phytochemicals to be obtained. On other hand, to recover bioactive substances like pectin or dietary fiber, high-temperature and drying methods need to be applied (Schieber et al., 2001, 2003).

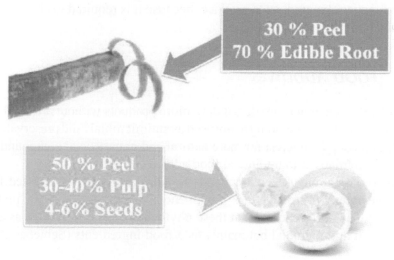

FIGURE 7.1 Example of by-product proportion in minimally processed fruits and vegetables.

7.4 CONSIDERATION FOR PRACTICAL USES OF THE DERIVED BY-PRODUCTS

Nowadays, by-products of plant food industry have been used as a source of bioactive and functional compounds that could be exploited as new field of food additive alternatives. However, for its application, it is necessary to know all the considerations that these additives need to have, for example, the exploration of by-products phytochemical composition, the security of being clear of toxic compounds derived from organic solvents used for extraction, and economic and sensorial impact. Various bioactive compounds found in plants appear to have a wide spectrum of bioactivities: anti-inflammatory, antioxidant, and antimicrobial, among others (Balasundram et al., 2006; Oreopoulou and Tzia, 2007).

It has been known that the addition of phytochemical derived from by-product can give some negative-flavor characteristic to some products such as the bitterness and astringency. Actually, as alternative, some research works report that these compounds can be included in functional foods applying an active packing, edible coatings, or encapsulating these compounds, retarding or masking these negatives properties (Silva et al., 2014).

In this sense, we know that these by-products contain a great source of bioactive compounds; however, it is important to think about the production of the functional ingredient to produce, because it is required to consider the cost and benefit of their obtaining.

7.4.1   FOOD ADDITIVES

Actually, the consumers are demanding more products without artificial and chemicals additives, including many used as antimicrobials and preservatives. For this reason, the interest for more natural and nonartificial compounds as potential alternatives to traditional food additives has increased.

The principal bioactive compounds found in minimally processed fruit and vegetables by-products are polyphenols, fibers, and a small protein content. It has been reported that these phytochemicals can be used as antioxidant and antimicrobial but mainly as a food ingredients (Schieber et al., 2001).

7.4.2   ANTIMICROBIAL USES

Some authors suggest that mainly the bioactive compounds present in plant-origin by-products have antimicrobial potential. Among them, flavonoids phenolic acids represent this property. Some mechanisms of the antimicrobial activity including a degradation of the microbial cell wall, targeting the cytoplasmic membrane, alterations proton, and effect on some primordial enzymes have been suggested. These compounds are capable of interacting with proteins causing changes in their structure and leakage compounds necessary for microorganisms (Cowan, 1999; Nohynek et al., 2006).

The citric industry produces large amounts of by-products from seeds and peels, oils rich in terpenes (limonene, terpineol, carvone, and others) (Marzouk, 2013). Different applications such as lemon extracts applied in active packing have also been performed to reduce the spoilage of dairy products and preserve its shelf life (Conte et al., 2007). In other works, extracts of pomegranate (seed and peel) present antibacterial potential against *Staphylococcus aureus* and *Bacillus cereus* applied in chicken products (Kanatt et al., 2010). Studies in vitro of peel of bergamot extract rich in eriodictyol, naringenin, and hesperidin had an inhibition of *Bacillus subtilis* (Mandalari et al., 2007). In this context, the plant-origin by-products are a promising source of antimicrobial compounds that offer a new commercial opportunity to the functional food industry. However, there is very scarce information research that needs immediate attention.

7.4.3 THICKENER USES

During the processing of some fruits like apples and citric fruits for fresh-cut products, peels represent between 50% and 65% of the total weight of the fruits and remain as the primary by-product. If not processed further, it becomes waste and can give rise to serious environmental pollution. It is well known that some of these wastes are disposed for use as livestock feed or used as fertilizing (Bocco et al., 1998; Tripodo et al., 2004); however, most of these are not given further use due to the quantities produced and the small amounts able to be processed in the immediate locality (Mandalari et al., 2006).

The main compound present in citrus fruit peels is pectin, one of the main components of dietary fiber, which is an important ingredient in the food industry as additive for their gelling, texturizer, emulsifier, stabilizer, and thickener properties (Masmoudi et al., 2008). The use of this polysaccharide and other derived by-products has brought with it a growing interest from ingredient companies because they are continually looking for cheaper but value-added ingredients (Garau et al., 2007).

7.4.4 NUTRACEUTICAL USES

There is a growing trend in the food industry to develop new food products that are both, easy to consume, with low preparation times but with high nutritional value, and that can be of interest for some portion of the population with specific dietary needs. An increasing percentage of the population presents different degrees of chronic diseases, like arterial hypertension, diabetes, cancer, arteriosclerosis, metabolic syndrome, among others which require a special diet. There is also another portion of the population with high regards for the quality of the diet they consume, especially in reference to health benefits associated to a diverse diet. To appeal to these potential consumers, who are usually willing to pay extra for a ready-to-eat meal or a highly nutritional product, the nutraceutical market is a promising alternative. Several studies are focusing in designing new products with healthy characteristics that differentiate them from other more traditional foods (Fig. 7.2).

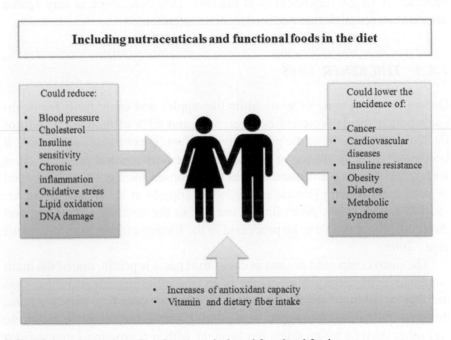

FIGURE 7.2   Health benefits of nutraceuticals and functional foods.

One very promising source to obtain compounds with potential application are the by-products resulting of the minimally processed vegetables industry (O'Shea et al., 2012). The amount of peels, seeds, pomace, pulp, discarded fruits, and vegetables that comprised these by-products is often discarded in the trash (with high environmental impact and the added cost to dispose it) or sold as animal feed at a very low cost (Angulo et al., 2012). Moreover, dietary guidelines advice to consume a diet rich in fruits and vegetables due to the amount of healthy compound they have. These organic residues are rich in nutritional compounds that can be extracted and added to other foods or used in the elaboration of nutraceutical products, therefore, increasing the revenue for the minimally processed vegetable producers, reducing disposal costs and diminishing the environmental impact. Most vegetables are seasonal that are processed into various products, which generates a significant amount of organic waste. For example, in mango fruit, peel constitutes about 20% of total fresh weight and it contains high amounts of several valuable compounds such as phenolic compounds, dietary fiber, vitamins, and several other micronutrients (Ajila et al., 2008). Total mango waste created as a consequence of the minimal processing can range from 35% to 60% of fresh weight (O'Shea et al., 2012). Several examples of the application of the by-products obtained from the minimally processed vegetables production are presented in Table 7.2.

TABLE 7.2 Applications of Commonly Used as Minimally Processed Vegetables By-products.

Produce	By-product	Source	Been added to	References
Orange	Peels	Dietary fiber	Yoghourt	O'Shea et al. (2012)
	Flavedo	Vitamin C	Meat pastes	Chau and Huang (2003)
	Pomace	Flavonoids	Dry fermented	Grigelmo-Miguel and
		Phenolic acids	sausages	Martín-Belloso (1999)
			Cookies	Fernández-López et al. (2008)
				Fernández-López et al. (2009)
				Larrea et al. (2005)
Mango	Peels	Tannins		Dorta et al. (2012)
	Seeds	Flavonoids		
		Protoanthocyanidins		
Peach	Kernels	Pectin	Jams	Grigelmo-Miguel and
	Peels	Dietary fiber	Muffins	Martín-Belloso (1999)
		Phenolic compounds		Kurz et al. (2008)
				Adil et al. (2007)
Apple	Skins	Phenolic compounds	Muffins	Rupasinghe et al. (2008)
	Pomace	Pectin		
Onion	Skins	Dietary fiber	Onion paste	Roldán et al. (2008)
	Roots	Flavonoids		Griffiths et al. (2002)
	Outer scales	Phenolic acids		Jaime et al. (2002)
				Benítez et al. (2011)
				Ng et al. (2000)
				Slimestad et al. (2007)

TABLE 7.2 *(Continued)*

Produce	By-product	Source	Been added to	References
Potato	Peels	Dietary fiber	Baked products	Abu-Ghannam and
	Pulp	Phenolic acids	Meat pastes	Crowley (2006)
		Anthocyanins		Kaack et al. (2006)
				Mattila and Hellström (2007)
Tomato	Peels	Dietary fiber	Fermented	Herrera et al. (2010)
	Pulp	Lycopene	sausage	Calvo et al. (2008)
		Carotene	Hamburgers	García et al. (2009)
			Barley-based extruded snacks	Altan et al. (2008)
Carrot	Peels	Phenolic acids	Juices	Chantaro et al. (2008)
		Dietary fiber	Candies	Gonçalves et al. (2010)
		Anthocyanins		Sun et al. (2009)
		Carotenes		Stoll et al. (2003)

7.4.5   DIETARY FIBER

Dietary fiber (DF) is one of the most wanted nutritional compounds in the food industry, mostly due to the dual benefit as a functional ingredient and for the added health benefits associated with its intake. DF is known to reduce cholesterol, diabetes, coronary heart disease, and easy constipation (Telrandhe et al., 2012). The interest in DF also depends on its high antioxidant capacity (Mildner-Szkudlarz et al., 2013). Several chemical compounds are classified as DF, mostly consists of cellulose, hemicellulose, lignin, pectin, β-glucans, and gums that are resistant to hydrolysis by human digestive enzymes (Chantaro et al., 2008). Several investigators have reviewed the importance of DF in recent years in the diet and its effects in the human metabolism (Brownlee, 2014; Foschia et al., 2013; Mudgil and Barak, 2013; Phillips, 2013).

DF has two categories, insoluble dietary fiber (IDF) and soluble dietary fiber (SDF). IDF mainly acts on the intestinal tract to produce mechanical peristalsis, while SDF can also influence available carbohydrate and lipid metabolism. Therefore, the proportion of SDF in DF is an important factor influencing the physiological function of DF. Composition of the obtained DF extracts depends on both the vegetable and processing technologies used in the extraction (Chantaro et al., 2008). Thermal processing, such as blanching or drying can inactivate enzymes, partial denaturalization of proteins, and degradation of some types of SDF.

In the past, DF was often included in the diet by consumption of whole cereals, like wheat, corn, or rice. However, in recent years, it was discovered that fruits and vegetables are also important sources of DF, especially from the by-products resulting of minimally processing (Ayala-Zavala et al., 2011).

Nowadays, it is known that fruit DF has a higher nutritive value than DF derived from cereals because it also contains significant amounts of other bioactive compounds such as polyphenols and carotenoids (Mildner-Szkudlarz et al., 2013). For example, apple, citrus fruits, *Brassica* vegetables by-products have been used as DF sources (Chantaro et al., 2008). Apple pomace, composed of peel and pulp, has a higher percentage of SDF than most cereals, with the added benefit of pectin content. DF constitutes in apple peels up to 0.91% of its fresh weight, while IDF and SDF fractions constitute 0.46% and 0.43% of the fresh weight, respectively (Gorinstein et al., 2001). Pectin can act as a stabilizer in foods, gelling and thickening agent, in addition to being a health-enhancing polymer due to its ability to reduce cholesterol and delay gastric emptying (O'Shea et al., 2012). Most of the DF present in apple is found in its peel (Angulo et al., 2012), in addition to the significant levels of calcium, zinc, iron copper, and magnesium, which can also increase the nutritional value of the apple by-products.

Another example of high level of waste production as a result of minimal processing is mango. Mango peel has been reported to contain high levels of DF, composed of 19% of SDF and 32% of IDF (O'Shea et al., 2012; Vergara-Valencia et al., 2007). Chau and Huang (2003) found an elevated DF content on orange peels, being the IDF predominant fraction, providing several health benefits, such as intestinal regulation and increase stool volume which in time reduces constipation and prevents colon cancer. Orange peels are also rich in pectic polysaccharides and cellulose as part of DF. Vegetables by-products are also important sources of DF, for example, in carrot peels; it was found a high amount of DF, 63.6% on dry weight, with 50.1% being the IDF fraction (Chau et al., 2004).

After being discovered that DF from vegetables has more health benefits than DF obtained from cereals, several cereal-based products are being supplemented with numerous forms of DF obtained from by-products of fruits or vegetables. For example, enrichment of baked products with extracts from by-products of fruits and vegetables has been studied. Mango peel is rich in DF, with a very good proportion of IDF and SDF, and biscuits prepared with DF from mango peel provide an 8.8% increase in IDF content and 5.4% increase in SDF content (Ajila et al., 2008). As a consequence, enrichment of cereal-based bakery products with mango DF enhances their nutritional quality (O'Shea et al., 2012).

7.4.6 ANTIOXIDANTS

Intake of dietary fiber and phytochemicals, such as polyphenols, carotenoids, tocopherols and ascorbic acid, have been related to the maintenance of health and protection from diseases such as cancer, cardiovascular diseases, and many other degenerative diseases (Ajila et al., 2008). The metabolic effect of polyphenols such as phenolic acids, anthocyanins, and flavonoids have been often linked with the reduction of the incidence of cardiovascular diseases, different types of cancer, reduction of systolic pressure, and lipid oxidation (O'Shea et al., 2012). As a consequence, an increase in the amount of fruits and vegetables included in the diet is encouraged. By-products obtained from the minimal processing of those fruits and vegetables are rich in these valuable nutrients but are often discarded or subutilized as animal feed. Extracting the nutrients present in the by-products would increase its value, obtaining phenolic compounds of natural sources that can be added to other foods or nutraceuticals. Several fruits and vegetables wastes had been studied and many of them had been found to be rich in different phenolic compounds that could be potentially used in nutraceuticals. For example, mango peel was found to elevated amounts of phenolic acids (110 mg/g dry mass [DM]), anthocyanins (565 mg/g DM), and β-carotene (436 mg/g DM). Peach peels have been found to have higher phenolic content than pulp, especially β-carotene and β-crytoxanthin (Gil et al., 2002). Carrot peels are also an important source of carotenes and anthocyanins (Chantaro et al., 2008), which have them often associated to eyesight enhancement. Carrot peels have the advantage of not having kernels or seeds, like in fruit by-products which simplifies the process of extraction and subsequent addition in the production of nutraceuticals.

Orange peels are one of the most important sources of phenolic compounds among the by-products of minimally processed fruits (even more if juice residues are included with them). Hesperidin is the main flavonoid present in orange peels, accounting for more than 50% of the total phenolic content, among others like hydroxycinnamic and ferulic acid (Fernández-López et al., 2009). Orange peel is both considered a functional ingredient and a nutraceutical ingredient, which can be added to several other food products, like meat pastes, baked goods, and yogurt (Fernández-López et al., 2008; Larrea et al., 2005; Sendra et al., 2010).

Several phytochemicals of potential use in nutraceuticals have been found in apple pomace and its seeds, mainly phenolic acids, such as chlorogenic, protocatechuic, and caffeic acid, and flavanols and flavonols (García et al., 2009; Schieber et al., 2003). These compounds are associated to reduce cancer-cell proliferation, lipid-oxidation decrease, and cholesterol reduction. Quercetin is one of the main flavonoids found in apple peel, often linked to reducing incidence of breast cancer and leukemia (O'Shea et al., 2012).

Some studies have suggested that inclusion of fruits and vegetables extracts as an ingredient in nutraceuticals can improve the nutritive properties of such products, enhancing the health benefits to the consumer. For example, apple pomace extracts rich in several antioxidant compounds has been used to enhance the nutritional value of several food products, like sausages, jams, muffins, and many other baked products (Henríquez et al., 2010; Rupasinghe et al., 2008). Stoll et al. (2003) added carrot peel hydrolyzate to carrot juice, enhancing its antioxidant capacity without affecting the stability of the final beverage. Tomato peel is also a very important source of antioxidant capacity, mainly due to its high content on lycopene. Calvo et al. (2008) added tomato peel powder to sausages, increasing the amount of carotenes present in the product. On the other hand, García et al. (2009) added lycopene extract from tomato peels to hamburgers improving its nutritional content but an undesired orange color was observed in the meat. More research should be done in the promising field of adding antioxidant compounds obtained from by-products of fruits and vegetables in foods to enhance its nutritional value, especially to combine its functional properties with the health-promoting properties.

7.4.7   VITAMINS

Orange and other citric fruits are the main source of vitamin C among fruits and vegetables; therefore, orange by-products are an important source of these important health nutrients.

Mango peels, when ripened, are rich in vitamins C and E (Ajila et al., 2007). These phytochemicals are very important due to their prevention of DNA oxidative damage, cancer prevention, and scavenging of free radicals activity.

7.5   FUTURE NEEDS AND CONCLUDING REMARKS

As it was stated in the present chapter, there is a very promising source of food ingredients in the by-product obtained from the minimal processing industry. These organic residues are rich in dietary fiber, phenolic compound with high antioxidant capacity, minerals, vitamins, and several other micronutrients with known functional properties and health benefits. There is also the added benefit of reducing costs and environmental impact. Nowadays, consumers are more concerned about the quality of the food they consume and are more insistent in the sustainability of the processed, utilized into obtaining food. As a consequence, new types of food should be developed to include these concerns, in addition to the optimization of the whole production (both from the economic point of view and from the environmental one). In the next years, research should focus on the extraction of compounds present in the by-products that have functional and/or nutritional importance to be used as ingredients in more elaborated food products.

Technologies should be developed to optimize the extraction process of the micronutrients found in these organic residues, as well as the continuance on the medical research in collaboration with the food technologist to improve the bioavailability and bioaccessibility of these nutrients obtained from the by-products of fruits and vegetables. Their extraction and subsequent application should always have the aim of increasing the nutritional value of the diet with maintaining the sensorial quality of the product by exploiting their functional properties as well.

KEYWORDS

- by-products
- food processing
- dietary fiber
- food industry
- bioactive compounds

REFERENCES

Abu-Ghannam, N.; Crowley, H. The Effect of Low Temperature Blanching on the Texture of Whole Processed New Potatoes. *J. Food Eng.* **2006,** *74* (3), 335–344.

Adil, I. H.; Cetin, H.; Yener, M.; Bayındırlı, A. Subcritical (Carbon Dioxide + Ethanol) Extraction of Polyphenols from Apple and Peach Pomaces, and Determination of the Antioxidant Activities of the Extracts. *J. Supercrit. Fluids* **2007,** *43* (1), 55–63.

Ajila, C.; Leelavathi, K.; Rao, U. P. Improvement of Dietary Fiber Content and Antioxidant Properties in Soft Dough Biscuits with the Incorporation of Mango Peel Powder. *J. Cereal Sci.* **2008,** 48 (2), 319–326.

Ajila, C.; Naidu, K.; Bhat, S.; Rao, U. P. Bioactive Compounds and Antioxidant Potential of Mango Peel Extract. *Food Chem.* **2007,** *105* (3), 982–988.

Altan, A.; McCarthy, K. L.; Maskan, M. Evaluation of Snack Foods from Barley–Tomato Pomace Blends by Extrusion Processing. *J. Food Eng.* **2008,** *84* (2), 231–242.

Angulo, J.; Mahecha, L.; Yepes, S. A.; Yepes, A. M.; Bustamante, G.; Jaramillo, H.; Valencia, E.; Villamil, T.; Gallo, J. Nutritional Evaluation of Fruit and Vegetable Waste as Feedstuff for Diets of Lactating Holstein Cows. *J. Environ. Manage.* **2012,** *95*, S210–S214.

Ayala-Zavala, J.; Vega-Vega, V.; Rosas-Domínguez, C.; Palafox-Carlos, H.; Villa-Rodriguez, J.; Siddiqui, M. W.; Dávila-Aviña, J.; González-Aguilar, G. Agro-industrial Potential of Exotic Fruit Byproducts as a Source of Food Additives. *Food Res. Int.* **2011,** *44* (7), 1866–1874.

Balasundram, N.; Sundram, K.; Samman, S. Phenolic Compounds in Plants and Agri-industrial By-products: Antioxidant Activity, Occurrence, and Potential Uses. *Food Chem.* **2006,** *99* (1), 191–203.

Benítez, V.; Mollá, E.; Martín-Cabrejas, M. A.; Aguilera, Y.; López-Andréu, F. J.; Cools, K.; Terry, L. A.; Esteban, R. M. Characterization of Industrial Onion Wastes (*Allium cepa* L.), Dietary Fibre and Bioactive Compounds. *Plant Foods Hum. Nutr.* **2011,** *66* (1), 48–57.

Bocco, A.; Cuvelier, M.-E.; Richard, H.; Berset, C. Antioxidant Activity and Phenolic Composition of Citrus Peel and Seed Extracts. *J. Agric. Food Chem.* **1998,** *46* (6), 2123–2129.

Brownlee, I. The Impact of Dietary Fibre Intake on the Physiology and Health of the Stomach and Upper Gastrointestinal Tract. *Bioact. Carbohydr. Diet. Fib.* **2014,** *4* (2), 155–169.

Calvo, M.; García, M.; Selgas, M. Dry Fermented Sausages Enriched with Lycopene from Tomato Peel. *Meat Sci.* **2008,** *80* (2), 167–172.

Conte, A.; Scrocco, C.; Sinigaglia, M.; Del Nobile, M. Innovative Active Packaging Systems to Prolong the Shelf Life of Mozzarella Cheese. *J. Dairy Sci.* **2007,** *90* (5), 2126–2131.

Cowan, M. M. Plant Products as Antimicrobial Agents. *Clin. Microbiol. Rev.* **1999,** *12* (4), 564–582.

Chantaro, P.; Devahastin, S.; Chiewchan, N. Production of Antioxidant High Dietary Fiber Powder from Carrot Peels. *LWT—Food Sci. Technol.* **2008,** *41* (10), 1987–1994.

Chau, C.-F.; Chen, C.-H.; Lee, M.-H. Comparison of the Characteristics, Functional Properties, and In Vitro Hypoglycemic Effects of Various Carrot Insoluble Fiber-Rich Fractions. *LWT—Food Sci. Technol.* **2004,** *37* (2), 155–160.

Chau, C.-F.; Huang, Y.-L. Comparison of the Chemical Composition and Physicochemical Properties of Different Fibers Prepared from the Peel of *Citrus sinensis* L. cv. Liucheng. *J. Agric. Food Chem.* **2003,** *51* (9), 2615–2618.

Dorta, E.; Lobo, M. G.; González, M. Using Drying Treatments to Stabilise Mango Peel and Seed: Effect on Antioxidant Activity. *LWT—Food Sci. Technol.* **2012,** *45* (2), 261–268.

Fernández-López, J.; Sendra, E.; Sayas-Barberá, E.; Navarro, C.; Pérez-Alvarez, J. Physico-chemical and Microbiological Profiles of "Salchichón" (Spanish Dry-Fermented Sausage) Enriched with Orange Fiber. *Meat Sci.* **2008,** *80* (2), 410–417.

Fernández-López, J.; Sendra-Nadal, E.; Navarro, C.; Sayas, E.; Viuda-Martos, M.; Alvarez, J. A. P. Storage Stability of a High Dietary Fibre Powder from Orange By-products. *Int. J. Food Sci. Technol.* **2009,** *44* (4), 748–756.

Foschia, M.; Peressini, D.; Sensidoni, A.; Brennan, C. S. The Effects of Dietary Fibre Addition on the Quality of Common Cereal Products. *J. Cereal Sci.* **2013,** *58* (2), 216–227.

Garau, M. C.; Simal, S.; Rossello, C.; Femenia, A. Effect of Air-Drying Temperature on Physico-chemical Properties of Dietary Fibre and Antioxidant Capacity of Orange (*Citrus aurantium* v. Canoneta) By-products. *Food Chem.* **2007,** *104* (3), 1014–1024.

García, M. L.; Calvo, M. M.; Selgas, M. D. Beef Hamburgers Enriched in Lycopene Using Dry Tomato Peel as an Ingredient. *Meat Sci.* **2009,** *83* (1), 45–49.

Gil, M. I.; Tomás-Barberán, F. A.; Hess-Pierce, B.; Kader, A. A. Antioxidant Capacities, Phenolic Compounds, Carotenoids, and Vitamin C Contents of Nectarine, Peach, and Plum Cultivars from California. *J. Agric. Food Chem.* **2002,** *50* (17), 4976–4982.

Gonçalves, E.; Pinheiro, J.; Abreu, M.; Brandão, T.; Silva, C. L. Carrot (*Daucus carota* L.) Peroxidase Inactivation, Phenolic Content and Physical Changes Kinetics Due to Blanching. *J. Food Eng.* **2010,** *97* (4), 574–581.

Gorinstein, S.; Zachwieja, Z.; Folta, M.; Barton, H.; Piotrowicz, J.; Zemser, M.; Weisz, M.; Trakhtenberg, S.; Màrtín-Belloso, O. Comparative Contents of Dietary Fiber, Total Pheno-lics, and Minerals in Persimmons and Apples. *J. Agric. Food Chem.* **2001,** *49* (2), 952–957.

Griffiths, G.; Trueman, L.; Crowther, T.; Thomas, B.; Smith, B. Onions—A Global Benefit to Health. *Phytother. Res.* **2002,** *16* (7), 603–615.

Grigelmo-Miguel, N.; Martín-Belloso, O. Comparison of Dietary Fibre from By-products of Processing Fruits and Greens and from Cereals. *LWT—Food Sci. Technol.* **1999,** *32* (8), 503–508.

Henríquez, C.; Speisky, H.; Chiffelle, I.; Valenzuela, T.; Araya, M.; Simpson, R.; Almonacid, S. Development of an Ingredient Containing Apple Peel, as a Source of Polyphenols and Dietary Fiber. *J. Food Sci.* **2010,** *75* (6), H172–H181.

Herrera, P. G.; Sánchez-Mata, M.; Cámara, M. Nutritional Characterization of Tomato Fiber as a Useful Ingredient for Food Industry. *Innov. Food Sci. Emerg. Technol.* **2010,** *11* (4), 707–711.

IBIS. *IBIS World Industry Report Global Fruit & Vegetables Processing*, 2015.

Jaime, L.; Mollá, E.; Fernández, A.; Martín-Cabrejas, M. A.; López-Andréu, F. J.; Esteban, R. M. Structural Carbohydrate Differences and Potential Source of Dietary Fiber of Onion (*Allium cepa* L.) Tissues. *J. Agric. Food Chem.* **2002,** *50* (1), 122–128.

Kaack, K.; Pedersen, L.; Laerke, H. N.; Meyer, A. New Potato Fibre for Improvement of Texture and Colour of Wheat Bread. *Eur. Food Res. Technol.* **2006,** *224* (2), 199–207.

Kanatt, S. R.; Chander, R.; Sharma, A. Antioxidant and Antimicrobial Activity of Pome-granate Peel Extract Improves the Shelf Life of Chicken Products. *Int. J. Food Sci. Technol.* **2010,** *45* (2), 216–222.

Kurz, C.; Carle, R.; Schieber, A. Characterisation of Cell Wall Polysaccharide Profiles of Apricots (*Prunus armeniaca* L.), Peaches (*Prunus persica* L.), and Pumpkins (*Cucurbita* sp.) for the Evaluation of Fruit Product Authenticity. *Food Chem.* **2008,** *106* (1), 421–430.

Larrauri, J. A.; Rupérez, P.; Saura-Calixto, F. Effect of Drying Temperature on the Stability of Polyphenols and Antioxidant Activity of Red Grape Pomace Peels. *J. Agric. Food Chem.* **1997,** *45* (4), 1390–1393.

Larrea, M.; Chang, Y.; Martinez-Bustos, F. Some Functional Properties of Extruded Orange Pulp and Its Effect on the Quality of Cookies. *LWT—Food Sci. Technol.* **2005**, *38* (3), 213–220.

Lowe, E. D.; Buckmaster, D. R. Dewatering Makes Big Difference in Compost Strategies. *Biocycle* **1995**, *36*, 78–82.

Mandalari, G.; Bennett, R.; Bisignano, G.; Trombetta, D.; Saija, A.; Faulds, C.; Gasson, M.; Narbad, A. Antimicrobial Activity of Flavonoids Extracted from Bergamot (*Citrus bergamia* Risso) Peel, a Byproduct of the Essential Oil Industry. *J. Appl. Microbiol.* **2007**, *103* (6), 2056–2064.

Mandalari, G.; Bennett, R. N.; Bisignano, G.; Saija, A.; Dugo, G.; Lo Curto, R. B.; Faulds, C. B.; Waldron, K. W. Characterization of Flavonoids and Pectins from Bergamot (*Citrus bergamia* Risso) Peel, a Major Byproduct of Essential Oil Extraction. *J. Agric. Food Chem.* **2006**, *54* (1), 197–203. DOI:10.1021/jf051847n.

Marzouk, B. Characterization of Bioactive Compounds in Tunisian Bitter Orange (*Citrus aurantium* L.) Peel and Juice and Determination of their Antioxidant Activities. *BioMed Res. Int.* **2013**, *2013*. http://dx.doi.org/10.1155/2013/345415.

Masmoudi, M.; Besbes, S.; Chaabouni, M.; Robert, C.; Paquot, M.; Blecker, C.; Attia, H. Optimization of Pectin Extraction from Lemon By-product with Acidified Date Juice Using Response Surface Methodology. *Carbohydr. Polym.* **2008**, *74* (2), 185–192.

Mattila, P.; Hellström, J. Phenolic Acids in Potatoes, Vegetables, and Some of their Products. *J. Food Compos. Anal.* **2007**, *20* (3), 152–160.

Mildner-Szkudlarz, S.; Bajerska, J.; Zawirska-Wojtasiak, R.; Górecka, D. White Grape Pomace as a Source of Dietary Fibre and Polyphenols and its Effect on Physical and Nutraceutical Characteristics of Wheat Biscuits. *J. Sci. Food Agric.* **2013**, *93* (2), 389–395.

Mudgil, D.; Barak, S. Composition, Properties and Health Benefits of Indigestible Carbohydrate Polymers as Dietary Fiber: A Review. *Int. J. Biol. Macromol.* **2013**, *61*, 1–6.

Ng, A.; Parker, M. L.; .; Parr, A. J.; Saunders, P. K.; Smith, A. C.; Waldron, K. W. Physicochemical Characteristics of Onion (*Allium cepa* L.) Tissues. *J. Agric. Food Chem.* **2000**, *48* (11), 5612–5617.

Nohynek, L. J.; Alakomi, H.-L.; Kähkönen, M. P.; Heinonen, M.; Helander, I. M.; Oksman-Caldentey, K.-M.; Puupponen-Pimiä, R. H. Berry Phenolics: Antimicrobial Properties and Mechanisms of Action against Severe Human Pathogens. *Nutr. Cancer* **2006**, *54* (1), 18–32.

O'Shea, N.; Arendt, E. K.; Gallagher, E. Dietary Fibre and Phytochemical Characteristics of Fruit and Vegetable By-products and their Recent Applications as Novel Ingredients in Food Products. *Innov. Food Sci. Emerg. Technol.* **2012**, *16*, 1–10.

Oreopoulou, V.; Tzia, C. Utilization of Plant By-products for the Recovery of Proteins, Dietary Fibers, Antioxidants, and Colorants. In *Utilization of By-products and Treatment of Waste in the Food Industry*; Springer: Berlin-Heidelberg, 2007; pp 209–232.

Phillips, G. O. Dietary Fibre: A Chemical Category or a Health Ingredient? *Bioact. Carbohydr. Diet. Fib.* **2013**, *1* (1), 3–9.

Rico, D.; Martin-Diana, A. B.; Barat, J.; Barry-Ryan, C. Extending and Measuring the Quality of Fresh-Cut Fruit and Vegetables: A Review. *Trends Food Sci. Technol.* **2007**, *18* (7), 373–386.

Roldán, E.; Sánchez-Moreno, C.; de Ancos, B.; Cano, M. P. Characterisation of Onion (*Allium cepa* L.) By-products as Food Ingredients with Antioxidant and Antibrowning Properties. *Food Chem.* **2008**, *108* (3), 907–916.

Rupasinghe, H. V.; Wang, L.; Huber, G. M.; Pitts, N. L. Effect of Baking on Dietary Fibre and Phenolics of Muffins Incorporated with Apple Skin Powder. *Food Chem.* **2008**, *107* (3), 1217–1224.

Schieber, A.; Hilt, P.; Streker, P.; Endreß, H.-U.; Rentschler, C.; Carle, R. A New Process for the Combined Recovery of Pectin and Phenolic Compounds from Apple Pomace. *Innov. Food Sci. Emerg. Technol.* **2003,** *4* (1), 99–107.

Schieber, A.; Stintzing, F.; Carle, R. By-products of Plant Food Processing as a Source of Functional Compounds—Recent Developments. *Trends Food Sci. Technol.* **2001,** *12* (11), 401–413.

Sendra, E.; Kuri, V.; Fernandez-Lopez, J.; Sayas-Barbera, E.; Navarro, C.; Perez-Alvarez, J. Viscoelastic Properties of Orange Fiber Enriched Yogurt as a Function of Fiber Dose, Size and Thermal Treatment. *LWT—Food Sci. Technol.* **2010,** *43* (4), 708–714.

Silva, L. M.; Hill, L. E.; Figueiredo, E.; Gomes, C. L. Delivery of Phytochemicals of Tropical Fruit By-Products Using Poly(DL-Lactide-*co*-Glycolide) (PLGA) Nanoparticles: Synthesis, Characterization, and Antimicrobial Activity. *Food Chem.* **2014,** *165*, 362–370.

Slimestad, R.; Fossen, T.; Vågen, I. M. Onions: A Source of Unique Dietary Flavonoids. *J. Agric. Food Chem.* **2007,** *55* (25), 10067–10080.

Sruamsiri, S.; Silman, P. Nutritive Value and Nutrient Digestibility of Ensiled Mango By-products. *Maejo Int. J. Sci. Technol.* **2009,** *3* (3), 371–378.

Stoll, T.; Schweiggert, U.; Schieber, A.; Carle, R. Application of Hydrolyzed Carrot Pomace as a Functional Food Ingredient to Beverages. *Food, Agric. Environ.* **2003,** *1* (2), 88–92.

Sun, T.; Simon, P. W.; Tanumihardjo, S. A. Antioxidant Phytochemicals and Antioxidant Capacity of Biofortified Carrots (*Daucus carota* L.) of Various Colors. *J. Agric. Food Chem.* **2009,** *57* (10), 4142–4147.

Szebiotko, K. Utilization of Agroindustrial By-products and Crop Residues by Monogastric Species in Europe. *FAO Animal Production and Health Paper (FAO)*, 1985.

Telrandhe, U.; Kurmi, R.; Uplanchiwar, V.; Mansoori, M.; Raj, V.; Jain, K. Nutraceuticals—A Phenomenal Resource in Modern Medicine. *Int. J. Univ. Pharm. Life Sci.* **2012,** *2* (1), 179–195.

Tripodo, M. M.; Lanuzza, F.; Micali, G.; Coppolino, R.; Nucita, F. Citrus Waste Recovery: A New Environmentally Friendly Procedure to Obtain Animal Feed. Bioresour. Technol. **2004,** *91* (2), 111–115.

Vergara-Valencia, N.; Granados-Pérez, E.; Agama-Acevedo, E.; Tovar, J.; Ruales, J.; Bello-Pérez, L. A. Fibre Concentrate from Mango Fruit: Characterization, Associated Antioxidant Capacity and Application as a Bakery Product Ingredient. *LWT—Food Sci. Technol.* **2007,** *40* (4), 722–729.

CHAPTER 8

BY-PRODUCTS OF THE NUT AND PEANUT AGRO-INDUSTRY AS SOURCES OF PHYTOCHEMICALS SUITABLE FOR THE NUTRACEUTICAL AND FOOD INDUSTRIES

ALMA ANGELICA VAZQUEZ-FLORES[1],
JOSÉ ALBERTO NÚÑEZ-GASTÉLUM[1], EMILIO ALVAREZ-PARRILLA[1],
ABRAHAM WALL-MEDRANO[2], JOAQUÍN RODRIGO-GARCÍA[2],
JESÚS FERNANDO AYALA-ZAVALA[3],
GUSTAVO ADOLFO GONZÁLEZ-AGUILAR[3], and
LAURA ALEJANDRA DE LA ROSA[1*]

[1]Departamento de Ciencias Químico-Biológicas, Instituto de Ciencias Biomédicas, Universidad Autónoma de Ciudad Juárez, Anillo Envolvente del PRONAF y Estocolmo s/n, Ciudad Juárez, Chihuahua, CP 322310, México

[2]Departamento de Ciencias de la Salud, Instituto de Ciencias Biomédicas, Universidad Autónoma de Ciudad Juárez, Anillo Envolvente del PRONAF y Estocolmo s/n, Ciudad Juárez, Chihuahua, CP 322310, México

[3]Centro de Investigación en Alimentación y Desarrollo, A. C. (CIAD, AC), Carretera a la Victoria Km. 0.6, La Victoria, Hermosillo, Sonora, CP 83000, México

*Corresponding author. E-mail: ldelaros@uacj.mx

CONTENTS

ABSTRACT

Nut by-products are the materials discarded from the peeling process of edible peanuts and tree nuts such as almonds, pistachio, pecans, and hazelnuts. These by-products constitute more than 50% of the total weight generated by the nut industry and represent a large amount of solid waste to handle by the producers. Currently, few useful strategies, with low or nule-added value, exist for disposal of these by-products. Some authors point out that there are more ambitious applications for their uses. They are an excellent source of phenolic compounds and dietary fiber with potential applications, not only as nutraceuticals or functional food ingredients to protect human health but also to enhance shelf life or quality of several foodstuffs when used as natural additives. The aim of this chapter is to review the nutraceutical potential of phytochemicals from nut by-products, their physiological effects in human health, and their use in the food industry.

8.1 INTRODUCTION

The term "nuts" refers to the dry and hard seeds that some trees develop; these include almond (*Prunus amygdalus*), pecan (*Carya illinoinensis*), pistachios (*Pistacia vera*), hazelnuts (*Corylus avellana*), walnuts (*Juglans regia*), among others (Brown, 2003). Peanuts (*Arachis hypogaea*) are not nuts from a botanical point of view; however, their chemical composition and nutrient profile (Prior and Gu, 2005) is very similar to that of tree nuts and therefore are included in the present chapter. Nuts have been part of the human diet for thousands of years; however, they are not a basic diet food group (Pimenta, 1992), since they are popularly regarded as "fatty foods" due to their high fat and caloric content (Vadivel and Biesalski, 2011). A lack of correct information has contributed to misunderstand the impact of nut consumption on body weight; nevertheless, some epidemiologic studies have shown that inclusion of nuts in a diet does not induce weight gain (Vadivel and Biesalski, 2011). The perception on this food has gradually changed and the average intake has risen specially after 2003, when the United States Food and Drug Administration made the first statement suggesting that consumption of 1.5 Oz (42.5 g) of nuts a day, as a part of a diet low in saturated fat and cholesterol, can reduce the risk of suffering from heart diseases (Brown, 2003). This statement may have contributed to the increase of amount of nuts consumed on average by the North American consumer for up to 45% since 1995 (Buzby and Pollack, 2008). Moreover, as nuts have been now regarded as "functional foods," that is foods that, in addition to their nutritive value, provide nonnutrient ingredients that promote health and reduce disease-risk in humans; scientific interest in these nonnutrient health-promoting ingredients has also risen (Bolling et al., 2011). Of special interest to scientists and producers has been the fact that some of these health-promoting or bioactive ingredients are highly concentrated in the shells and skins of nuts (Wijeratne et al., 2006; de la Rosa et al., 2011; O'Shea et al., 2012), which are usually discarded as waste products during nut commercialization, so these by-products are rich sources of health-promoting natural compounds with several biological and chemical properties, potentially useful for the food and health industries (O'Shea et al., 2012).

The purpose of this chapter is to provide a general overview of the trends that researches are taking on the isolation of bioactive compounds from by-products of the nut industry and their biological activities, as well as their in vitro and in vivo experiments to prove their functional or bioactive properties. Especial attention is given to the animal studies which suggest protective activity of by-product extracts against different kinds of disease. Potential applications of some extracts as food preservatives or additives capable of maintaining food quality and increasing its healthy properties are also discussed, considering, when possible, the doses that would be required for their actual biological activity.

8.2   NUT BY-PRODUCTS

As a consequence of the current increase in the consumption of nut products, the nut industry faces the issue of by-product accumulation. By-products are derived from the inedible portion of nuts: leaves, husks, plant bark, and most of all, shells which are harvested with the dried fruits and released in the peeling process. These by-products are a concern for the industry due to the lack of dumping sites appropriate to avoid the spread of insects and unwanted wildlife in areas nearby its confinement, becoming, therefore, an environmental problem (Suppadit et al., 2012; Zhao et al., 2012). According to Food and Agriculture Organization of the United Nations (Nations, 2013), the production of tree nuts was about 14 million tons in 2013. Approximately, 50–60% of the whole nut production of the nut industry drifts into devalued by-products through the peeling process, and, in the best case, is used as a resource to obtain heat energy (Chen et al., 2010), applied as fertilizer on agricultural crops, production of activated carbon filters (Esfahlan et al., 2010), or as dietary supplement for barnyard animal food (Chen et al., 2010). Nevertheless, the post-crop utilization of this by-product is not sufficiently profitable, and development of new applications that lead to a transformation and to a greater usefulness of these products is desirable.

Nowadays, consumers are searching for new and better foods that not only fulfill the daily requirements of nutrients for the body but also provide some bioactive ingredients to promote and protect their health, the so-called functional foods, that may exert health benefits beyond basic nutrition by the presence of phytochemical compounds, which are not properly nutrients (e.g., polyphenolic compounds). Functional foods can contribute to improve physiological, physical, and mental health and the functional food industry has been steadily growing in the last decade. Closely linked to the design and production of novel functional foods, commercialization, and use of nutraceuticals (i.e., pharmaceuticals obtained from food) or health supplements, which are constituted by the isolated bioactive principles of functional foods and presented to the consumers as capsules, or other medical form, has also risen (Lau et al., 2012). It has been proposed that by-products from the nut and other agricultural commodities can be used as sources of bioactive components, exploitable by the functional food and nutraceutical industry, adding economic value to these by-products (O'Shea et al., 2012).

Almond, pecan nut, hazelnut, pistachio, and peanut by-products are the most studied as sources of nutraceuticals and bioactive ingredients. The amount and type of nut by-product varies according to the nut processed. Some nuts such as almonds, hazelnuts, and peanuts possess not only the hard and outer shell but also an inner thin layer, which we identify as a skin, that surrounds the kernel. In other nuts, such as pecan and pistachios, the skin is not discarded due to the difficulty to separate it from the edible kernel, and or consumers' preference, so they are consumed as part of the kernel (Grijalva-Contreras et al., 2011). There is strong scientific evidence for the presence of high concentrations of bioactive compounds in the shells or skins of peanut (Taha et al., 2012), pecan nut (Villarreal-Lozoya et al., 2007), pistachio (Fabani et al., 2013), hazelnut (Alasalvar et al., 2006; Shahidi et al., 2007), and almond (Mandalari et al., 2010a; Tsujita et al., 2013), all of which are promising sources of bioactive compounds for the functional foods industry. However, to take full advantage of their properties, it is first necessary to effectively isolate and characterize their main bioactive ingredients, and to test their biological activities in several systems from in vitro chemical tests to cell and animal models to finally use them in human trials. Also, if the by-product-derived component is designed to be used as a food ingredient, both its effectiveness and safety should be evaluated, as well as its effects on the sensory attributes of the finished food product.

8.3   BIOACTIVE COMPONENTS OF NUT BY-PRODUCTS

The edible portion of nuts has been, in general, well characterized on their nutritional quality and bioactive components, due to the fact that it is part of the human diet. Some authors point out that the constituents of this portion are a reflection of what exists in the outermost surrounding layers but up to 10 times more concentrated in the latter (Villarreal-Lozoya et al., 2007; Prado et al., 2009) because they act as an protective barriers for the seed which constitutes the edible portion. However, nutshells and skins are not rich in all of the compounds of the fruit; this is especially evident by their lack of lipids and their scant proteins (Zhao et al., 2012). In this way, the main shell and skin phytochemicals are dietary fiber and polyphenolic compounds (Table 8.1).

Phenolic compounds are secondary metabolites synthesized by plants; they are characterized by the presence of at least one aromatic group with at least one hydroxyl moiety in its structure. In plant tissues, phenolic compounds act as defensive mechanisms against herbivores, producing a bitter taste or inducing digestive enzyme inhibition (Tsujita et al., 2013; He et al., 2008; Sarnoski et al., 2012). As part of the human diet, the main property for which phenolic compounds are recognized is for their antioxidant activity. It is well described that the beneficial effect of antioxidants is through the protection of cells against free radicals caused by metabolism, smoking, inflammatory processes, and others (Chen et al., 2010), although many other mechanisms of action of antioxidant polyphenols are recognized (Wang et al., 2014).

Phenolic compounds are ubiquitous in plants and include several types of chemical structures. Many good reviews on the structure of phenolic compounds exist in the scientific literature (Soto-Vaca et al., 2012). Table 8.1 shows major classes of phenolic compounds, and, in some cases, main individual polyphenols found in by-products of hazelnut, almond, peanut, pecan, and pistachios. From the analysis of this table, it is evident that condensed tannins, as well as their precursors: catechin, epicatechin, and epigallocatechin are major components of almost all nut by-products. Other polyphenols include the flavonolsquercetin and kaempherol and their glycosides (Pereira et al., 2007; Montella et al., 2013). However, most of the studies have only reported total phenolic compounds, usually determined by the method of Folin–Ciocalteu, without identifying individual components. This has some advantages, as a total phenolic extract is simple to obtain, total phenolic compounds are easy to quantify, and some applications can be found for this kind of extracts. Nevertheless, further characterization and purification of bioactive principles may be needed for other uses, as will be discussed in the following sections.

TABLE 8.1 Phenolic Compounds and Dietary Fiber in By-products of the Nut Industry.

Source	Type of phytochemical	References
Almond skin	Catechin (90 µg/g)	Esfahlan et al. (2010)
	Epicatechin (36 µg/g)	
	Condensed tannins (75 µg/g)	
	Insoluble dietary fiber (40%)	Xu et al. (2012)
	Insoluble dietary fiber (4–8%)	Mandalari et al. (2010b)
Hazelnut skin	Catechin (250 mg/g)	Montella et al. (2013)
	Quercetin (85.9 mg/g)	
	Kaempherol (35.6 mg/g)	
	Condensed tannins (2.94 mg CE/g)	Alasalvar et al. (2006)
	Insoluble dietary fiber (52.7%)	Montella et al. (2013)
Hazelnut hard shell	Total phenols (211.1 mg CE/g)	Shahidi et al. (2007)
Hazelnut green cover	Total phenols (134.7 mg/CE g)	Shahidi et al. (2007)
Peanut skin	Epigallochatequin, catechin, and epicatechin	Francisco and Resurreccion (2008)
	Total phenolic compounds (149–169 mg GAE/g)	Taha et al. (2012)
Peanut shell	Condensed tannins (17%)	Ma et al. (2014)
	Insoluble dietary fiber (55%)	Ma et al. (2014)
	Insoluble dietary fiber (60–70%)	Zhao et al. (2012)
Pecan shell	Condensed tannins (310–460 mg CE/g)	de la Rosa et al. (2011)
	Insoluble dietary fiber (44–49%)	do Prado et al. (2013)
Pistachio skin	Flavonoids (70 mg CE/g) and condensed tannins (27.5 mg CyE/g)	Tomaino et al. (2010)
Walnut leaves	Quercetin-3-galactoside, 3-caffeoylquinic acid, and quercetin-3-xyloside	Pereira et al. (2007)

CE, Catechin equivalents; *CyE*, cyanidin equivalents, *GAE*, gallic acid equivalents.

Another important phytochemical found in nut by-products is dietary fiber, which is defined as carbohydrate polymers with 10 or more interconnected monomer units that fail to be endogenously hydrolyzed by enzymes in the small intestine (Kendall et al., 2010). Dietary fiber consists mainly of pectin, cellulose, hemicellulose, and lignin; nevertheless, the most abundant constituents in nutshells are the insoluble fibers lignin and hemicellulose, which comprise at least a 40% of the total shell. Table 8.1, shows the amount of dietary insoluble fiber found in different nut by-products. Peanut shells and skins contain the largest fiber concentration with 70% and 55% of insoluble fiber, respectively. On the other hand, hazelnut skin shows the lowest fiber value. Neither fiber nor phenolic compounds represent an essential dietary nutrient, but actually provide biological benefits to the consumer by reducing the risk of diabetes mellitus 2, cardiovascular diseases, and colon cancer, among others (O'Shea et al., 2012).

It is complicated to establish a comparison between by-products, to select the best source of bioactive ingredients, since studies have determined these compounds under different extraction conditions, which affect yields and quantifications especially for phenolic compounds. For instance, acetone extraction has been reported as the best solvent for the extraction of phenolic compounds from almond and hazelnut shells and skins (Alasalvar et al., 2006; Contini et al., 2008), due to the presence of high molecular weight phenolic compounds (condensed tannins) in these materials. The authors explain this due to the ability of acetone to break the bonds between polyphenols and cell walls (dietary fiber) from the matrix (Mandalari et al., 2010a). However, other authors have pointed out that some polyphenols may be too tightly bound to the cell walls or have molecular weights too high to be soluble in solvents such as acetone, so other extraction conditions may be required to obtain this "nonextractable" phenolic compounds (Taha et al., 2012).

It is advisable to extend the research and find the best extraction conditions for other nut products whose production and commercialization is of regional interest in different parts of the world, and whose by-products may be an economical source of dietary fiber or phenolic compounds, noncaloric functional ingredients and capable of providing added physiological benefits (O'Shea et al., 2012).

8.4   PROPERTIES OF BIOACTIVE COMPOUNDS FROM NUT BY-PRODUCTS

The biological activities that have been found in phenolic compounds and dietary fiber from nut by-products have been divided into two major groups. The first group includes those that have a direct impact on the health of living organisms, or are applicable to improve any physiological condition, such as antioxidant (Garrido et al., 2007), anti-inflammatory (Zhao et al., 2012), anticancer (Villarreal-Lozoya et al., 2007), neuroprotective (Reckziegel et al., 2011; Trevizol et al., 2011), antimicrobial (Rajaei et al., 2009), weight-control (McDougall et al., 2009; Gonçalves et al., 2011; Kimura et al., 2011), antidiabetes (Zhao et al., 2012), antimutagenic (Rajaei et al., 2009), and cardioprotective activities (Tamura et al., 2012). The second group is integrated by those activities which are suitable to improve food processing such as antioxidant properties that can increase the shelf life of food products.

Antioxidant activity of phenolic compounds occurs by donation of a hydrogen atom from the phenol hydroxyl group to the free radical in the food or biological matrix, avoiding the oxidation processes on lipids, protein, and DNA molecules. In vivo, the antioxidants are also capable to inhibit enzymes such as, lipoxygenase, responsible for the oxidative changes in cell lipids (do Prado et al., 2013). It is known that nuts have high fat content, and that is why it is believed that antioxidant compounds are abundant in the outer layers that surround the seed, since they protect and prevent the lipid oxidation of these seeds. Several in vitro studies have demonstrated the antioxidant activity against lipid oxidation of most nut by-products (Villarreal-Lozoya et al., 2007; Prado et al., 2009; do Prado et al., 2013).

The antioxidant activity is probably one of the most commonly studied biological activities of nut by-products. Table 8.2 summarized the main studies carried out on polyphenolic compounds extracted from nut by-products and their antioxidant activity. DPPH and ABTS^{+} radical scavenging techniques are certainly the most commonly used assays. Results in this table show high antioxidant activity indicating that nut by-products are promising economical sources of natural antioxidant compounds. Nevertheless, it is important to point out that the lack of a unique method to express the antioxidant capacity hinders the comparison between studies. Furthermore, many of these studies are in vitro systems, so more information about in vivo assays, both in living and food systems, is needed.

On the other hand, natural products and their derivatives are important sources of new anti-infectious drugs and many nut by-products have demonstrated these properties. Antimicrobial activity has been shown from the lipid fraction (triterpenoids and phytosterols) of the leaves and bark from the pecan tree, against the bacteria *Mycobacterium tuberculosis H37Rv*. A dose of 125 mg/mL was enough to inhibit 50–80% of the bacterial growth in vitro (Cruz-Vega et al., 2008). Also, a study of our research group found that a polyphenolic extract of pecan shell showed antimicrobial activity against Gram-positive and Gram-negative bacteria with minimal inhibitory concentrations in the range 15–24 mg/mL and minimal bactericidal concentrations of 18–27 mg/mL (Fig. 8.1).

TABLE 8.2   Antioxidant Activity of Phenolic Compounds Extracted from By-products of the Nut Industry.

Source	Active compound	Antioxidant activity	References
Hazelnut skin	Total phenolic compounds and total tannins	DPPH EC_{50} = 143 µg/mL DPPH 99.5% radical inhibition	Contini et al. (2008)
	Total phenolic compounds and total tannins		Alasalvar et al. (2006)
Hazelnut shell	Total phenolic compounds and total tannins	DPPH 74% radical inhibition ABTS 157.6 µmol TE/g	Xu et al. (2012)
Hazelnut shell	Total phenolic compounds and total tannins	DPPH EC_{50} = 2.28 mg/mL	Contini et al. (2008)
Hazelnut green cover	Total phenolic compounds	DPPH 99.5% radical inhibition	Shahidi et al. (2007)
Peanut shell	Poliphenolic-rich purified extract	DPPH 95% radical inhibition, 80% β-carotene bleaching	Zhang et al. (2013)
Pecan nut shell	Phenolic compounds and condensed tannins	ABTS 644.2 µmol TE/g, DPPH 720 µmol TE/g, ORAC 227 µmol TE/g	de la Rosa et al. (2011)
Pecan nut shell	Phenolic compounds and condensed tannins	ABTS 2228 µmol TE/g, DPPH 529.4 µmol TE/g	Reckziegel et al. (2011)
Pistachio shell	Flavonoids	DPPH EC_{50} = 2.53 µg/mL, β-carotene EC_{50} = 7.85 µg/mL	Rajaei et al. (2010)
Walnut husk	Polyphenolic compounds	DPPH 60% radical inhibition	Fernández-Agulló et al. (2013)

The green husk of walnut (an outer layer than the shell) was also evaluated as a natural antimicrobial agent. Pathogenic Gram-positive (*Bacillus cereus*, *Bacillus subtilis*, *Staphyloccocus aureus*, *Staphyloccocus epidermis*) and Gram-negative bacteria (*Escherichia coli*, *Pseudomonas aeruginosa*) were sensible to the alcoholic extract, rich in phenolic compounds, from this by-product. The major effect was over the Gram positive, with a minimum inhibitory concentration of 20 mg/mL for *B. cereus*, and 50 mg/mL for *B. subtilis*, *S. aureus*, and *S. epidermis*. Thus, walnut by-products like the green hulk not only could be a rich source of compounds that could be applied in food technology, against food-poisoning bacteria, but also in the pharmaceutical industry for the treatment of pathogenic human bacteria (Fernández-Agulló et al., 2013). Almond skin phenolic extracts have antimicrobial activity over Gram-positive and Gram-negative bacteria. *Salmonella enterica*, *Listeria monocytogenes*, and *S. aureus* were mostly affected by the phenolic extracts of these by-products in doses ranging from 250 to 500 µg/mL. Authors

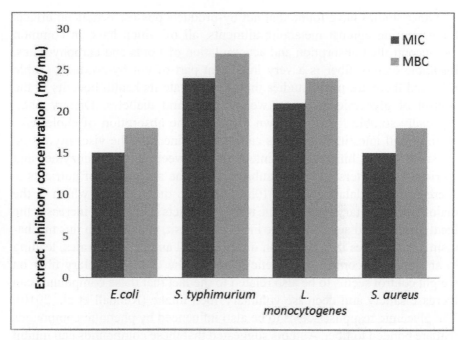

FIGURE 8.1 Minimal inhibitory concentrations (MIC) and minimal bactericidal concentrations (MIB) of a pecan shell acetonic extract against different Gram-negative (*Escherichia coli, Salmonella typhimurim*) and Gram-positive (*Listeria monocytogenes, Staphyloccocus aureus*) bacteria (unpublished results).

suggested that this activity was due to the presence of antioxidants acting synergistically among them, specifically epicatechin, protocatechuic acid, and naringenin (Mandalari et al., 2010a). Pistachio by-products have also been tested as antibacterial agents in vitro. Shell extracts in a concentration of 1 mg/mL showed inhibitory activity against *S. aureus, Escherichia coli, Salmonella thyphi, Navia intermedia, P. aeruginosa, B. cereus*, and *Candida albicans*. This activity is attributed to the presence of flavonoid compounds acting on the enzyme DNA gyrase of bacteria (Rajaei et al., 2009). Another mechanism that explains the antimicrobial effect of this type of extracts is the presence of phenolic compounds that inhibit the adhesion of bacteria to the urethra walls, in such a way that is not possible for the bacteria to maintain a bacteria–host relation (Hisano et al., 2012; Krueger et al., 2013). Considering the above, we can suggest that nut by-products are economically and easily accessible sources of antimicrobial compounds, allowing us to explore alternative pharmaceutical applications on these by-products.

Other studies have found that nut by-products possess beneficial effects for protecting against metabolic ailments, all of which have in common the uncontrolled absorption and accumulation of lipids and carbohydrates. Insoluble dietary fiber is a very important part of nut by-products (Table 8.1), and there are many studies that demonstrate its health benefits in the control of glycemic response, weight gain, and diabetes. Dietary fiber, especially soluble fiber, is known to impair the absorption of cholesterol on the small intestine, due to its capability to increase the viscosity of the intestinal lumen, hindering the interaction between digestive enzymes and nutrient transporters with their substrates, so the absorption of nutrients is decreased (Mandalari et al., 2010b). Insoluble dietary fiber, which is the major type of dietary fiber in some nut by-products (Table 8.1), increases the fecal material and accelerates the intestinal transit, and, through this mechanism, also reduces the absorption of cholesterol and carbohydrates, leading to an improved control of diabetic patients. The effect of dietary fiber on weight control seems to be also related to the fact that these compounds can increase satiety and decrease voluntary food intake (Kendall et al., 2010). The glycemic response seems to be also influenced by phenolic compounds that are bonded to fiber. Authors suggested that these compounds can inhibit carbohydrate absorption and increase insulin production, improving the control of glucose in blood. Nonetheless, these studies were made with fiber and phenolic compounds obtained from edible parts of fruit products (Devalaraja et al., 2011). There is a scarcity of investigation about the use of fiber and phenolic compounds from nut by-products in the control of these health issues. One study found that phenolic compounds obtained from peanut by-products showed hipocholesterolemic activity in an animal model, where additions of a polyphenol extract in the drinking water of Wistar rats increased the amount of cholesterol in feces. Authors suggested that polyphenols are capable of reducing the cholesterol solubility at the small intestine level, by inhibiting the formation of bile micelles (Tamura et al., 2012).

Other nut by-products may also be used as sources of bioactive ingredients for the development of drugs that prevent the absorption of nutrients with high caloric content, therefore helping to control the obesity pandemic. Inhibition of enzymes involved in fat and carbohydrate absorption, significantly reduces the caloric content of a diet (Gonçalves et al., 2011; Kimura et al., 2011). A peanut shell extract demonstrated inhibitory activity of pancreatic lipase, avoiding excessive weight gain on rats fed on a high lipid diet (de la Garza et al., 2011). A mechanism that explains this phenomenon lies on the ability of the phenolic compounds to complex proteins, generating conformational changes in lipase through hydrophobic interactions of the aromatic groups from both molecules (McDougall et al., 2009). A study on almond skins showed that polymeric phenolic compounds from this product, mainly condensed tannins based in flavan 3-ol monomers, inhibited human α amylase with an IC_{50} = 2.2 µg/mL, thus showing a potential for slowing carbohydrate absorption and lowering postprandial hyperglycemia (Tsujita et al., 2013). These two studies show the presence of bioactive compounds with inhibitory lipase and amylase activity, in nut by-products, and suggest the possibility for developing effective natural anti-obesity drugs from these sources.

Several extracts from nut by-products have been tested to prevent growth on tumor cells of several types of cancer. Pecan shell extracts showed anti-proliferative activity in colon cancer cell lines (Villarreal-Lozoya et al., 2007). Phenolic compound extracts from peanut skins inhibited the growth of colon, cervix, and liver tumoral cell lines in vitro. Colon cells showed more sensibility to this compounds with an IC_{50} = 10.9 µg/mL, followed by cervix cancer cells (IC_{50} = 12.6 µg/mL) and liver cancer cells (IC_{50} = 19.3 µg/mL). This effect is attributed to the phenolic compounds like cate-chin, epicatechin, gallic acid, and protocatechuic acid already known for their anticarcinogenic properties (Taha et al., 2012). Nevertheless, in vivo research is required to know the real extent of these antitumoral effects.

There are some reports showing that extracts of nut by-products are effective in reducing the symptoms of certain neurological disorders in animal models, suggesting its possible use as a raw material for the extraction of nutraceutical compounds which may be included in commercial drug formulations (Reckziegel et al., 2011) found that pecan shell aqueous extracts with high antioxidant activity also minimized the anxiety induced by cigarette abstinence in rats that had been exposed to the smoke of at least six cigarettes daily for 3 weeks. The authors concluded that the antioxidant attributes of the extract prevented superoxide radical formation in the brain, protecting the cell membranes, rich in polyunsaturated acids. In a related study, an aqueous extract from the pecan shell showed a positive effect in the prevention of dyskinesia induced by prolonged treatment with antipsychotics in rats (Trevizol et al., 2011). This study also showed a partial recovery of the disease for those animals that already had symptoms and were treated with the pecan shell extract. Authors proposed that the protective effect of the shell extract was also by controlling the oxidative stress generated in the brain area (Trevizol et al., 2011). Further studies on the beneficial effects of nut by-products on the central nervous system are necessary to explore the mechanisms for their actions and the possibility of extracting and purifying the active compounds.

Despite all the information on the different biological activities of extracts obtained from nut by-products, it is necessary to carry out toxicological evaluations to assess the safety of these materials for human use. This type of study has been performed for the aqueous extract of pecan shells, and authors found no acute or subacute toxicity of the pecan shell aqueous extract in doses up to 2000 mg/kg, single dose (acute test), or 100 mg/kg for 28 days (subacute test). The same extract was not mutagenic in *Salmonella typhimurium* strains (Porto et al., 2013).

In summary, although several studies show that nut by-products are rich in bioactive ingredients with many potential health effects, and some extracts possess antioxidant, antimicrobial, anticholesterolemic, anti-obesity, antiproliferative, and neuroprotective activities, more studies are needed to provide strong evidence for the nutraceutical and pharmaceutical industries to consider their use. These studies should include a complete characterization of the by-products or their extracts, which is sometimes difficult due to the high content of polymeric compounds (fiber and tannins), more in vivo animal studies, including safety assessments of the extracts, and finally human studies with carefully characterized extracts or purified active ingredients.

8.5   POTENTIAL USES OF NUT BY-PRODUCTS

An important option for the use of nut by-products is their addition, either as raw materials or in different grades of purification, to human foods and animal feeds, as sources of bioactive ingredients and also to extend the product's shelf life and quality attributes. Some by-products from the wine industry are already processed and added in foodstuffs to enrich the original product and improve their functional quality (Taha et al., 2012; Chouchouli et al., 2013). In other cases, isolated bioactive compounds from the natural matrix may be transformed into nutraceutical products.

The addition of insoluble fiber or phenolic compounds to processed foods is not only done to impact on the consumer's health but also to improve their organoleptic conditions or prolonging its shelf life (Mudgil and Barak, 2013). In the latter case, phenolic compounds and insoluble dietary fiber are added as food additives. In this context, both bioactive compounds have different functions when added to processed foods: first, as "functional food enhancers," and second, as food preservatives and additives. However, until now, phenolic compounds or insoluble dietary fiber isolated from nut by-products have been applied only on a few food products, increasing their functionality or as preservatives (Stintzing et al., 2001).

On recent studies, the skin that covers peanuts was added in powder during the elaboration of peanut butter. Authors pointed out that addition of 20% peanut peel to the peanut butter increased by more than 50% the antioxidant activity of the final product, with only slight modifications in color, and no significant changes in the taste nor smell (Hathorn and Sanders, 2012). Another application for peanut skins is on flaxseed oil products, where the addition of peanut skins can extend the shelf life of flaxseed oil, due to its phenolic portion and antioxidant capacity, delaying the presence of unwanted odors and flavors that come after an oil oxidation process (Taha et al., 2012).

In a different study researchers have combined a beverage, the Espresso coffee, with hazelnut skins, which are one of the richest edible sources of antioxidant compounds (Rio et al., 2011), and found a greater in vitro antioxidant capacity (DPPH and FRAP techniques) of the new beverage, as well as a greater antioxidant activity in vivo on serum from the rats which were fed with this beverage (Contini et al., 2012). Some authors have pointed out that hazelnut skin could be also a good prebiotic ingredient due to its high content of dietary fiber. Its prebiotic activity was tested with two strains of bifidobacteria, where the addition of 0.01% of hazelnut skin, increased bacterial growth (Montella et al., 2013). Almond skin has also shown prebiotic activity on beneficial bacteria of the human gut, a condition that favors a better balance in the intestinal flora. Authors explained this phenomenon due to the ability of the colonic bacteria, to ferment the dietary fiber to obtain energy. Consequently, populations of bifidobacteria and rectal coccoids were increased to a sufficient level to prevent growth of other pathogenic bacteria in the intestine (Mandalari et al., 2010b). Is important to point up that the scarcity of studies about the technological applications of nut by-products may be due to the lack of evidence that ensure the extracts are safe to humans. Therefore, safety-assessment studies of carefully characterized extracts are much needed. The effect of the addition of nut by-products or their extracts to the sensory attributes of the food products should also be considered, although, when low concentrations are effectively used, their impact on sensory quality is usually insignificant (Hathorn and Sanders, 2012).

Finally the potential use of nut by-products, their extracts, or active ingredients in animal feeds is also a promising field of study and future applications.

8.6   CONCLUSION

Currently, several countries, especially in the developed world, are going through an "epidemiological transition," with a lower incidence of infectious diseases and an increase of degenerative and chronic disease like cardiovascular and metabolic disorders, diabetes, and cancer. By-products of the nut industry contain high amounts of active compounds that can significantly contribute to control these diseases over the next years. Therefore, the development of studies demonstrating the best methods for their extraction, purification, and characterization as well as their safe use in the production of functional foods or nutraceutical formulations is of great importance. In this way, if we can ensure that all by-products from this industry, are safely processed, applied, and effectively used, they will increase their economic value and become a new source of profit for the agro-industry. On the other hand, it is important to consider that in spite of the biological applications that have been described about dietary fiber and phenolic compounds, their optimal doses, in enriched functional foods or nutraceutical supplements, must be determined to develop their maximal functionality, while making sure that its addition will not generate unwanted changes in food, which can be linked to the consumers' rejection. We noted in this chapter, the scarcity of studies on the safety and effectiveness of extracts from by-products and even on the detailed identification of the active principles, which currently delays the full use of these products, in any of the areas suggested by this chapter.

ACKNOWLEDGMENTS

Funding of this work by Mexican's National Council of Science and Technology, CONACYT (projects CB-2001-01-167164 and ALFANUTRA Functional Foods Network) and PROMEP (Red de Uso de Subproductos de la IndustriaAgroalimentaria) is gratefully acknowledged. A. A. Vazquez-Flores also acknowledges CONACYT for her Ph.D. scholarship (362130).

KEYWORDS

- **nuts**
- **health-promoting natural compounds**
- **husks**
- **peeling process**
- **bioactive ingredients**

REFERENCES

Alasalvar, C.; Karamać, M.; Amarowicz, R.; Shahidi, F. Antioxidant and Antiradical Activities in Extracts of Hazelnut Kernel (*Corylus avellana* L.) and Hazelnut Green Leafy Cover. *J. Agric. Food Chem.* **2006,** *54* (13), 4826–4832.

Bolling, B. W.; Chen, Y. C.; McKay, D. L.; Blumberg, J. B. Tree Nut Phytochemicals: Composition, Antioxidant Capacity, Bioactivity, Impact Factors. A Systematic Review of Almonds, Brazils, Cashews, Hazelnuts, Macadamias, Pecans, Pine Nuts, Pistachios and Walnuts. *Nutr. Res. Rev.* **2011,** *24* (02), 244–275.

Brown, D. FDA Considers Health Claim for Nuts. *J. Am. Diet. Assoc.* **2003,** *103* (4), 426.

Buzby, J. C.; Pollack, S. L. Almonds Lead Increase in Tree Nut Consumption. *Amber Waves* **2008**.

Chen, P.; Yanling, C.; Shaobo, D.; Xiangyang, L. Utilization of Almond Residues. *Int. J. Agric. Biol. Eng.* **2010,** *3* (4), 1–18.

Chouchouli, V.; Kalogeropoulos, N.; Konteles, S. J.; Karvela, E.; Makris, D. P.; Karathanos, V. T. Fortification of Yoghurts with Grape (*Vitis vinifera*) Seed Extracts. *LWT—Food Sci. Technol.* **2013,** *53* (2), 522–529.

Contini, M.; Baccelloni, S.; Frangipane, M.; Merendino, N.; Massantini, R. Increasing Espresso Coffee Brew Antioxidant Capacity Using Phenolic Extract Recovered from Hazelnut Skin Waste. *J. Funct. Foods* **2012,** *4* (1), 137–146.

Contini, M.; Baccelloni, S.; Massantini, R.; Anelli, G. Extraction of Natural Antioxidants from Hazelnut (*Corylus avellana* L.) Shell and Skin Wastes by Long Maceration at Room Temperature. *Food Chem.* **2008,** *110* (3), 659–669.

Cruz-Vega, D.; Verde-Star, M.; Salinas-González, N.; Rosales-Hernández, B.; Estrada-García, I.; Mendez-Aragón, P.; Carranza-Rosales, P.; González-Garza, M.; Castro-Garza, J. Antimycobacterial Activity of *Juglans regia, Juglans mollis, Carya illinoinensis* and *Bocconia frutescens. Phytother. Res.* **2008,** *22* (4), 557–559.

de la Garza, A. L.; Milagro, F. I.; Boque, N.; Campión, J.; Martinez, J. A. Natural Inhibitors of Pancreatic Lipase as New Players in Obesity Treatment. *Planta Med.* **2011,** *77* (8), 773–785.

de la Rosa, L. A.; Alvarez-Parrilla, E.; Shahidi, F. Phenolic Compounds and Antioxidant Activity of Kernels and Shells of Mexican Pecan (*Carya illinoinensis*). *J. Agric. Food Chem.* **2011,** *59* (1), 152–162.

Devalaraja, S.; Jain, S.; Yadav, H. Exotic fruits as Therapeutic Complements for Diabetes, Obesity and Metabolic Syndrome. *Food Res. Int.* **2011,** *44* (7), 1856–1865.

do Prado, A.; Manion, B. A.; Seetharaman, K.; Deschamps, F.; Arellano, D.; Block, J. Relationship between Antioxidant Properties and Chemical Composition of the Oil and the Shell of Pecan Nuts [*Carya illinoinensis* (Wangenh) C. Koch]. *Ind. Crops Prod.* **2013,** *45*, 64–73.

Esfahlan, A.; Jamei, R.; Esfahlan, R. The Importance of Almond (*Prunus amygdalus* L.) and its By-products. *Food Chem.* **2010,** *120* (2), 349–360.

Fabani, M. P.; Luna, L.; Baroni, M. V.; Monferran, M. V.; Ighani, M.; Tapia, A.; Wunderlin, D. A.; Feresin, G. Pistachio (*Pistacia vera* var Kerman) from Argentinean Cultivars. A Natural Product with Potential to Improve Human Health. *J. Funct. Foods* **2013,** *5* (3), 1347–1356.

Fernández-Agulló, A.; Pereira, E.; Freire; Valentão, P.; Andrade, P. B.; González-Álvarez, J.; Pereira, J. A. Influence of Solvent on the Antioxidant and Antimicrobial Properties of Walnut (*Juglans regia* L.) Green Husk Extracts. *Ind. Crops Prod.* **2013,** *42*, 126–132.

Francisco, M. L.; Resurreccion A. V. A. Functional Components in Peanuts. *Crit. Rev. Food Sci. Nutr.* **2008,** *48* (8), 715–746.

Garrido, I.; Juan, M.; Josefina, M.; Gómez-Cordovés, C. Extracción de antioxidantes a partir de subproductos del procesado de la almendra. *Grasas Aceites* **2007,** *58* (2), 130–135.

Gonçalves, R.; Mateus, N.; de Freitas, V. Inhibition of α-Amylase Activity by Condensed Tannins. *Food Chem.* **2011,** *125* (2), 665–672.

Grijalva-Contreras, R.; Martínez-Díaz, G.; Macías-Duarte, R.; Robles-Contreras, F. Effect of Ethephon on Almond Bloom Delay, Yield, and Nut Quality under Warm Climate Conditions in Northwestern Mexico. *Chilean J. Agric. Res.* **2011,** *71* (1), 34–38.

Hathorn, C. S.; Sanders, T. H. Flavor and Antioxidant Capacity of Peanut Paste and Peanut Butter Supplemented with Peanut Skins. *J. Food Sci.* **2012,** *77* (11), S407–S411.

He, F.; Pan, Q.-H.; Shi, Y.; Duan, C.-Q. Biosynthesis and Genetic Regulation of Proanthocyanidins in Plants. *Molecules (Basel, Switzerland)* **2008,** *13* (10), 2674–2703.

Hisano, M.; Bruschini, H.; Nicodemo, A.; Srougi, M. Cranberries and Lower Urinary Tract Infection Prevention. *Clinics* **2012,** *67* (6), 661–667.

Kendall, C.; Esfahani, A.; Jenkins, D. The Link between Dietary Fibre and Human Health. *Food Hydrocolloids* **2010,** *24* (1), 42–48.

Kimura, H.; Ogawa, S.; Akihiro, T.; Yokota, K. Structural Analysis of A-type or B-type Highly Polymeric Proanthocyanidins by Thiolytic Degradation and the Implication in their Inhibitory Effects on Pancreatic Lipase. *J. Chromatogr. A* **2011,** *1218* (42), 7704–7712.

Krueger, C. G.; Reed, J. D.; Feliciano, R. P.; Howell, A. B. Quantifying and Characterizing Proanthocyanidins in Cranberries in Relation to Urinary Tract Health. *Anal. Bioanal. Chem.* **2013,** *405* (13), 4385–4395.

Lau, T.-C.; Chan, M.-W.; Tan, H.-P.; Kwek, C.-L. Functional Food: A Growing Trend among the Health Conscious. *Asian Soc. Sci.* **2012,** *9* (1). DOI:10.5539/ass.v9n1p198.

Ma, Y.; Kerr, W.; Swanson, R.; Hargrove, J.; Pegg, R. Peanut Skins-Fortified Peanut Butters: Effect of Processing on the Phenolics Content, Fibre Content and Antioxidant Activity. *Food Chem.* **2014,** *145*, 883–891.

Mandalari, G.; Bisignano, C.; D'Arrigo, M.; Ginestra, G.; Arena, A.; Tomaino, A.; Wickham, J. Antimicrobial Potential of Polyphenols Extracted from Almond Skins. *Lett. Appl. Microbiol.* **2010a,** *51* (1), 83–89.

Mandalari, G.; Faulks, R. M.; Bisignano, C.; Waldron, K. W.; Narbad, A.; Wickham, M. In Vitro Evaluation of the Prebiotic Properties of Almond Skins (*Amygdalus communis* L.). *FEMS Microbiol. Lett.* **2010b,** *304* (2), 116–122.

McDougall, G. J.; Kulkarni, N. N.; Stewart, D. Berry Polyphenols Inhibit Pancreatic Lipase Activity In Vitro. *Food Chem.* **2009,** *115* (1), 193–199.

Montella, R.; Coïsson, J.; Travaglia, F.; Locatelli, M.; Malfa, P.; Martelli, A.; Arlorio, M. Bioactive Compounds from Hazelnut Skin (*Corylus avellana* L.): Effects on *Lactobacillus plantarum* P17630 and *Lactobacillus crispatus* P17631. *J. Funct. Foods* **2013,** *5* (1), 306–315.

Mudgil, D.; Barak, S. Composition, Properties and Health Benefits of Indigestible Carbohydrate Polymers as Dietary Fiber: A Review. *Int. J. Biol. Macromol.* **2013,** *61*, 1–6.

Nations, F. A. A. O. O. T. U. FAO Statistical Yearbook 2013. In *World food and Agriculture*; FAO: Rome, 2013; p 307 [Online].

O'Shea, N.; Arendt, E. K.; Gallagher, E. Dietary Fibre and Phytochemical Characteristics of Fruit and Vegetable By-products and their Recent Applications as Novel Ingredients in Food Products. *Innov. Food Sci. Emerg. Technol.* **2012,** *16*, 1–10.

Pereira, J.; Oliveira, I.; Sousa, A.; Valentão, P.; Andrade, P. B.; Ferreira, I. C. F. R.; Ferreres, F.; Bento, A.; Seabra, R.; Estevinho, L. Walnut (*Juglans regia* L.) Leaves: Phenolic Compounds, Antibacterial Activity and Antioxidant Potential of Different Cultivars. *Food Chem. Toxicol.* **2007**, *45* (11), 2287–2295.

Pimenta, J. *Arq. Bras. Cardiol.* **1992**, *58* (5), 343–344 [original article].

Porto, L.; da Silva, J.; de Ferraz, A.; Corrêa, D.; dos Santos, M.; Porto, C.; Picada, J. Evaluation of Acute and Subacute Toxicity and Mutagenic Activity of the Aqueous Extract of Pecan Shells [*Carya illinoinensis* (Wangenh.) K. Koch]. *Food Chem. Toxicol.* **2013**, *59*, 579–585.

Prado, A. C. P.; Aragão, A. M.; Fett, R. Phenolic Compounds and Antioxidant Activity of Pecan [*Carya illinoinensis* (Wangenh.) C. Koch] Kernel Cake Extracts. *Grasas Aceites (Sevilla, Spain)* **2009**, *60* (5), 458–467.

Prior, R. L.; Gu, L. Occurrence and Biological Significance of Proanthocyanidins in the American Diet. *Phytochemistry* **2005**, *66* (18), 2264–2280.

Rajaei, A.; Barzegar, M.; Mobarez, A.; Sahari, M.; Esfahani, Z. Antioxidant, Anti-microbial and Antimutagenicity Activities of Pistachio (*Pistachia vera*) Green Hull Extract. *Food Chem. Toxicol.* **2009**, *48* (1), 107–112.

Reckziegel, P.; Boufleur, N.; Barcelos, R. C. S.; Benvegnú, D. M.; Pase, C. S., Muller, L. G.; Teixeira, A. M.; Zanella, R.; Prado, A. C. P.; Fett, R.; Block, J. M.; Burger, M. E. Oxidative Stress and Anxiety-Like Symptoms Related to Withdrawal of Passive Cigarette Smoke in Mice: Beneficial Effects of Pecan Nut Shells Extract, a By-product of the Nut Industry. *Ecotoxicol. Environ. Saf.* **2011**, *74* (6), 1770–1778.

Rio, D.; Calani, L.; Dall'Asta, M.; Brighenti, F. Polyphenolic Composition of Hazelnut Skin. *J. Agric. Food Chem.* **2011**, *59* (18), 9935–9941.

Sarnoski, P. J.; Johnson, J. V.; Reed, K. A.; Tanko, J. M.; O'Keefe, S. F. Separation and Characterisation of Proanthocyanidins in Virginia Type Peanut Skins by LC–MSn. *Food Chem.* **2012**, *131* (3), 927–939.

Shahidi, F.; Alasalvar, C.; Liyana-Pathirana, C. M. Antioxidant Phytochemicals in Hazelnut Kernel (*Corylus avellana* L.) and Hazelnut Byproducts. *J. Agric. Food Chem.* **2007**, *55* (4), 1212–1220.

Soto-Vaca, A.; Gutierrez, A.; Losso, J. N.; Xu, Z.; Finley, J. W. Evolution of Phenolic Compounds from Color and Flavor Problems to Health Benefits. *J. Agric. Food Chem.* **2012**, *60* (27), 6658–6677.

Stintzing, F. C.; Schieber, A.; Carle, R. Phytochemical and Nutritional Significance of Cactus Pear. *Eur. Food Res. Technol.* **2001**, *212* (4), 396–407.

Suppadit, T.; Kitikoon, V.; Phubphol, A.; Neumnoi, P. Effect of Quail Litter Biochar on Productivity of Four New Physic Nut Varieties Planted in Cadmium-Contaminated Soil. *Chilean J. Agric. Res.* **2012**, *72* (1), 125–132.

Taha, F. S.; Wagdy, S. M.; Singer, F. A. Comparison between Antioxidant Activities of Phenolic Extracts from Different Parts of Peanuts. *Life Sci. J.* **2012**, *9*, 207–215.

Tamura, T.; Inoue, N.; Shimizu-Ibuka, A.; Tadaishi, M.; Takita, T.; Arai, S.; Mura, K. Serum Cholesterol Reduction by Feeding a High-Cholesterol Diet Containing a Lower-Molecular-Weight Polyphenol Fraction from Peanut Skin. *Biosci., Biotechnol. Biochem.* **2012**, *76* (4), 834–837.

Tomaino, A.; Martorana, M.; Arcoraci, T.; Monteleone, D.; Giovinazzo, C.; Saija, A. Antioxidant Activity and Phenolic Profile of Pistachio (*Pistacia vera* L., Variety Bronte) Seeds and Skins. *Biochimie* **2010**, *92*, 1115–1122.

Trevizol, F.; Benvegnú, D. M.; Barcelos, R. C. S.; Pase, C. S.; Segat, H. J.; Dias, V.; Dolci, G. S.; Boufleur, N.; Reckziegel, P.; Bürger, M. E. Comparative Study between two Animal Models of Extrapyramidal Movement Disorders: Prevention and Reversion by Pecan Nut Shell Aqueous Extract. *Behav. Brain Res.* **2011**, *221* (1), 13–18.

Tsujita, T.; Shintani, T.; Sato, H. α-Amylase Inhibitory Activity from Nut Seed Skin Polyphenols. 1. Purification and Characterization of Almond Seed Skin Polyphenols. *J. Agric. Food Chem.* **2013**, *61* (19), 4570–4576.

Vadivel, V.; Biesalski, H. K. Contribution of Phenolic Compounds to the Antioxidant Potential and Type II Diabetes Related Enzyme Inhibition Properties of *Pongamia pinnata* L. Pierre Seeds. *Process Biochem.* **2011**, *46* (10), 1973–1980.

Villarreal-Lozoya, J. E.; Lombardini, L.; Cisneros-Zevallos, L. Phytochemical Constituents and Antioxidant Capacity of Different Pecan [*Carya illinoinensis* (Wangenh.) K. Koch] Cultivars. *Food Chem.* **2007**, *102* (4), 1241–1249.

Wang, L.; Gao, S.; Jiang, W.; Luo, C.; Xu, M.; Bohlin, L.; Rosendahl, M.; Huang, W. Antioxidative Dietary Compounds Modulate Gene Expression Associated with Apoptosis, DNA Repair, Inhibition of Cell Proliferation and Migration. *Int. J. Mol. Sci.* **2014**, *15* (9), 16226–16245.

Wijeratne, S. S. K.; Abou, M. M. A.; Shahidi, F. Antioxidant Polyphenols in Almond and its Coproducts. *J. Agric. Food Chem.* **2006**, *54*, 312–318.

Xu, Y.; Edwars, N.; Sismour, N.; Parry, J.; Hanna, M. A.; Li, H. Nutritional Composition and Antioxidant Activity in Hazelnut Shells from US-Grown Cultivars. *Int. J. Food Sci. Technol.* **2012**, *47*, 940–946.

Zhang, G.; Hu, M.; He, L.; Fu, P.; Wang, L.; Zhou, J. Optimization of Microwave-Assisted Enzymatic Extraction of Polyphenol from Waste Peanut Shells and Evaluation of its Antioxidant and Antibacterial Activities In Vitro. *Food Bioprod. Process.* **2013**, *91*, 158–168.

Zhao, X.; Chen, J.; Du, F. Potential Use of Peanut By-products in Food Processing: A Review. *J. Food Sci. Technol.* **2012**, *49* (5), 521–529.

CHAPTER 9

BY-PRODUCTS FROM COFFEE PROCESSING

LAURA A. CONTRERAS-ANGULO, JACQUELINE RUIZ-CANIZALES, ALEXIS EMUS-MEDINA, and JOSÉ BASILIO HEREDIA*

Centro de Investigación en Alimentación y Desarrollo A.C. Carretera a El Dorado km 5.5 Campo El Diez, 80110 Culiacán, Sinaloa, México

Corresponding author. E-mail: jbheredia@ciad.mx

CONTENTS

ABSTRACT

Coffee is one of the most significant foods in the world. Therefore, the coffee by-products (pulp, silverskin, and husk) offer great opportunities to be used as substrate for bioprocesses, as feed for animals, organic fertilizer, organics acids, thickener, flavors and aroma compounds, productions of molds and yeast, in the growth of larvae and worms but mainly have a potential use as a source of bioactive compounds. These compounds may improve health, and their use as ingredients of functional foods leads to give added value and to reduce contamination. Coffee wastes have potential to innovate strategies applied in industrial scale processes.

9.1 TYPES OF PROCESSING AND STATISTICS IN PRODUCTION

In agriculture, coffee is one of the most significant foods in the world, after oil; it is the second most important because for decades, it has been internationally marketed. Coffee is produced in 10 countries according to the International Coffee Organization (ICO), among the main producers Guatemala, Uganda, Mexico, Honduras, India, Ethiopia, Indonesia, Colombia, Vietnam, and Brazil are found. The latter is the most important producer worldwide totaling 2,720,520 t of coffee beans in 2014. The coffee world production has increased by 73% from 1980 to 2014. In June 2015, 9.69 million bags were exported (60 kg each), compared to 2014, being *Coffea arabica* (67.31 million bags) and *Coffea canephora* (43.12 million bags) the two most commercialized varieties of coffee until June 2015 (Davis et al., 2011; ICO, 2015).

Coffee is one of the oldest cultures in the world; it is native to Africa, where it was originally cultivated in the province of Kaffa in Ethiopia. This culture extends throughout the tropics with more than 70 species of the genus *Coffea* of the family Rubiaceae, subfamily Cinchonoideae and coffeae tribe (Clifford et al., 1989). Only three species have commercial importance; *C. arabica*, *C. canephora* Pierre, and Coffea *liberica* Hiern (Farah, 2012; Murthy and Madhava Naidu, 2012; Pohlan and Janssens, 2012). Nonetheless, 80% of global production is identified as Arabian (*C. arabica*) and 20% as Robusta (*C. canephora*), both species differ in several aspects like the climate in which they grow, physical appearance, chemical composition, and type of drink made with roasted grains (Pohlan and Janssens, 2012; Padmapriya et al., 2013).

Coffee is a perennial that has a prominent vertical stem with shallow roots; the plant produces a berry that is used for the preparation of coffee, which is a beverage with a very particular aroma and flavor (Pohlan and Janssens, 2012). The berry has two cavities, each containing an oval and flattened by a grain faces. These grains are covered with two thin layers, which are removed before roasting. The raw coffee is roasted to obtain an exquisite aroma and flavor, as well as to intensify the color. The morphological structure of the coffee consists of an outer covering (exocarp), which determines the color of the fruit. Inside, there are different layers: mesocarp known as pulp (it is a gum rich in sugars), the endocarp called parchment, the yellow layer covering the grain known as silverskin, the endosperm, known as green coffee, which are toasted to prepare different types of coffee drinks and derived products (Suárez Agudelo, 2012; Fig. 9.1).

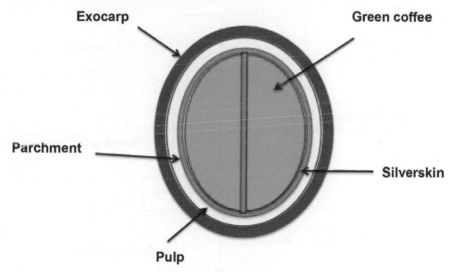

FIGURE 9.1 Morphological structure of coffee bean.

Traditionally, the coffee plant is cultivated only for the berry, which can be separated by the processing method: dry or wet (Chu, 2012; Padmapriya et al., 2013; Habtamu, 2014). Depending on the used method to separate the mesocarp from the endocarp is how the quality of the coffee would be. Also, the processing time can affect the desired quality of grains by losing different compounds found in waste (pulp, hulls, silverskin, and spent coffee) (Fig. 9.2). Dry processing is performed immediately after harvesting the fruit of coffee. This method is considered as GRAS (generally recognized as safe) and is used when you want to generate a final drink with sweet aroma. In this process, the berries are exposed to sunlight for a period of time until moisture reaches 10–12%. This is accomplished by placing the berries in dry floor at a depth of no more than 40 mm, with frequent movements to prevent fermentation or discoloration. Once the fruit dries, it is cleaned and peeled, removing skin and flesh, leaving mucilage on the seeds. Then the defective beans are removed to obtain a drink of high quality (Toci et al., 2006; Farah, 2012). The disadvantages of the dry process are that the product is exposed to dust, dirt, and climatic conditions, besides being a slow process that lasts up to four weeks (Padmapriya et al., 2013). This process can also be used as rotating equipment, horizontal or vertical dryer where the process can be continuous.

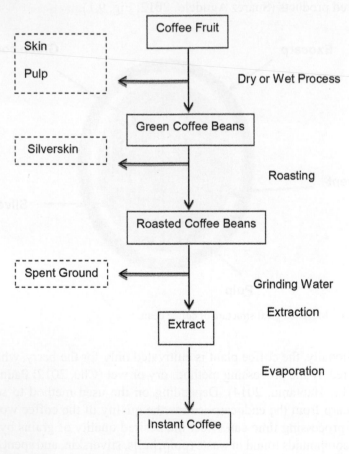

FIGURE 9.2   Schematic representation of the coffee product generation (Mussatto and Teixeira, 2010).

The method of wet processing is the most used in the world and involves mechanical and biological operations to separate the grain or seed exocarp (skin), mesocarp (mucilaginous pulp), and endocarp (parchment) (Clark, 1985). This process is more complex than dry processing, but excellent quality coffee is produced. The used bean must be mature (to facilitate pulping), usually selected by flotation. Pulping should be done immediately after harvest (not more than 24 h), using a pulping machine (drum pulper, disk pulper, vertical spiral drum pulper). At this stage, the pulp which accounts for about 40% of the weight of the fresh fruit and nowadays is scarcely used (causing serious pollution problems), is separated from the skin and parchment. The mucilaginous layer must be removed by fermentation (optimum temperature 30–35°C) with or without water; this is done several times with some pectolytic enzymes. After a period of 6–40 h in which the acidity (pH 4.5) is increased, the seed is washed thoroughly. Afterward, coffee is sun or hot air dried, or by a mixture of both. At this point, samples must be carefully handled to prevent moisture re-absorption. After peeling is done to remove the dried parchment, silverskin is also removed (Pandey et al., 2000; Hicks, 2001; Murthy, 2011). Once the coffee bean is dried, other processes are performed as polished, roasted, and decaffeinated.

Roasting is the process by which it gives the flavor, aroma, and color desired to coffee. Green coffee beans have an undesirable flavor; this is considered the most important and delicate step because by roasting grain also turns brown or even black, and fragility increases to facilitate the milling process (Baggenstoss et al., 2008). Changes in chemical composition that occur during this process are dependent on time and temperature, but features such as species, variety, and origin also influence the organoleptic quality of the grain (Ruosi et al., 2012; Mussato et al., 2011).

In roasted, there are hundreds of chemical transformations, such as polyphenolic compounds in a complex mixture of products of the Maillard reaction and organic compounds formed from pyrolysis (Czerny et al., 1999; Sacchetti et al., 2009). Changes also occur in sulfur compounds by oxidation, thermal degradation, hydrolysis, and increased vanillin (Kumazawa and Masuda, 2003). During roasting, there are also changes such as moisture loss, color changes, volume, mass, shape, pH, density, and volatile, which occur while the CO_2 is generated (Hernández et al., 2008).

Grains are roasted at 200 and 300°C which is done with constant stirring that allows to develop the taste of coffee; at this stage, the grain begins to absorb heat and takes a brown color. Furthermore, there is a change of pH (5.7–6.0 to 4.9–5.5), due to the formation and volatilization of the acetic acid from the breakdown of sugar, acids formed by decarboxylation of the rearrangement of sugar molecules, and decarboxylation of chlorogenic acid. These changes are responsible for the pleasant aroma (over 800 volatile compounds). The main product in this process is the silverskin (Balzer, 2001; Farah et al., 2005, 2006; Napolitano et al., 2007; López-Galilea et al., 2006).

Once roasted coffee beans must be quickly cooled to stop the exothermic reactions and prevent excessive roasting, which could reduce the quality of the product. Unground roasted coffee stored for 30 days can begin to lose flavor due to air contact. Then beans are milled and packed, and sometimes they are packaged as whole grains by using vacuum-sealed bags for marketing (Baggenstoss et al., 2008).

Besides roasting, preparation is another significant factor for consumers. Making coffee is a heterophase ranging from soft pure solution to an emulsion (drip filter coffee, boiled coffee Nordic, Turkish-style coffee, espresso, and cappuccino). The processing of coffee is an art and science involves several stages, each with a different purpose. So to obtain high-quality coffee is imperative that all steps are carried out according to the recommended procedures (Murthy and Madhava Naidu, 2012).

To obtain instant coffee, extraction is required at 175°C under pressurized conditions. This is performed after roasting and grinding, to extract soluble solids volatiles giving aroma and flavor. This should be performed by evaporation concentration to increase the soluble materials. The concentrated extract is then dried to produce the instant coffee and the residue obtained is called spent ground coffee (Mussato and Teixeria, 2014).

9.2 CHARACTERIZATION OF THE DERIVED BY-PRODUCTS

In recent years, food waste and its residues have been a concern. This is generated in the entire food supply chain from harvesting, storage, transport, processing, and consumers. Among the main causes of losses or waste generation pests, handling damage, loss of quality, inadequate process, separation of materials for different purposes, and inadequate storage practices can be named (Liu et al., 2013). According to statistics from the Food and Agriculture Organization of the United Nations (FAO), about 1.3 billion metric tons of food losses are estimated throughout the supply chain (Gustavsson et al., 2011). Other sources suggest that about 61 million tons of food wastes are generated each year in the United States, while in Europe 90 million tons are reported (GMA, 2012; Girotto et al., 2015). This affects the economic sector, which by 2008 in the United States alone the calculated losses were of $165.6 billion (Buzby and Hyman, 2012). It also reported that approximately 42% of the waste is generated by consumers and 39% by the food industry. As a result, waste has become an environmental and economic problem faced by many countries (Mirabella et al., 2014). Furthermore, loss of food has a social approach, as has the amount of waste associated with the global hunger and malnutrition. Therefore, the organization of the United Nations Food and Agriculture (FAO), a mandate to reduce food loss. From that initiative, there are programs currently in operation (Parfitt et al., 2010). Therefore, efforts have increased the number of strategies to resolve the situation and fate of food waste, which will contribute to the preservation of natural resources as well as to save financial resources involved in the production of surplus (Girotto et al., 2015).

As mentioned above, coffee consumption is of great importance worldwide, being one of the most consumed beverages and found in a variety of products. However, the dry, wet, and toasted process used to obtain coffee can generate a considerable amount of waste closely associated to a contamination problem in producing countries (Bonilla-Hermosa et al., 2014). That is highly related to inadequate or nonexistent generating process for removal of water and land pollution.

Depending on the type of process of coffee berry, many different industrial waste or by-products can be obtained. The main ones are the skin, pulp, mucilage, and parchment (Esquivel and Jiménez, 2012). In the dry process, the obtained by-products are coffee husks, skin, pulp, and peel (Brand et al., 2001). Similarly, in the semidry process, the seed pulp covering is removed, and the mucilage that covers the grain is washed before being dried in the sun. The water used to remove the mucilage is known as coffee wastewater contains high concentrations and organic pollutants (Bonilla-Hermosa et al., 2014). Finally, the wet process allows recovery of the pulp and skin in a fraction, mucilage, and the parchment among others (Esquivel and Jiménez, 2012).

Coffee pulp accounts for about 30% of the coffee fruit weight and is the main by-product produced during processing (Bhoite et al., 2013). The coffee pulp is essentially rich in sugars, proteins, minerals, and also contains significant amounts of tannins, polyphenols, and caffeine, considered toxic in nature (Pandey et al., 2000). The pulp contains about 76.7% moisture, 10% protein, and 21% crude fiber. Among the major organic compounds, we may mention pectin (6.5%), reducing sugars (12.4%), caffeine (1.3%), tannins (1.8–8.56%), caffeic acid (1.6%), and chlorogenic acid (2.6%), while major minerals are K, Ca, P, Zn, Cu, Mn, and B. It also contains amino acids such as lysine, valine, isoleucine, leucine, and glutamic acid among others

(Bressani et al., 1972; Braham and Bressani, 1979). Also, it can be considered a major source of antioxidant compounds, as has been reported chlorogenic acid as a major component (42.2%), followed by epicatechin (21.6%, acid isochlorogenic I, II, and II with 5.7%, 19.3%, and 4.4%, respectively), catechin (2.2%), rutin (2.1%), protocatechuic acid (1.6%), and ferulic acid (1%), among other compounds (Ramirez-Martinez, 1988).

Similarly, the coffee husk is another important product, mainly obtained from the dry process. Although its components are similar to the coffee pulp, it presents differences due to the obtaining process (Pandey et al., 2000). It has high carbohydrate content (58–85%), which is expected by the amount of pulp and reported containing 8–11% protein, 3–7% minerals, 0.5–3% lipids, 1% caffeine, and 5% tannins. The reducing sugar content was observed in 14% (Adams and Dougan, 1987; Franca and Oliveira, 2009). It also has high fiber content (30.8%) (Brand et al., 2001). Also, it could be considered a source of phenolic compounds, especially anthocyanins (Prata and Oliveira, 2007).

Another by-product obtained from roasted coffee is the silverskin, which is covering the coffee bean. The silverskin of the coffee is low in fat (2.2%) and reducing sugars (0.21%); on the other hand, it has a high protein content (18.6%) and ash (7%) suggesting a good source of minerals. Also, it is proper to emphasize the total dietary fiber (DF) content (62.2%) and soluble fiber (8.7%), so it can be considered as a functional ingredient for the development of products with high fiber content. It also has a high antioxidant capacity, which can be compared with tomatoes, peaches, and apples, antioxidant capacity is related to its content of chlorogenic acid and caffeic acid, although these acids are less than in roasted coffee (Borrelli et al., 2004).

Similarly, coffee consumed in industries, restaurants, cafes, and homes is produced by tons each day. The traditional way of use of this waste could be as fertilizer; however, this practice has been limited. Currently, various investigations suggest the use of coffee spent for producing animal feed, biofuel, and coffee extract as a source of antioxidant compounds. The latter was recently highlighted, since research has found a significant amount of caffeoylquinic acid, mainly monocaffeoylquinic acid (3-CQA, 4-CQA, 5-CQA) and dicaffeoylquinic acid (3,4-diCQA, 3,5-diCQA, 4,5-diCQA), which are in concentrations ranging from 2.91 up to 25.62 mg/g (Bravo et al., 2012). Other authors have reported a high content of phenolic compounds (18 mg GAE/g) and an appropriate antioxidant activity (Mussatto et al., 2011a). Likewise, it has been reported that the content of ferulic acid, p-coumaric acid, sinapic acid, and 4-hydroxybenzoic acid (Monente et al., 2015). Flavonoids are also reported between 2.11 and 8.03 mg QE/g, and caffeine in a range of 5.99–11.50 mg/g (Panusa et al., 2013).

Coffee parchment is a lesser amount; this residue is a strong fiber that covers the seeds of coffee and keeps it separated. It is obtained with the peel and pulp in the dry process, while the wet process is obtained after drying and shelling by separate steps. There is little information on its composition separately, but there are reports that demonstrate a high α-cellulose content (40–49%), hemicellulose (25–32%), and lignin (33–35%), so they have been considered for the use of building particleboard (Beliz et al., 2009; Bekalo and Reinhardt, 2010; Esquivel and Jiménez, 2012).

9.3 CONSIDERATION FOR PRACTICAL USES OF THE DERIVED BY-PRODUCTS

Among the main considerations for using coffee residues, in addition to the quantities produced, are generating an annual temporal and regional distribution, moisture content, storage and preservation, the technology used in its process and chemical composition and to determine the use of these by-products (Rathinavelu and Graziosi, 2005; García, 2009).

The coffee pulp is one of the residues obtained during the processing of coffee. It is a mucilaginous fiber material and constitutes approximately 43% of the fresh weight of the coffee bean. The pulp has high humidity and is rich in carbohydrates, protein, fiber, minerals, potassium, tannins, caffeine, and polyphenols (Pandey et al., 2000; Murthy and Madhava Naidu, 2012). Depending on their chemical composition, it can be used as a dietary supplement for animal nutrition based on the premium organic compounds present in pulp material. Also, as organic fertilizer for improving soil structure, and as key ingredient for the production of molds and yeasts in the growth of larvae and worms, including for combustion (Rathinavelu and Graziosi, 2005; García, 2009). The pulp solids contain only a small part of soil nutrients; still can be a source of humus and organic carbon as long as the pulp is quickly processed for this purpose. The advantage of using this residue is the odor-generated earth is stable and does not attract flies (Rathinavelu and Graziosi, 2005). Similarly, the coffee pulp can be used as direct fuel gas combustion systems or to produce electricity (Valencia and Franco, 2010).

Pandey et al. (2000) reported for coffee pulp 50% carbohydrate, 10% protein, 18% fiber, and 2.5% fat, characteristics considered to be used in the preparation of supplements in cattle feed. This residue may replace up to 20% of the concentrated in dairy cattle feeding without adverse reactions. Also, in pig diets, it is possible to replace corn by dehydrated coffee pulp (16% maximum per serving) without any negative effect on weight and feed conversion (Rathinavelu and Graziosi, 2005). It has also been used to produce molasses, finding that adding to pig diets by 30%, it has shown successful results as compared to molasses from sugarcane (Buitrago et al., 1968).

Furthermore, due to their anthocyanin content, the pulp can be used as a natural pigment (Murthy and Madhava Naidu, 2012). Likewise, various authors mention that according to the sugar content and pectin, it may be feasible to use as a substrate for the production of pectic enzymes (Trejo-Hernandez et al., 1991; Antier et al., 1993; Kashyap et al., 2001; Couto and Sanromán, 2006).

Caffeine of one of the components found in the pulp is an alkaloid that gets good market prices. However, this component is undesirable in animal nutrition, so its extraction is for addition to other foods and beverages to enhance flavor, or as a stimulant drug. Other compounds are soluble sugars and tannins, which can be extracted from the pulp for industrial purposes like caffeine (Braham and Bressani, 1979; Horvat et al., 2005).

Mucilage represents between 15% and 20% coffee bean and is composed of water, nonreducing and reducing sugars, cellulose, and also contains small quantities of protein, fat, ash, and some minerals such as K, Ca, Mg, and P. Mucilage has several applications in animal diets, plant nutrients, and diverse industries (Sarasty-Zambrano, 2012). Due to its content of pectic substances, it is feasible to obtain pectins; its sugar content can be used in the production of honey and ethyl alcohol and can produce methane gas from the anaerobic fermentation (Calle, 1977).

The coffee husk residue is perhaps the most productive, where for every ton of coffee beans produced, about a ton of husk is generated during coffee processing (Saenger et al., 2001). The residue of coffee husk have a high cellulose and hemicellulose content, which is used for commodities like wood, so it may have similar applications, such as a production of particle board (Mussatto and Teixeira, 2010). Oliveira et al. (2008) evaluated coffee husk as bioabsorbable material to remove heavy metals from aqueous solutions, finding good absorption at low ion concentration. However, an increase in the concentration showed a decrease in its bioabsorbability efficiency (Pandey et al., 2000).

Depending on their physical and chemical characteristics (carbon, hydrogen, sulfur, oxygen, and moisture), the coffee husk can be considered for high its calorific value, making it an excellent fuel, which can be used in furnaces of the same machines for drying coffee. Also, it can be a good substrate for the growth of edible fungi (Basidiomycetes); other uses may be as a soil conditioner (García, 2009). It serves as a substrate to produce citric acid from *Aspergillus niger* (Shankaranand and Losane, 1999) in a fermentation system and as the carbon source in the production of gibberellins from strains of *Gibeberella fujikuroi* (Machado et al., 2002). In the cosmetics industry it can be used as a source of antioxidants, based on their total content of phenolics and flavonoids, also contains anthocyanins, which can be an important source of natural colorant (Prata and Oliviera, 2007; Rodrigues et al., 2014).

Silverskin is the integument that covers the coffee bean has a high content of DF and a high antioxidant capacity, it is low in fat and carbohydrates that make it can be used as a functional ingredient (Borrelli et al., 2004). This residue has been evaluated for biotechnologically phenolic compounds or in the production of enzymes and ethanol. In fermentation systems, about 20 g/L of total sugars can be obtained (xylose, arabinose, galactose, and mannose), which serve as a culture medium in fermentation processes (Mussato and Teixeira, 2014). Silverskin was utilized as a nutrient source of *Aspergillus japonicus* in the production of β-fructofuranosidase and fructooligosaccharides (Mussatto and Teixeira, 2010).

Based on the phenolic acids found in silverskin, it can be used as a promissory source for nutritional supplements and may represent an innovative and functional ingredient, which may increase the antioxidant capacity if added in food products (Bresciani et al., 2014). Silverskin coffee is a residue that can be incorporated in the development of pancakes, bread, biscuits, and snacks because their properties could be an alternative in the food industry (Mussatto et al., 2011).

Spent coffee is rich in polysaccharides with a low amount of soluble solids. Because of its nutritional limitation is not considered for food, so can be used in fuel burn, in addition to being used as compost in gardens. This can also extract oil and biodiesel become a high yield (Kondamudi et al., 2008). Spent coffee can produce biodiesel esters, 47.1% saturated and 51.2% unsaturated. Besides, biodiesel quality was evaluated according the ASTM standards, showing results within normal limits except for carbon residue (Haile, 2014). In a study by Adi and Noor (2008), it was found that this waste could be decomposed by vermicomposting, using the earthworm *Lumbricus rubellus*. On the other hand, it can be used to affect the development of *Aedes aegypi* early, at concentrations of 50 mg/mL (Laranja et al., 2003).

It can be used as feed for ruminant animals such as pigs, chickens, and rabbits, although because of the high lignin content, it may be limited for use as food. However, this peculiarity makes it useful for making fuel pellets also by its content of cellulose is considered to produce paper, with a high calorie content (5000 kcal/kg), and therefore affording the property to be used as fuel in boilers (Mussatto et al., 2011). The charcoal from spent coffee can be used as an adsorbent to remove the acid colorant orange 7 (Nakamura et al., 2003).

9.4 FOOD ADDITIVES

9.4.1 ANTIMICROBIAL USES

Several authors suggest that some compounds in coffee have antimicrobial properties. Among them, phenolic compounds, these secondary metabolites in plants, mainly represented by flavonoids and phenolic acids. There are different mechanisms by which exhibit antimicrobial activity and degradation of the cell wall, damage to the cytoplasmic membrane, leakage of cell contents, coagulation of cytoplasm, alterations in the proton motive force, electron flux, and activity of conveyors. Phenolic compounds are capable of binding with proteins by hydrophobic interactions with phospholipids or react components of the cell membrane, these interactions cause changes in the composition of fatty acids, precipitate proteins, and leakage compounds necessary for microorganisms and increases the susceptibility to damage UV (Sikkema et al., 1995; Cowan, 1999; Nychas and Tassou, 2014).

Daglia et al. (1994) identified several compounds in roasted coffee as potent indicators of antimicrobial activity, among them 5-caffeoylquinic, caffeic acid, nicotinic acid, trigonelline, and 5-(hydroxymethyl) furfuraldehyde. Likewise, they found that Gram-positive bacteria are generally more sensitive than those of Gram-negative (Daglia et al., 1994).

Among the characteristic compounds of residues of coffee, chlorogenic acid (caffeoylquinic acid) has been investigated, finding an antimicrobial effect against *Listeria monocytogenes* and *Salmonella enteritidis* in concentrations of 500 mg/mL (Shen et al., 2014). Other authors report a change in morphology of the cell wall of *Salmonella choleraesuis* and *Escherichia coli* O157:H7 in the presence of chlorogenic acid (15 ppm), causing damage to the cell wall and the loss of matter within the bacterium (Cetin-Karaca and Newman, 2015). Other potential antimicrobial compounds found in coffee residues are protocatechuic acid and caffeic acid, since it has been reported that can inhibit *E. coli*, *Klebsiella pneumoniae*, and *Bacillus cereus* concentrations from 0.3 mg/mL (Aziz et al., 1997).

Similarly, Almeida et al. (2006) has shown that coffee has antibacterial properties against several strains of Enterobacteria, due to compounds such as caffeine, chlorogenic acid, protocatechuic acid, caffeic acid trigonelline. It also showed an effect against *Serratia marcescens*, *Enterobacter cloacae*, and against *Legionella pneumophila*, responsible for respiratory diseases (Dogasaki et al., 2002).

Melanoidins are other compounds present in roasted coffee as described with antimicrobial activity. They are present in the soluble fraction of coffee and are generated from Maillard reactions or caramelization of carbohydrates (Daglia et al., 1994). Rufian-Henares and de la Cueva (2009) describe the mechanisms by which melanoidins have antimicrobial properties and found that at low concentrations of melanoidins exert bacteriostatic activity by chelating iron (Fe^{3+}), so decreases the virulence of pathogens. Likewise, when melanoidins are in high concentrations can act Mg^{2+} cations, from outside the cell, promotes the rupture of the cell membrane.

Meanwhile, coffee spent and coffee melanoidins were studied, showing antimicrobial activity against *Staphylococcus aureus* and *E. coli*, but CS showed negligible antimicrobial effect. However, the combination of coffee melanoidins and coffee spent different significantly and coffee melanoidins alone were effective only against *S. aureus* (Jiménez-Zamora et al., 2015).

9.4.2 ANTIOXIDANT USES

The preservation of food has been instrumental in maintaining safe food available for consumption. For this reason, the food industry have developed different technologies to avoid or reduce oxidation, these range from modified atmosphere storage temperatures, hermetic packaging to the addition of antioxidant compounds (Ibáñez et al., 2003; Ayala-Zavala et al., 2012).

Antioxidants are natural or synthetic substances that may act by different mechanisms: reducing agents, sequestering free radicals, chelation, and through synergism (Pokorný, 1991; Pokorný et al., 2001).

Both synthetic and natural food retard oxidation. Among the synthetic phenolics are butylated hydroxyanisole, butylated hydroxytoluene, *tert*-butylhydroquinone, besides sulfites and gallates. On the other hand, among natural antioxidants are included tocopherols, ascorbic acid, carotenoids, and plant-derived phenolic compounds (Madhavi et al., 1995; Honglian et al., 2001).

The use of synthetic antioxidants in recent years have been replaced by naturally occurring, and it is believed that synthetic cause or promote negative health effects (Egea et al., 2010), which is why the trend has been used those from spices, herbs, fruits, and even waste that may have antioxidant capacity, as in the case of waste from coffee processing (Moure et al., 2001).

Process residues of coffee are an important source of antioxidants; the main compounds found are polyphenols, chlorogenic acid, and tannins (Table 9.1). Its content varies according to the type of waste (Murthy and Madhava Naidu, 2012). Little information is generated in function to the use of waste as a coffee additive. However, there is a study of Yoo et al. (2011) where they evaluated the addition of residues of coffee to chocolate to see the effect and determined their antioxidant capacity and content of flavonoids, as well as its sensory acceptability; the addition of coffee waste was performed at different concentrations (0, 10, 20, 30 40 g of residue of coffee). The results showed an increase in the content of total flavonoids and their antioxidant capacity (DPPH scavenging activity %), depending on the concentration of added waste, finding no negative effect on the sensory acceptability.

TABLE 9.1 Antioxidant Compounds in Coffee Residues.

Parameter (%)	Pulp	Husk	Silverskin	Spent
Total polyphenols[a]	1.48	1.22	1.32	1.02
Tannins	3.0	5.0	0.02	0.02
Chlorogenic acid	2.4	2.5	3.0	2.3

[a]% w/w GA.

In the pulp of coffee (*arabica*), four kinds of polyphenol flavan-3-ols, hydroxycinnamic acids, flavonols, and anthocyanidins were found (Ramirez-Coronel et al., 2004). The identified phenolic acids in coffee pulp were chlorogenic (5-caffeoylquinic acid), epicatechin, 3,4-dicaffeoylquinic acid, 3,5-dicaffeoylquinic acid, 4,5-dicaffeoylquinic acid, catechin, rutin, ferulic acid, and protocatechin (Ramirez-Martinez, 1988). According to Rodríguez-Durán et al. (2014), phenolic acids found in coffee pulp chlorogenic acid is the most abundant in the soluble fraction, followed by caffeic acid in the bound fraction. Other acids such as p-coumaric and ferulic found in both fractions lesser amount. Also, other compounds are soluble tannins, mainly procyanidins, and fewer insoluble (Farah and Donangelo, 2006).

Coffee husk contains anthocyanins mainly cyanidin 3-rutinoside as antioxidant compound. However, being rich in cellulose, hemicellulose, and lignin, these fiber components can be attached to and provide phenolic antioxidant (1.84 mmol Trolox equivalent/100 g dw) (Murthy and Madhava Naidu, 2012).

On the other silverskin side of coffee, there is one of the residues attached to DF (antioxidant–fiber complex) that show a strong antioxidant activity due to their content of chlorogenic acid and caffeic, as well as products of the Maillard reaction, melanoidins. These compounds are polymers produced during roasting and its content increases according to the intensity of the heat treatment, and their ability to inhibit peroxidation (Borrelli et al., 2002, 2004). The melanoidins present in silverskin can give flavor, aroma, color, and antioxidant properties (Moreira et al., 2012). According to these characteristics, Martinez-Saez et al. (2014) formulation of a drink extract silverskin from *arabica* and Robusta coffee, getting a drink with high antioxidant and sensorial acceptable potentially.

Residues spent coffee have an important action on the stability of food lipids, by the presence of phenolic acids (chlorogenic acid and caffeic acid), these compounds have good potential for use as a natural antioxidant (Yen et al., 2005). Both other free and bound phenolic acids were quantified in this residue finding caffeoylquinic, dicaffeoylquinic, caffeic, ferulic, ρ-coumaric, sinapic, and 4-hydroxybenzoic acids (Monente et al., 2015). Meanwhile, Panusa et al. (2013) evaluated the antioxidant activity of coffee spent in the two most important varieties of coffee (*arabica* and *robusta*), finding that the content of phenolic compounds is higher in the Robusta variety. Likewise, Bravo et al. (2012) reports that waste from spent coffee has an important content of caffeoylquinic acids (6.55–13.24 mg/g), and an antioxidant capacity ranging from 46% to 102.3% maximum, depending on the origin of the waste, concluding it can be a good source of bioactive compounds hydrophilic nature.

The compounds present in the waste of processed coffee can be a good source of antioxidants, if their bioactive compounds are extracted to be used as an additive, as in the case of some phenolic compounds such as chlorogenic acid. The latest has the property of retarding oxidation in food, keeping a fresh flavor for a longer period and is widely used in the food industry to prevent microbial growth (Andrés-Lacueva et al., 2010; Zhangjiajie Hengxing Biotechnology Co. Ltd., 2015). It has also been used as an additive in foods such as fish, emulsions, pate, and lard because of its antioxidant properties (Medina et al., 2009; Almajano et al., 2007). Polyphenols have antioxidant action because of its ability to scavenging free radicals, metal chelator, and inactivation of hemoproteins (Laranjinha et al., 1995; Kortenska-Kancheva et al., 2005; Iglesias et al., 2012).

In the same context, anthocyanins are other compounds that are found in coffee residues, these are responsible for the color of some fruits and flowers. Belonging to the family of flavonoids, anthocyanins are glycosides, that is, in its structure have joined a simple sugar such as glucose, galactose, rhamnose, and arabinose or a sugar compound as rutinose and sambubiose (Hartati et al., 2012). Its basic structure is the flavylium ion (2-phenyl-benzopyrilium), comprising two aromatic groups, one benzopropyl, and a phenolic ring; flavylium generally functions as the cation (Badui Dergal, 2006). These compounds are used in the food industry mainly as colorants, but can provide protection to own food components (vitamins) of factors such as light, and helps keep them for longer periods because that protects them from oxidation, its ability to sequester free radicals (Ortíz et al., 2011; Hartati et al., 2012).

Tannins consist of a large group of compounds having various structures that have the ability to bind and precipitate proteins. These are divided into condensed and hydrolyzable (Vermerris and Nicholson, 2006). The condensed tannins also called proanthocyanidins found in cranberries many products, seeds, bark, and other plants. These compounds may be found in the coffee pulp, primarily epicatechin, which represent over 90% of the units of proanthocyanidins. In a study, in fish muscle, proanthocyanidins at 25 ppm were applied, and it was found that proanthocyanidins were able to narrow the peroxide formation, and with an increase in concentration, there is an increase of the antioxidant effect (Iglesias et al., 2012).

9.4.3 THICKENER USES

Stabilizers, thickeners, and gelling agents are obtained from a wide range of compounds known as hydrocolloids. In general, substances or mixtures of substances maintained homogeneous products made of two or more immiscible phases, preventing their separation (Badui Dergal, 2006). The global market valued in the period 2007–2008 was set for these products at $4.2 billion in the United States, highlighting the use of starch, gelatin, pectin, and carrageenan. Current trends in the development of these products focus on convenience factors, natural, quality, and cost nutrition (Imeson, 2011).

Currently, the galactomannans are used as thickeners, emulsifiers, and stabilizers microencapsulating in the food, pharmaceutical, and cosmetic (Busch et al., 2015). Generally, galactomannans are present in the endosperm of various plants, their structure is composed of β-(1–4)-D-mannan backbone with single D-galactose branches linked α-(1–6) and the proportions mannose and glucose (M/G) depend on the extraction matrix (Kök et al., 1999). Main galactomannans used in industry are guar gum (M/G 2:1), tara gum (M/G 3:1), and locus vean gum (M/G 3.5:1) (Cerqueira et al., 2009).

Coffee residues could be used as a possible source of these products. It has been reported that about 50% of the coffee bean comprises polysaccharides; their main polymers are arabinogalactan, mannan or galactomannan, and cellulose. Coffee polysaccharides are responsible for the retention of volatile compounds, the foam stability, and viscosity of the body or coffee (Arya and Rao, 2007), while its major monosaccharides are mannose (22.4%), galactose (12.4%), glucose (8.7%), arabinose (4.0%), rhamnose (0.3%), and xylose (0.2%) (Bradbury and Halliday, 1990). Similarly, Avallone et al. (2001) reported arabinose, galactose, xylose, rhamnose, fucose, and glucose as major monosaccharides.

In that sense, in spent coffee, Ballesteros et al. (2015) reported polysaccharides thermoset and galactose (60.27%) as the dominant sugar, followed by arabinose (19.93%), sugar (15.37%), and mannose (4.43%). Moreover, Mussatto et al. (2011c) found 43.8% of mannose, galactose 30.4%, 19% glucose, and 3.8% arabinose. Other studies showed similar results with a 57% mannose, followed by 26% galactose, glucose 11%, and 6% arabinose (Simões et al., 2009).

Galactomannans represent about 70% of polysaccharides in the coffee infusion and with type II arabinogalactan as its main components. Galactomannans of green and roasted coffee are made by β-(1–4)-linked D-mannopyranosyl residues substituted at O-6 by single a-D-galactopyranosyl residues (Navarini et al., 1999). Similarly, arabinogalactan are made by β1–3-linked galactan main chain with frequent arabinose and galactose residue containing side chains, whereas the mannan resembles cellulose in that it has a linear β1–4-linked structure (Bradbury and Halliday, 1990).

Therefore, the characteristics and composition of coffee make it useful for the extraction of polysaccharides to be used as a thickener in the food industry; however, it needs further investigation of the rheological properties of each of the coffee products.

9.4.4 SENSORIAL IMPACT

In general, the production of flavors for the food industry is carried out by chemical synthesis. However, current trends are heading to the natural production of aromatic compounds by microbial biosynthesis or bioconversion (Welsh et al., 1989). Coffee residues could be used as substrates in such processes. The solid-state fermentation allows the use of industrial waste for treating solid waste and production of enzymes, organic acids, and secondary metabolites. The coffee husk has been used as a substrate of *Ceratocystis fimbriata*, a fungus that produces a large amount of flavors like pineapple, peach, banana, citrus, and pink (Lanza et al., 1976) getting good results in developing a strong smell of pineapple with total volatile of 6.58 and 5.24 mmol/L g for 20 and 35% glucose, respectively. Furthermore, banana flavors were presented by adding leucine (Soares et al., 2000). Others, used coffee pulp and coffee husks, finding a maximum total volatile (28 mmol/L g) after 72 h increased production compounds were ethyl acetate, ethanol, and acetaldehyde with 84.7%, 7.6%, and 2.0%, respectively (Medeiros et al., 2003).

Another application for the use of waste coffee could be obtaining dyes. Natural dyes are in high demand in the food industry, mainly in the United States, European Union, and Japan. It is necessary to consider the characteristics such as of color stability, price, pH, heat, and light sensitivities (Chattopadhyay et al., 2008). In this context, Prata and Oliveira (2007), they suggest using coffee husk as a possible source of anthocyanins and application of natural dyes for food, finding dominant anthocyanin cyadinin-3-rutinoside, further delphinidin, cyanidin, and malvidin. Similarly, it has extracted dye dehydrated coffee pulp with cafes husk, wherein the color obtained was attributed to the content of flavonoids and tannins (Diaz-Giron and Elias-Marroquin, 2009).

The aroma of coffee is characterized as complex as it encompasses a lot of reactions and routes as Maillard reactions, Strecker degradation, disruption of sulfur amino acids, proline and hydroxyproline rupture, degradation trigonelline, quinic acid degradation, degradation of pigments and lipids, among others. The volatiles kinds of roasted coffee are sulfur compounds, pyrazines, pyridines, pyrroles, oxazole, furans, aldehydes, ketones, and phenols. However, despite being a large number of compounds involved in the aroma, various researchers suggest about 20 as relevant coffee aroma (Buffo and Cardelli-Freire, 2004). Coffee residues mainly roasted, characterized by maintaining some of these compounds. In the coffee silverskin, they have reported a variety of compounds such as proteins (3.3–15.7 g/100 g), carbohydrates (2.3–12.1 g/100 g), and phenolic compounds (0.6–3.6 g/100 g) (Narita and Inouye, 2014). Similarly, coffee husk presents phenolic compounds, proteins, fibers, minerals, carbohydrates, and caffeine (Brand et al., 2001). For its part, spent coffee contains carbohydrates, caffeine, browcolors compounds derived from Maillard reactions, minerals, and phenolic compounds (Campos-Vega et al., 2015). Previously described in detail is the composition of coffee waste, which could be a source for the extraction of compounds of interest.

9.4.5 NUTRACEUTICAL USES

Prevalence of chronic degenerative diseases has influenced eating habits. Nutraceutical and functional food market increased every year because consumers prefer products made with natural ingredients, which offer health benefits. According to Stephen DeFelice, a nutraceutical is a food or part of food that exert health benefits, including prevention or treatment of some diseases. A functional food is a product that provides proteins, lipids, carbohydrates, vitamins, minerals, and other substances that improve health such as DF, antioxidants, and prebiotics (Kalra, 2003; Espín et al., 2007). Coffee is one of the most popular beverages in the world and its increasing production generates more than 30 million tons of by-products that cause pollution problems (Rojas et al., 2002). By-products of coffee processing (silverskin, husk, pulp, and spent coffee ground [SCG]) are good sources of protein, carbohydrates, lipids, fiber, and phenolic compounds, making them suitable to produce nutraceutical and functional foods (Esquivel and Jimenez, 2012; Narita and Inouye, 2014). Nonetheless, it's necessary to continue studying their real biological properties on in vivo models, determine secondary effects, estimate lethal doses, as well as, create an international regulation of production and marketing of this type of products.

9.4.5.1 FIBERS

DF consists of plant nondigestible carbohydrates and lignin that are completely or partially fermented in the large intestine, which is divided into soluble and insoluble. Soluble fiber (pectin, inulin, oligofructose, etc.) can formed gels in the intestine and control lipid and glucose levels; insoluble fiber (cellulose, lignin) absorbs water but does not dissolve it and helps to excrete feces and controls carbohydrate-absorption time. Ingestion of DF contributes with health benefits, preventing diabetes, hyperlipidemia, and cardiovascular disorders (Buff, 2002; Borrelli et al., 2004; Narita and Inouye, 2014; Campos-Vega et al., 2015).

Coffee silverskin (CS) is a source of DF that contains 54 and 9 g/100 g of insoluble and soluble DF, respectively (Pourfarzad et al., 2013); this content is higher than other DF sources such as apples, cabbages, broccoli, carrot, wheat bran, and potato. CS fiber is composed of 30% lignin, 17.8% glucan, 4.7% xylan, 2% arabinan, 3.8% galactan, and 2.6% mannan (Narita and Inouye, 2014), and it has the potential to be used as a functional ingredient for fiber-rich food production because it has a low content of fats and reducing carbohydrates; it has a high amount of DF and a considerable antioxidant activity due to the linked antioxidant compounds to the fiber, such as Maillard products and phenolic acids (Borrelli et al., 2004).

SCG is composed by organic compounds such as lignin, hemicellulose, and other polysaccharides. Thereof some carbohydrates are extracted during coffee beverage preparation, some of them still linked with SCG matrix. Mannose, galactose, arabinose (from hemicellulose), and glucose (from cellulose) are the main sugar present in SCG fraction (Campos-Vega et al., 2015). Mussatto et al. (2011b) optimized the conditions of SCG hydrolysis to extract sugars at 163°C for 45 min, using 100 mg acid/g dry matter and 10 g/g liquid-to-solid ratio, reaching efficiencies of 100%, 77.4%, 89.5%, and 87.4% for galactan, mannan, arabinan, and hemicellulose, respectively. Arabinogalactan, galactomannan, and cellulose are the dominant polysaccharides in coffee beans; these polysaccharides can control blood glucose and insulin response. An in vivo study shown that galactomannans decrease liver and plasma cholesterol concentrations and inhibit hepatic cholesterol production (Evans et al., 1992). As mentioned, coffee by-products are a promising source of fiber to enrich foods and give them added value, mainly to exert benefits in human health.

9.4.5.2 ANTIOXIDANTS

Coffee by-products, spent grounds, and silverskin are source of antioxidants such as chlorogenic, caffeic and ferulic acid, melanoidines, anthocyanins, and lignans (Esquivel and Jiménez, 2012). Antioxidants are chemical compounds which can transfer an electron or a hydrogen atom to stabilize free radicals, reducing cellular oxidative stress. Presence of antioxidants in daily diet has been associated with prevention of chronic degenerative diseases such as diabetes, atherosclerosis, cardiovascular disease, and some types of cancer. As coffee consumption increases every year and generates millions of tons of organic residues, coffee by-products can be used to produce nutraceuticals and functional foods (Jiménez-Zamora et al., 2015).

Different investigation groups have identified and extracted bioactive compounds present in coffee by-products. Narita and Inouye (2012) applied subcritical water extraction to obtained caffeine and chlorogenic acid from

CS. Caffeine concentration was 4.4 mg/g and 1.5 mg/g for 5-caffeoylquinic acid, showing potential for its use as nutraceutical ingredient of food products. On the other hand, Bresciani et al. (2014) identified the main antioxidants in silverskin fraction as 3-caffeoylquinic acid, 5-caffeoylquinic acid, 5- and 4-feruloylquinic acid, corresponding to 74% of the total chlorogenic acids detected in the sample. The caffeine concentration was 10 mg/g, two times higher when compared to Narita and Inouye (2012). Costa et al. (2014) observed that hydroalcoholic extraction (50%w:50%e) at 40°C for 60 min showed good results for phenolics (302 mg GAE/L), flavonoids (83 mg ECE/L), tannins (0.43 mg TAE/L), and antioxidant activity (326 mg TE/L, 1791 mg SFE/L, DPPH and FRAP, respectively).

Also, spent ground coffee is considered as a source of antioxidants (21 mg GAE/g) and caffeine (1.8 mg/g) according to Zuorro and Lavecchia (2012) and Campos-Vega et al. (2015), respectively, which is higher compared to other agroindustrial wastes as carrot peel, grape pomace, and similar to kiwi and apple peel. Xu et al. (2015) optimized an extraction procedure to obtained phenolic compounds using response surface methodology, they estimated the phenolic content (88.34 mg GAE/g) and antioxidant activity (88.65 mmol TE/100 g and 38.28 mmol TE/100 g, ABTS and DPPH, respectively) at the optimized subcritical water conditions of 179°C, 36 min, and 14.08 g/L. Melanoidins have also been detected in this fraction, corresponding to 16% of dry matter whose composition has not yet been elucidated (Campos-Vega et al., 2015).

Coffee pulp is a source of chlorogenic acid, epicatechin, catechin, protocatechuic acid, and ferulic acid (Pandey et al., 2000). Coffee husk can be used as raw material to obtain anthocyanins; Prata and Oliveira (2007) quantified 19.2 mg of cyanindin-3-rutinoside/100 g. Anthocyanins are flavonoids present in grains, flowers, fruits, and vegetables that impart color and have been studied for their biological properties as antioxidant, anti-inflammatory, chemoprotective, vasoprotective, and hepatoprotective agent (Wang et al., 1997).

The higher antioxidant activity contribution in coffee by-products is mainly due to chlorogenic acid. Chlorogenic acid is an ester of caffeic and quinic acids, exhibit antimutagenic, anticarcinogenic, and antioxidant properties due to its capacity to quenched reactive species of oxygen and nitrogen (Kono et al., 1997). To show its bioactivity, chlorogenic acid must be absorbed. Olthof et al. (2001) found that 33% of chlorogenic acid were absorbed in small intestine of humans by probably two mechanisms, the first involves intact absorption of the molecule and second suggests hydrolysis of the compound to quinic and caffeic acid before absorption.

Hypotensive effect of chlorogenic acid has been evaluated in vivo. Twelve subjects ingested 140 mg of CGA daily for 12 weeks, showing a significant reduction of blood pressure compared with the placebo group ($n = 12$). The mechanism suggest that CGA scavenges superoxide anions, reducing oxidative inactivation of nitric oxide, which reacts with superoxide anions, producing peroxynitrite and reducing nitric oxide bioavailability in endothelial tissue (Thomas et al., 2001; Watanabe et al., 2006). Antioxidant activity of caffeine showed capacity to inhibit lipid peroxidation and oxidative DNA breakage by reduction of hydroxyl radicals, peroxides and singlet oxygen, considering it as a strong antioxidant as glutathione and ascorbic acid (Azam et al., 2003; Gotteland and de Pablo, 2007).

Melanoidins are bioactive molecules formed during roasting of coffee beans by Maillard reaction, which are formed through the reaction of reducing sugars with aminoacids, and give brown color to the roasted coffee (Bartel et al., 2015). Although their chemical composition is still unclear, their antioxidant mechanism is based on the ability for scavenging electrophilic species, oxygen radicals, and metal chelation (Delgado-Andrade et al., 2005; Rufián and Morales, 2007).

9.4.5.3 PREBIOTICS

A prebiotic is a nondigestible food compound that present selective stimulation of growth and/or activity of microbial species in the gut microbiota that confers health benefits to the host such as inulin, galactooligosaccharides, and oligofructose. To consider any food ingredient as a prebiotic, it has to resist gastric conditions and be unabsorbed in the small intestine but be fermented by large intestine microbiota and stimulates intestinal bacteria (Langlands et al., 2004; Al-Sheraji et al., 2013; Yasmin, et al., 2015). The prebiotic effect of coffee and coffee by-products is due to their nondigestible carbohydrates that modulate colonic microflora as well as other bowl functions as evacuation frequency (Vitaglione et al., 2012).

CS has prebiotic effect due to the presence of nondigestible polysaccharides such as arabinogalactans and galactomannans. An in vitro study demonstrated that concentration of bifidobacteria and lactobacilli, increased during hindgut fermentation because these microorganisms were capable to use CS as a source of carbon and nitrogen (Borrelli et al., 2004). Also, melanoidins could increase growth of total anaerobes, Bacteroides, Clostridia at 6 h of an in vitro digestion; meanwhile, a significant increase in Bifidobacteria was observed until 24 h of incubation (Ames et al., 1999).

Mannose, galactose, glucose, and arabinose are the main components of SCG, the high content of mannose suggests that galactomannans are the major polysaccharides present in SCG. These compounds showed prebiotic effect, causing health benefits in humans (Simões et al., 2013). Also, manno-oligosaccharides (MOS) can be obtained by thermal degradation of SCGs and increased Bifidobacterium and Eubacterium when subjects ingested 1 g/day of MOS for 2 weeks as well as improved frequency and days of defecation (Asano et al., 2004). Nondigestible polysaccharides promoted growth of Bifidobacterium by lowering pH via the production of short-chain fatty acids, and low pH decreased harmful bacteria growth, improve constipation by stimulation intestinal inner wall, and promote intestinal peristalsis (Asano et al., 2003; Umemura et al., 2004; Yasmin et al., 2015).

Coffee by-products present industrial potential as a raw material during prebiotic extraction; the polysaccharides obtained can be used as ingredient of food products, improving the taste, mouth feel, texture, and shelf-life and give a well-balanced nutritional composition. Prebiotics are used to enrich bakery, dairy, and chocolate products and their solubility allow them to be incorporated into drinks without any increase in caloric content (Al-Sheraji et al., 2013; Yasmin et al., 2015). The use of coffee by-product will reduce environmental pollution, while giving value added to various food products and will generate economic incomes for coffee producers.

9.5 OTHER HEALTH BENEFITS

Coffee and its by-products have different compounds that exert biological activity such as weight control, antihypertensive, anti-inflammatory, anticarcinogenic, antiallergic, and so on. Chlorogenic acid, the main phenolic acid in coffee, showed potential to prevent intestinal inflammation by inhibit of TNF-α and H_2O_2-induced IL-8 in vitro production (Shin et al., 2015), protects against HOCl-induced endothelial dysfunction and enhance endothelial cell survival following HOCl-induced oxidative damage (Jiang et al., 2015) and also protects neuronal cells from glutamate toxicity during ischemic stroke, offering an alternative treatment of vascular dementia and Alzheimer disease (Mikami and Yamazawa, 2015). MOS from coffee by-products showed significant reduction of weight, visceral, and subcutaneous adipose tissue in men by inhibition of liver lipogenesis due to propionic acid formation during intestinal fermentation. These results give information to use of MOS in weight management to reduce obesity in men (Salinardi et al., 2010; St-Onge et al., 2012).

Jawi et al. (2012) evaluated the antihypertensive effect and endothelial nitric oxide synthesis of anthocyanins on an in vivo model, observing a reduction in blood pressure and an increasing of eNOS expression (endothelial nitric oxide synthase). Ronceros et al. (2012) determined the effect in lowering serum lipids in a group of diabetic patients with dyslipidemia treated with anthocyanins. At the doses studied, there was a reduction in the amount of triglycerides, increase in LDH cholesterol and glucose was controlled in patients under study.

A study done by Bravo et al. (2013) demonstrated that Robusta SCG gave protection against oxidation and DNA damage in human cells (HeLa); during the evaluation, cell viability was not affected in a concentration of 1000 μg/mL. Spent coffee extracts significantly reduced the increase of ROS level and DNA strand breaks induced by H_2O_2 and also decreased photosensitizer-induced oxidative DNA damage after 24 h of exposure.

9.6 CONCLUDING REMARKS

Millions tons of agroindustrial wastes are generated every year and their management and final deposition only generate pollution. The coffee by-products (pulp, silverskin, and husk) offer potential opportunities to be used as substrate for bioprocesses, as feed for animals, organic fertilizer, organics acids, thickener, flavors and aroma compounds, and productions of molds and yeast, in the growth of larvae and worms, but mainly have a potential use as a source of bioactive compounds that improve health, their use as ingredients of functional foods lead to give added value and reduce contamination. Coffee wastes have potential to innovate strategies applied in industrial-scale processes.

KEYWORDS

- pulp
- hulls
- silverskin
- spent coffee
- fermentation

REFERENCES

Adams, M.; Dougan, J. *Waste Products: Coffee*; Springer: Netherlands, 1987; pp 257–291.

Adi, A. J.; Noor, Z. M. Waste Recycling: Utilization of Coffee Grounds and Kitchen Waste in Vermicomposting. *Bioresour. Technol.* **2009,** *100* (2), 1027–1030.

Almajano, M. P.; Carbo, R.; Delgado, M. E.; Gordon, M. H. Effect of pH on the Antimicrobial Activity and Oxidative Stability of Oil-in-Water Emulsions Containing Caffeic Acid. *J. Food Sci.* **2007,** *72* (5), C258–C263.

Almeida, A. A. P.; Farah, A.; Silva, D. A.; Nunan, E. A.; Glória, M. B. A. Antibacterial Activity of Coffee Extracts and Selected Coffee Chemical Compounds against Enterobacteria. *J. Agric. Food Chem.* **2006,** *54* (23), 8738–8743.

Al-Sheraji, S. H.; Ismail, A.; Manap, M. Y.; Mustafa, S.; Yusof, R. M.; Hassan, F. A. Prebiotics as Functional Foods: A Review. *J. Funct. Foods* **2013,** *5* (4), 1542–1553.

Ames, J. M.; Wynne, A.; Hofmann, A.; Plos, S.; Gibson, G. R. The Effect of a Model Melanoidin Mixture on Faecal Bacterial Populations in vitro. *Br. J. Nutr.* **1999,** *82* (06), 489–495.

Andrés-Lacueva, C.; Medina-Remon, A.; Llorach, R.; Urpi-Sarda, M.; Khan, N.; Chiva-Blanch, G.; Lamuela-Raventos, R. M. Phenolic Compounds: Chemistry and Occurrence in Fruits and Vegetables. *Fruit and Vegetable Phytochemicals. Chemistry, Nutritional Value and Stability*. Blackwell Publishing: Ames, IA, 2010; pp 53–80.

Antier, P.; Minjares, A.; Roussos, S.; Viniegragonzalez, G. New Approach for Selecting Pectinase Producing Mutants of *Aspergillus niger* Well Adapted to Solid State Fermentation. *Biotechnol. Adv.* **1993,** *11* (3), 429–440.

Arya, M.; Rao, L. J. M. An Impression of Coffee Carbohydrates. *Crit. Rev. Food Sci. Nutr* **2007,** *47* (1), 51–67.

Asano, I.; Hamaguchi, K.; Fujii, S.; Lino, H. *In Vitro* Digestibility and Fermentation of Mannooligosaccharides from Coffee Mannan. *Food Sci. Technol. Res.* **2003,** *9* (1), 62–66.

Asano, I.; Umemura, M.; Fujii, S.; Hoshino, H.; Lino, H. Effects of Mannooligosaccharides from Coffee Mannan on Fecal Microflora and Defecation in Healthy Volunteers. *Food Sci. Technol. Res.* **2004,** *10* (1), 93–97.

Avallone, S.; Guiraud, J. P.; Guyot, B.; Olguin, E.; Brillouet, J. M. Fate of Mucilage Cell Wall Polysaccharides during Coffee Fermentation. *J. Agric. Food Chem.* **2001,** *49* (11), 5556–5559.

Ayala-Zavala, J. F.; Vega-Vega, V.; Rosas-Domínguez, C.; Palafox-Carlos, H.; Villa-Rodriguez, J. A.; Siddiqui, M. W. and González-Aguilar, G. A. Agro-industrial Potential of Exotic Fruit Byproducts as a Source of Food Additives. *Food Res. Int.* **2011,** *44* (7), 1866–1874.

Azam, S.; Hadi, N.; Khan, N. U.; Hadi, S. M. Antioxidant and Prooxidant Properties of Caffeine, Theobromine and Xanthine. *Med. Sci. Monit.* **2003,** *9* (9), BR325–BR330.

Aziz, N.; Farag, S.; Mousa, L.; Abo-Zaid, M. Comparative Antibacterial and Antifungal Effects of Some Phenolic Compounds. *Microbios* **1997,** *93* (374), 43–54.

Badui Dergal, S. *Química de los alimentos*. Alhambra Mexicana: México, 2006; pp 507–543.

Baggenstoss, J.; Poisson, L.; Kaegi, R.; Perren, R.; Escher, F. Coffee Roasting and Aroma Formation: Application of Different Time–Temperature Conditions. *J. Agric. Food Chem.* **2008,** *56* (14), 5836–5846.

Ballesteros, L. F.; Cerqueira, M. A.; Teixeira, J. A.; Mussatto, S. I. Characterization of Polysaccharides Extracted from Spent Coffee Grounds by Alkali Pretreatment. *Carbohydr. Polym.* **2015,** *127*, 347–354.

Balzer, H. H. Acids in Coffee. In: *Coffee Recent Development*. Blackwell Science: London, UK. 2001; pp 18–32.

Bartel, C.; Mesías, M.; Morales, F. J. Investigation on the Extractability of Melanoidins in Portioned Espresso Coffee. *Food Res. Int.* **2015**, *67*, 356–365.

Bekalo, S. A.; Reinhardt, H. W. Fibers of Coffee Husk and Hulls for the Production of Particleboard. *Mater. Struct.* **2010**, *43* (8), 1049–1060.

Beliz, H.; Grosch, W.; Schieberle, P. *Food Chemistry*; Springer-Verlag: Berlin-Heidelberg, 2009; pp 1–998.

Bhoite, R. N.; Navya, P.; Murthy, P. S. Statistical Optimization, Partial Purification, and Characterization of Coffee Pulp β-Glucosidase and Its Application in Ethanol Production. *Food Sci. Biotechnol.* **2013**, *22* (1), 205–212.

Bonilla-Hermosa, V. A.; Duarte, W. F.; Schwan, R. F. Utilization of Coffee By-products Obtained from Semi-washed Process for Production of Value-Added Compounds. *Bioresour. Technol.* **2014**, *166*, 142–150.

Borrelli, R. C.; Esposito, F.; Napolitano, A.; Ritieni, A.; Fogliano, V. Characterization of a New Potential Functional Ingredient: Coffee Silverskin. *J. Agric. Food Chem.* **2004**, *52* (5), 1338–1343.

Borrelli, R. C.; Visconti, A.; Mennella, C.; Anese, M.; Fogliano, V. Chemical Characterization and Antioxidant Properties of Coffee Melanoidins. *J. Agric. Food Chem.* **2002**, *50* (22), 6527–6533.

Bradbury, A. G.; Halliday, D. J. Chemical Structures of Green Coffee Bean Polysaccharides. *J. Agric. Food Chem.* **1990**, *38* (2), 389–392.

Braham, J. E.; Bressani, R. *Coffee Pulp: Composition, Technology, and Utilization*. International Development Research Center: Guatemala, 1979; pp 1–89.

Brand, D.; Pandey, A.; Rodriguez-Leon, J. A.; Roussos, S.; Brand, I.; Soccol, C. R. Packed Bed Column Fermenter and Kinetic Modeling for Upgrading the Nutritional Quality of Coffee Husk in Solid-State Fermentation. *Biotechnol. Progress* **2001**, *17* (6), 1065–1070.

Bravo, J.; Arbillaga, L.; de Peña, M. P.; Cid, C. Antioxidant and Genoprotective Effects of Spent Coffee Extracts in Human Cells. *Food Chem. Toxicol.* **2013**, *60*, 397–403.

Bravo, J.; Juániz, I.; Monente, C.; Caemmerer, B.; Kroh, L. W.; De Peña, M. P.; Cid, C. N. Evaluation of Spent Coffee Obtained from the Most Common Coffeemakers as a Source of Hydrophilic Bioactive Compounds. *J. Agric. Food Chem.* **2012**, *60* (51), 12565–12573.

Bresciani, L.; Calani, L.; Bruni, R.; Brighenti, F.; Del Rio, D. Phenolic Composition, Caffeine Content and Antioxidant Capacity of Coffee Silverskin. *Food Res. Int.* **2014**, *61*, 196–201.

Bressani, R.; Estrada, E.; Jarquin, R. Pulpa y pergamino de café. Composición química y contenido de aminoácidos de la proteína de la pulpa. *Turrialba* **1972**, *22* (3), 299–304.

Buff, S. Trimming the Fat, Cutting the Cholesterol. *The Good Fat, Bad Fat Counter*; Macmillan: New York, NY, 2002; pp 35–37.

Buffo, R. A.; Cardelli-Freire, C. Coffee Flavour: An Overview. *Flavour Fragrance J.* **2004**, *19* (2), 99–104.

Buitrago, J.; Gallo, J. T.; Corzo, M.; Calle, N. Evaluation of Coffee Molasses in the Diet for Growing Hogs. Instituto Colombiano Agropecuario, Bogota, Colombia, Memorias ALPA **1968**, *3*, 159.

Busch, V. M.; Kolender, A. A.; Santagapita, P. R.; Buera, M. P. Vinal Gum, a Galactomannan from *Prosopis ruscifolia* Seeds: Physicochemical Characterization. *Food Hydrocolloids* **2015**, *51*, 495–502.

Buzby, J. C.; Hyman, J. Total and Per Capita Value of Food Loss in the United States. *Food Policy* **2012**, *37* (5), 561–570.

Calle, V. H. Subproductos del café. *Bol. Téc.*; *Cenicafé* **1977**, *6*, 13.

Campos-Vega, R.; Loarca-Piña, G.; Vergara-Castañeda, H.; Oomah, B. D. Spent Coffee Grounds: A Review on Current Research and Future Prospects. *Trends Food Sci. Technol.* **2015**, *45*, 24–36.

Cerqueira, M. A.; Pinheiro, A. C.; Souza, B. W.; Lima, Á. M.; Ribeiro, C.; Miranda, C.; Gonçalves, M. P. Extraction, Purification and Characterization of Galactomannans from Non-Traditional Sources. *Carbohydr. Polym.* **2009**, *75* (3), 408–414.

Cetin-Karaca, H.; Newman, M. C. Antimicrobial Efficacy of Plant Phenolic Compounds against *Salmonella* and *Escherichia coli*. *Food Biosci.* **2015**, *11*, 8–16.

Chattopadhyay, P.; Chatterjee, S.; Sen, S. K. Biotechnological Potential of Natural Food Grade Biocolorants. *Afr. J. Biotechnol.* **2008**, *7* (17), 2972–2985.

Chu, Y. F. *Coffee: Emerging Health Effects and Disease Prevention*; Wiley-Blackwell: Ames, IA, 2012; pp 1–319.

Clark, R. J. *Coffee: Botany, Chemistry and Production of Beans and Beverage*; Clifford, M. N., Willson, K. C., Eds.; Croom Helm: London, UK, 1985; pp 230–250.

Clifford, M. N.; Williams, T.; Bridson, D. Chlorogenic Acids and Caffeine as Possible Taxonomic Criteria in *Coffea* and *Psillanthus*. *Phytochemistry* **1989**, *28*, 829–38.

Costa, A. S.; Alves, R. C.; Vinha, A. F.; Barreira, S. V.; Nunes, M. A.; Cunha, L. M.; Oliveira, M. B. P. Optimization of Antioxidants Extraction from Coffee Silverskin, a Roasting By-product, Having in View a Sustainable Process. *Ind. Crops Prod.* **2014**, *53*, 350–357.

Couto, S. R.; Sanromán, M. A. Application of Solid-State Fermentation to Food Industry—A Review. *J. Food Eng.* **2006**, *76* (3), 291–302.

Cowan, M. M. Plant Products as Antimicrobial Agents. *Clin. Microbiol. Rev.* **1999**, *12* (4), 564–582.

Czerny, M.; Mayer, F.; Grosch, W. Sensory Study on the Character Impact Odorants of Roasted *Arabica* Coffee. *J. Agric. Food Chem.* **1999**, *47*, 695–699.

Daglia, M.; Cuzzoni, M. T.; Dacarro, C. Antibacterial Activity of Coffee: Relationship between Biological Activity and Chemical Markers. *J. Agric. Food Chem.* **1994**, *42* (10), 2270–2272.

Davis, A. P.; Tosh, J.; Ruch, N.; Fay, M. Growing Coffee: On the Bases of Plastid and Nuclear DNA Sequences: Implication for the Size, Morphology, Distribution and Evolutionary History of Coffea. *Bot. J. Linn. Soc.* **2011**, *167*, 357–377.

Delgado-Andrade, C.; Rufián-Henares, J. A.; Morales, F. J. Assessing the Antioxidant Activity of Melanoidins from Coffee Brews by Different Antioxidant Methods. *J. Agric. Food Chem.* **2005**, *53* (20), 7832–7836.

Diaz-Giron, M. Y.; Elias-Marroquin, G. N. *Propuesta para la obtención de un colorante natural a partir de la pulpa seca de café (Tesis de Licenciatura)*; Universidad de El Salvador: San Salvador, El Salvador, 2009.

Dogasaki, C.; Shindo, T.; Furuhata, K.; Fukuyama, M. Identification of Chemical Structure of Antibacterial Components against *Legionella pneumophila* in a Coffee Beverage. *Yakugaku Zasshi: J. Pharm. Soc. Jpn.* **2002**, *122* (7), 487–494.

Egea, I.; Sánchez-Bel, P.; Romojaro, F.; Pretel, M. T. Six Edible Wild Fruits as Potential Antioxidant Additives or Nutritional Supplements. *Plant Foods Hum. Nutr.* **2010**, *65* (2), 121–129.

Espín, J. C.; García-Conesa, M. T.; Tomás-Barberán, F. A. Nutraceuticals: Facts and Fiction. *Phytochemistry* **2007**, *68* (22), 2986–3008.

Esquivel, P.; Jiménez, V. M. Functional Properties of Coffee and Coffee By-products. *Food Res. Int.* **2012**, *46* (2), 488–495.

Evans, A. J.; Hood, R. L.; Oakenfull, D. G.; Sidhu, G. S. Relationship between Structure and Function of Dietary Fibre: A Comparative Study of the Effects of Three Galactomannans on Cholesterol Metabolism in the Rat. *Br. J. Nutr.* **1992**, *68* (01), 217–229.

Farah, A. Coffee Constituents. In *Coffee: Emerging Health Effects and Disease Prevention*; Chu, Y.-F.; Ed.; Wiley-Blackwell: Ames, IA, 2012, pp 21–50.

Farah, A.; Donangelo, C. M. Phenolic Compounds in Coffee. *Braz. J. Plant Physiol.* **2006**, *18* (1), 23–36.

Farah, A.; de Paulis, T.; Moreira, D. P.; Trugo, L. C.; Martin, P. R. Chlorogenic Acids and Lactones in Regular and Water-Decaffeinated *arabica* Coffees. *J. Agric. Food Chem.* **2006**, *54*, 374–381.

Farah, A.; de Paulis, T.; Trugo, L. C.; Martin, P. R. Effect of Roasting on the Formation of Chlorogenic Acid Lactones in Coffee. *J. Agric. Food Chem.* **2005**, *53*, 1505–1513.

Franca, A. S.; Oliveira, L. S. Coffee Processing Solid Wastes: Current Uses and Future Perspectives. *Agric. Issues Pol. Ser.* **2009**, *15*, 171–205.

García, P. J. I. *Evaluación del rendimiento de extracción de pectinas en aguas mieles de beneficiado de café procedentes de desmucilaginado mecánico.* Tesis Licenciatura, Universidad de El Salvador: San Salvador, El Salvador, 2009.

Girotto, F.; Alibardi, L.; Cossu, R. Food Waste Generation and Industrial Uses: A Review. *Waste Manage.* **2015**, *45*, 32–41.

GMA (Grocery Manufactures Association). *Food Waste: Tier 1 Assessment*, 2012. http://www.foodwastealliance.org/wp-content/uploads/2013/06/FWRA_BSR_Tier1_FINAL.pdf (retrieved 02 Sept., 2015).

Gotteland, M.; de Pablo, S. Algunas verdades sobre el café. *Rev. Chil. Nutr.* **2007**, *34* (2), 105–115.

Gustavsson, J.; Cederberg, C.; Sonesson, U.; van Otterdijk, R.; Meybeck, A. *Global Food Losses and Food Waste*; Food and Agriculture Organization of the United Nations, 2011.

Habtamu, L. D. A Critical Review on Feed Value of Coffee Waste for Livestock Feeding. *World J. Biol. Biol. Sci.* **2014**, *2* (5), 072–086.

Haile, M. Integrated Volarization of Spent Coffee Grounds to Biofuels. *Biofuel Res. J.* **2014**, *1* (2), 65–69.

Hartati, I.; Riwayati, I.; Kurniasari, L. Potential Production of Food Colorant from Coffee Pulp. *Pros. Sem. Nacl. Sains Teknol. Fakult. Tek.* **2012**, *1* (1), 66–71.

Hernández, J. A.; Heyd, B.; Trystram, G. On-line Assessment of Brightness and Surface Kinetics During Coffee Roasting. *J. Food Eng.* **2008**, *87*, 314–322.

Hicks, P. A. *Postharvest Processing and Quality Assurance for Speciality/Organic Coffee Products*. The first Asian Regional Round-Table on Sustainable, Organic and Speciality Coffee Production, Processing and Marketing; FAO: Bangkok, Thailand, 26–28, 2001.

Honglian, S. Introducing Natural Antioxidants. *Antioxidants in Food: Practical Applications*; CRC Press: Boca Raton, FL, 2001; pp 147–158.

Horvat, E. A. G.; Grela, C. A.; Latapie, K. I. D.; Morales, D. Y. Influencia de la ingesta de cafeina en estudiantes de 6 año de la facultad de medicina de la Universidad Nacional del Nordeste. *Rev. Posgrad. VI Cát. 4 Med.* **2005**, *141*, 1–6.

Ibáñez, F.; Torre, P.; Irigoyen, A. Aditivos alimentários. *Rev. divul. N + D Nutr. Diet.* **2003**, *1* (8), 1–9.

ICO. *Estadisticas del comercio*. International Coffee Organization, 2015. http://www.ico.org (retrieved Sept., 2015).

Iglesias, J.; Pazos, M.; Torres, J. L.; Medina, I. Antioxidant Mechanism of Grape Pocyanidins in Muscle Tissues: Redox Interactions with Endogenous Ascorbic Acid and α-Tocopherol. *Food Chem.* **2012**, *134* (4), 1767–1774.

Imeson, A. *Food Stabilisers, Thickeners and Gelling Agents*; John Wiley & Sons: London, 2011; pp 1–368.

Jawi, I. M.; Sutirta-Yasa, I. W. P.; Suprapta, D. N.; Mahendra, A. N. Antihypertensive Effect and eNOS Expressions in NaCl-Induced Hypertensive Rats Treated with Purple Sweet Potato. *Univ. J. Med. Dentistry* **2012,** *1* (9), 102–107.

Jiang, R.; Hodgson, J. M.; Mas, E.; Croft, K. D.; Ward, N. C. Chlorogenic Acid Improves *Ex Vivo* Vessel Function and Protects Endothelial Cells against HOCl-Induced Oxidative Damage, via Increased Production of Nitric Oxide and Induction of Hmox-1. *J. Nutr. Biochem.* **2016.** DOI:10.1016/j.jnutbio.2015.08.017.

Jiménez-Zamora, A.; Pastoriza, S.; Rufián-Henares, J. A. Revalorization of Coffee By-products. Prebiotic, Antimicrobial and Antioxidant Properties. *LWT—Food Sci. Technol.* **2015,** *61* (1), 12–18.

Kalra, E. K. Nutraceutical—Definition and Introduction. *AAPS Pharm. Sci.* **2003,** *5* (3), 27–28.

Kashyap, D. R.; Vohra, P. K.; Chopra, S.; Tewari, R. Applications of Pectinases in the Commercial Sector: A Review. *Bioresour. Technol.* **2001,** *77* (3), 215–227.

Kök, M. S.; Hill, S. E.; Mitchell, J. R. Viscosity of Galactomannans during High Temperature Processing: Influence of Degradation and Solubilisation. *Food Hydrocolloids* **1999,** *13* (6), 535–542.

Kondamudi, N.; Mohapatra, S. K.; Misra, M. Spent Coffee Grounds as a Versatile Source of Green Energy. *J. Agric. Food Chem.* **2008,** *56* (24), 11757–11760.

Kono, Y.; Kobayashi, K.; Tagawa, S.; Adachi, K.; Ueda, A.; Sawa, Y.; Shibata, H. Antioxidant Activity of Polyphenolics in Diets: Rate Constants of Reactions of Chlorogenic Acid and Caffeic Acid with Reactive Species of Oxygen and Nitrogen. *Biochim. Biophys. Acta (BBA)—Gen. Subj.* **1997,** *1335* (3), 335–342.

Kortenska-Kancheva, V. D.; Yanishlieva, N. V.; Kyoseva, K. S.; Boneva, M. I.; Totzeva, I. R. Antioxidant Activity of Cinnamic Acid Derivatives in Presence of Fatty Alcohol in Lard Autoxidation. *Riv. Ital. Sost. Grasse* **2005,** *82* (2), 87–92.

Kumazawa, K.; Masuda, H. Investigation of the Change in the Flavor of a Coffee Drink during Heat Processing. *J. Agric. Food Chem.* **2003,** *51,* 2674–2678.

Langlands, S. J.; Hopkins, M. J.; Coleman, N.; Cummings, J. H. Prebiotic Carbohydrates Modify the Mucosa Associated Microflora of the Human Large Bowel. *Gut* **2004,** *53* (11), 1610–1616.

Lanza, E.; Ko, K. H.; Palmer, J. K. Aroma Production by Cultures of *Ceratocystis moniliformis. J. Agric. Food Chem.* **1976,** *24* (6), 1247–1250.

Laranja, A. T.; Manzatto, A. J.; Campos Bicudo, H. E. M. D. Effects of Caffeine and Used Coffee Grounds on Biological Features of *Aedes aegypti* (Diptera, Culicidae) and their Possible Use in Alternative Control. *Genet. Mol. Biol.* **2003,** *26* (4), 419–429.

Laranjinha, J.; Almeida, L.; Madeira, V. Reduction of Ferrylmyoglobin by Dietary Phenolic Acid Derivatives of Cinnamic Acid. *Free Rad. Biol. Med.* **1995,** *19* (3), 329–337.

Liu, J.; Lundqvist, J.; Weinberg, J.; Gustafsson, J. Food Losses and Waste in China and their Implication for Water and Land. *Environ. Sci. Technol.* **2013,** *47* (18), 10137–10144.

López-Galilea, I.; Fournier, N.; Cid, C.; Guichard, E. Changes in Headspace Volatile Concentrations of Coffee Brews Caused by the Roasting Process and the Brewing Procedure. *J. Agric. Food Chem.* **2006,** *54* (22), 8560–8566.

Machado, C. M.; Soccol, C. R.; de Oliveira, B. H.; Pandey, A. Gibberellic Acid Production by Solid-State Fermentation in Coffee Husk. *Appl. Biochem. Biotechnol.* **2002,** *102* (1–6), 179–191.

Madhavi, D. L.; Deshpande, S. S.; Salunkhe, D. K. Food Antioxidants: Technological: Toxicological and Health Perspectives. CRC Press: New York, NY, 1995; pp 1–493.

Martinez-Saez, N.; Ullate, M.; Martin-Cabrejas, M. A.; Martorell, P.; Genovés, S.; Ramon, D.; del Castillo, M. D. A Novel Antioxidant Beverage for Body Weight Control Based on Coffee Silverskin. *Food Chem.* **2014,** *150,* 227–234.

Medeiros, A. B. P.; Christen, P.; Roussos, S.; Gern, J. C.; Soccol, C. R. Coffee Residues as Substrates for Aroma Production by *Ceratocystis fimbriata* in Solid State Fermentation. *Braz. J. Microbiol.* **2003,** *34* (3), 245–248.

Medina, I.; González, M. J.; Iglesias, J.; Hedges, N. D. Effect of Hydroxycinnamic Acids on Lipid Oxidation and Protein Changes as well as Water Holding Capacity in Frozen Minced Horse Mackerel White Muscle. *Food Chem.* **2009,** *114* (3), 881–888.

Mikami, Y.; Yamazawa, T. Chlorogenic Acid, a Polyphenol in Coffee, Protects Neurons against Glutamate Neurotoxicity. *Life Sci.* **2015,** *139,* 69–74.

Mirabella, N.; Castellani, V.; Sala, S. Current Options for the Valorization of Food Manufacturing Waste: A Review. *J. Clean. Prod.* **2014,** *65,* 28–41.

Monente, C.; Ludwig, I. A.; Irigoyen, A.; De Peña, M. P.; Cid, C. Assessment of Total (Free and Bound) Phenolic Compounds in Spent Coffee Extracts. *J. Agric. Food Chem.* **2015,** *63* (17), 4327–4334.

Moreira, A. S. P.; Nunes, F. M.; Domingues, M. R.; Coimbra, M. A. Coffee Melanoidins: Structures, Mechanisms of Formation and Potential Health Impacts. *Food Funct.* **2012,** *3,* 903–915.

Moure, A.; Cruz, J. M.; Franco, D.; Domínguez, J. M.; Sineiro, J.; Domínguez, H.; Parajó, J. C. Natural Antioxidants from Residual Sources. *Food Chem.* **2001,** *72* (2), 145–171.

Murthy, P. S. *Biotechnological Approaches to Production of Bioactives from Coffee By-Products.* Thesis Doctoral, University of Mysore, India, 2011.

Murthy, P. S.; Madhava Naidu, M. Sustainable Management of Coffee Industry By-products and Value Addition—A Review. *Resour., Conserv. Recycl.* **2012,** *66,* 45–58.

Mussatto, S. I.; Teixeira, J. A. Increase in the Fructooligosaccharides Yield and Productivity by Solid-State Fermentation with *Aspergillus japonicus* Using Agro-industrial Residues as Support and Nutrient Source. *Biochem. Eng. J.* **2010,** *53* (1), 154–157.

Mussatto, S. I.; Ballesteros, L. F.; Martins, S.; Teixeira, J. A. Extraction of Antioxidant Phenolic Compounds from Spent Coffee Grounds. *Sep. Purif. Technol.* **2011a,** *83,* 173–179.

Mussatto, S. I.; Carneiro, L. M.; Silva, J. P.; Roberto, I. C.; Teixeira, J. A. A Study on Chemical Constituents and Sugars Extraction from Spent Coffee Grounds. *Carbohydr. Polym.* **2011b,** *83* (2), 368–374.

Mussatto, S. I.; Machado, E. M.; Martins, S.; Teixeira, J. A. Production, Composition, and Application of Coffee and Its Industrial Residues. *Food Bioprocess Technol.* **2011c,** *4* (5), 661–672.

Nakamura, T.; Tokimoto, T.; Tamura, T.; Kawasaki, N.; Tanada, S. Decolorization of Acidic Dye by Charcoal from Coffee Grounds. *J. Health Sci.* **2003,** *49* (6), 520–523.

Napolitano, A.; Fogliano, V.; Tafuri, A.; Ritieni, A. Natural Occurrence of Ochratoxin A and Antioxidant Activities of Green and Roasted Coffees and Corresponding Byproducts. *J. Agric. Food Chem.* **2007,** *55* (25), 10499–10504.

Narita, Y.; Inouye, K. High Antioxidant Activity of Coffee Silverskin Extracts Obtained by the Treatment of Coffee Silverskin with Subcritical Water. *Food Chem.* **2012,** *135* (3), 943–949.

Narita, Y.; Inouye, K. Review on Utilization and Composition of Coffee Silverskin. *Food Res. Int.* **2014,** *61,* 16–22.

Navarini, L.; Gilli, R.; Gombac, V.; Abatangelo, A.; Bosco, M.; Toffanin, R. Polysaccharides from Hot Water Extracts of Roasted *Coffea arabica* Beans: Isolation and Characterization. *Carbohydr. Polym.* **1999,** *40* (1), 71–81.

Nychas, G. J.; Tassou, C. Traditional Preservatives—Oils and Spices. In *Encyclopedia of Food Microbiology*; Elsevier: San Diego, CA. 2014; Vol 3, pp. 1717–1722.

Oliveira, L. S.; Franca, A. S.; Camargos, R. R.; Ferraz, V. P. Coffee Oil as a Potential Feedstock for Biodiesel Production. *Bioresour. Technol.* **2008,** *99* (8), 3244–3250.

Olthof, M. R.; Hollman, P. C.; Katan, M. B. Chlorogenic Acid and Caffeic Acid Are Absorbed in Humans. *J. Nutr.* **2001,** *131* (1), 66–71.

Ortíz, M. A.; Vargas, M. D. C. R.; Madinaveitia, R. G. C.; Velázquez, J. A. M. Propiedades funcionales de las antocianinas. *Rev. Cien. Biol. Salud* **2011,** *13* (2), 16–22.

Padmapriya, R.; Tharian, J. A.; Thirunalasundari, T. Coffee Waste Management—An Overview. *Int. J. Curr. Sci.* **2013,** *9*, 83–91.

Pandey, A.; Soccol, C. R.; Nigam, P.; Brand, D.; Mohan, R.; Roussos, S. Biotechnological Potential of Coffee Pulp and Coffee Husk for Bioprocesses. *Biochem. Eng. J.* **2000,** *6* (2), 153–162.

Panusa, A.; Zuorro, A.; Lavecchia, R.; Marrosu, G.; Petrucci, R. Recovery of Natural Antioxidants from Spent Coffee Grounds. *J. Agric. Food Chem.* **2013,** *61* (17), 4162–4168.

Parfitt, J.; Barthel, M.; Macnaughton, S. Food Waste within Food Supply Chains: Quantification and Potential for Change to 2050. *Philos. Trans. R. Soc. B: Biol. Sci.* **2010,** *365* (1554), 3065–3081.

Pohlan, H. A. J.; Janssens, M. J. Growth and Production of Coffee. *Soil, Plant Growth Crop Prod.* **2012,** *3*, 1–11.

Pokorný, J. Natural Antioxidants for Food Use. *Trends Food Sci. Technol.* **1991,** *2*, 223–227.

Pokorny, J.; Yanishlieva, N.; Gordon, M. H. *Antioxidants in Food. Practical Applications*; CRC Press: Boca Raton, FL, 2001; pp 1–387.

Pourfarzad, A.; Mahdavian-Mehr, H.; Sedaghat, N. Coffee Silverskin as a Source of Dietary Fiber in Bread-Making: Optimization of Chemical Treatment Using Response Surface Methodology. *LWT—Food Sci. Technol.* **2013,** *50* (2), 599–606.

Prata, E. R.; Oliveira, L. S. Fresh Coffee Husks as Potential Sources of Anthocyanins. *LWT—Food Sci. Technol.* **2007,** *40* (9), 1555–1560.

Ramírez-Coronel, M. A.; Marnet, N.; Kumar-Kolli, V. S.; Roussos, S.; Guyot, S.; Augur, C. Characterization and Estimation of Proanthocyanidins and Other Phenolicsin Coffee Pulp (*Coffea arabica*) by Thiolysis-High-Performance Liquid Chromatography. *J. Agric. Food Chem.* **2004,** *52*, 1344–1349.

Ramirez-Martinez, J. R. Phenolic Compounds in Coffee Pulp: Quantitative Determination by HPLC. *J. Sci. Food Agric.* **1988,** *43* (2), 135–144.

Rathinavelu, R.; Graziosi, G. Uso alternativo potencial de detritos e subprodutos do café. Organização Internacional do Café. Trieste (Italia): ICS-Unido, Parque da Ciência Padriciano: Costa Rica, 2005; pp 1–4.

Rodrigues, F.; Palmeira-de-Oliveira, A.; das Neves, J.; Sarmento, B.; Amaral, M. H.; Oliveira, M. B. P. Coffee Silverskin: A Possible Valuable Cosmetic Ingredient. *Pharm. Biol.* **2014,** *53* (3), 386–394.

Rodríguez-Durán, L. V.; Ramírez-Coronel, M. A.; Aranda-Delgado, E.; Nampoothiri, K. M.; Favela-Torres, E.; Aguilar, C. N.; Saucedo-Castañeda, G. Soluble and Bound Hydroxycinnamates in Coffee Pulp (*Coffea arabica*) from Seven Cultivars at Three Ripening Stages. *J. Agric. Food Chem.* **2014,** *62* (31), 7869–7876.

Rodriguez Valencia, N.; Zambrano Franco, A. Los subproductos del café: fuente de energia renovable. *Avance técnicos: Cenicafé* **2010**, *3*.

Rojas, J. U.; Verreth, J. A. J.; Van Weerd, J. H.; Huisman, E. A. Effect of Different Chemical Treatments on Nutritional and Antinutritional Properties of Coffee Pulp. *Anim. Feed Sci. Technol.* **2002**, *99* (1), 195–204.

Ronceros, G.; Ramos, W.; Arroyo, J.; Galarza, C.; Gutiérrez, E. L.; Ortega-Loayza, A. G.; Palma, L. Estudio comparativo del maíz morado (*Zea mays* L.) y simvastatina en la reducción de lípidos séricos de pacientes diabéticos normotensos con dislipidemia. *Anal. Fac. Med.* **2012**, *73* (2), 113–117.

Rufian-Henares, J. A.; de la Cueva, S. P. Antimicrobial Activity of Coffee Melanoidins: A Study of their Metal-Chelating Properties. *J. Agric. Food Chem.* **2009**, *57* (2), 432–438.

Rufián-Henares, J. A.; Morales, F. J. Functional Properties of Melanoidins: *In Vitro* Antioxidant, Antimicrobial and Antihypertensive Activities. *Food Res. Int.* **2007**, *40* (8), 995–1002.

Ruosi, M. R.; Cordero, C.; Cagliero, C.; Rubiolo, P.; Bicchi, C.; Sgorbini, B.; Liberto, E. A Further Tool to Monitor the Coffee Roasting Process: Aroma Composition and Chemical Indices. *J. Agric. Food Chem.* **2012**, *60* (45), 11283–11291.

Sacchetti, G.; Di Mattia, C.; Pittia, P.; Mastrocola, D. Effect of Roasting Degree, Equivalent Thermal Effect and Coffee Type on the Radical Scavenging Activity of Coffee Brews and their Phenolic Fraction. *J. Food Eng.* **2009**, *90*, 74–80.

Saenger, M.; Hartge, E. U.; Werther, J.; Ogada, T.; Siagi, Z. Combustion of Coffee Husks. *Renew. Energy* **2001**, *23* (1), 103–121.

Salinardi, T. C.; Rubin, K. H.; Black, R. M.; St-Onge, M. P. Coffee Mannooligosaccharides, Consumed as Part of a Free-Living, Weight-Maintaining Diet, Increase the Proportional Reduction in Body Volume in Overweight Men. *J. Nutr.* **2010**, *140* (11), 1943–1948.

Sarasty-Zambrano, D. J. Alternativas de tratamiento del mucílago residual producto del beneficiado de café. *Tesis Especialista en Química Ambiental*, Bucaramanga, Colombia, Universidad Industrial de Santander, 2012; 99 p.

Shankaranand, V. S.; Lonsane, B. K. Coffee Husk: An Inexpensive Substrate for Production of Citric Acid by *Aspergillus niger* in a Solid-State Fermentation System. *World J. Microbiol. Biotechnol.* **1994**, *10* (2), 165 168.

Shen, X.; Sun, X.; Xie, Q.; Liu, H.; Zhao, Y.; Pan, Y.; Wu, V. C. Antimicrobial Effect of Blueberry (*Vaccinium corymbosum* L.) Extracts against the Growth of *Listeria monocytogenes* and *Salmonella enteritidis*. *Food Control* **2014**, *35* (1), 159–165.

Shin, H. S.; Satsu, H.; Bae, M. J.; Zhao, Z.; Ogiwara, H.; Totsuka, M.; Shimizu, M. Anti-inflammatory Effect of Chlorogenic Acid on the IL-8 Production in Caco-2 Cells and the Dextran Sulphate Sodium-Induced Colitis Symptoms in C57BL/6 mice. *Food Chem.* **2015**, *168*, 167–175.

Sikkema, J.; De Bont, J.; Poolman, B. Mechanisms of Membrane Toxicity of Hydrocarbons. *Microbiol. Rev.* **1995**, *59* (2), 201–222.

Simões, J.; Madureira, P.; Nunes, F. M.; do Rosário Domingues, M.; Vilanova, M.; Coimbra, M. A. Immunostimulatory Properties of Coffee Mannans. *Mol. Nutr. Food Res.* **2009**, *53* (8), 1036–1043.

Simões, J.; Nunes, F. M.; Domingues, M. R.; Coimbra, M. A. Extractability and Structure of Spent Coffee Ground Polysaccharides by Roasting Pre-treatments. *Carbohydr. Polym.* **2013**, *97* (1), 81–89.

Soares, M.; Christen, P.; Pandey, A.; Soccol, C. R. Fruity Flavour Production by *Ceratocystis fimbriata* Grown on Coffee Husk in Solid-State Fermentation. *Process Biochem.* **2000**, *35* (8), 857–861.

St-Onge, M. P.; Salinardi, T.; Herron-Rubin, K.; Black, R. M. A Weight-Loss Diet Including Coffee-Derived Mannooligosaccharides Enhances Adipose Tissue Loss in Overweight Men but not Women. *Obesity* **2012,** *20* (2), 343–348.

Suárez Agudelo, J. M. *Aprovechamiento de los residuos sólidos provenientes del beneficio del café, en el municipio de Betania Antioquia: usos y aplicaciones.* Tesis Doctoral, Universidad Lasallista, Caldas Antioqua, 2012.

Thomas, G. D.; Zhang, W.; Victor, R. G. Nitric Oxide Deficiency as a Cause of Clinical Hypertension: Promising New Drug Targets for Refractory Hypertension. *J. Am. Med. Assoc.* **2001,** *285* (16), 2055–2057.

Toci, A.; Farah, A.; Trugo, L. C. Efeito do processo de descafeinação com diclorometano sobre a composição química dos cafés arábica e robusta antes e após a torração. *Quím. Nova* **2006,** *29* (5), 965–971.

Trejo-Hernandez, M. R.; Oriol, E.; Lopez-Canales, A.; Roussos, S.; Viniegra-González, G.; Raimbault, M. Pectic Enzymes Production by *Aspergillus niger* in Solid State Fermentation. *Micol. Neotrop. Apl.* **1991,** *4*, 49–62.

Umemura, M.; Fuji, S.; Asano, I.; Hoshino, H.; Lino, H. Effect of Coffee Mix Drink Containing Mannooligosaccharides from Coffee Mannan on Defecation and Fecal Microbiota in Healthy Volunteers. *Food Sci. Technol. Res.* **2004,** *10* (2), 195–198.

Vermerris, W.; Nicholson, R. *Phenolic Compound Biochemistry.* Springer: Florida. 2007; pp 1–267.

Vitaglione, P.; Fogliano, V.; Pellegrini, N. Coffee, Colon Function and Colorectal Cancer. *Food Funct.* **2012,** *3* (9), 916–922.

Watanabe, T.; Arai, Y.; Mitsui, Y.; Kusaura, T.; Okawa, W.; Kajihara, Y.; Saito, I. The Blood Pressure-Lowering Effect and Safety of Chlorogenic Acid from Green Coffee Bean Extract in Essential Hypertension. *Clin. Exp. Hypertens.* **2006,** *28* (5), 439–449.

Welsh, F. W.; Murray, W. D.; Williams, R. E.; Katz, I. Microbiological and Enzymatic Production of Flavor and Fragrance Chemicals. *Crit. Rev. Biotechnol.* **1989,** *9* (2), 105–169.

Wang, H.; Cao, G.; Prior, R. L. Oxygen Radical Absorbing Capacity of Anthocyanins. *J. Agric. Food Chem.* **1997,** *45* (2), 304–309.

Xu, H.; Wang, W.; Liu, X.; Yuan, F.; Gao, Y. Antioxidative Phenolics Obtained from Spent Coffee Grounds (*Coffea arabica* L.) by Subcritical Water Extraction. *Ind. Crops Prod.* **2015,** *76*, 946–954.

Yasmin, A.; Butt, M. S.; Afzaal, M.; van Baak, M.; Nadeem, M. T.; Shahid, M. Z. Prebiotics, Gut Microbiota And Metabolic Risks: Unveiling the Relationship. *J. Funct. Foods* **2015,** *17*, 189–201.

Yen, W. J.; Wang, B. S.; Chang, L. W.; Duh, P. D. Antioxidant Properties of Roasted Coffee Residues. *J. Agric. Food Chem.* **2005,** *53* (7), 2658–2663.

Yoo, K. M.; Song, M. R.; Ji, E. J. Preparation and Sensory Characteristics of Chocolate with Added Coffee Waste. *Korean J. Food Nutr.* **2011,** *24* (1), 111–116.

Zhangjiajie Hengxing Biotechnology Co. Ltd. http://www.china-zjjhxbio.com/news_detail_ en/id/4.html (revisado 20 Sept. 2015).

Zuorro, A.; Lavecchia, R. Spent Coffee Grounds as a Valuable Source of Phenolic Compounds and Bioenergy. *J. Clean. Prod.* **2012,** *34*, 49–56.

CHAPTER 10

DIETARY FIBER AND LYCOPENE FROM TOMATO PROCESSING

DANIEL LIRA-MORALES, MAGALY B. MONTOYA-ROJO, NANCY VARELA-BOJÓRQUEZ, MIRIAN GONZÁLEZ-AYÓN, ROSABEL VÉLEZ-DE LA ROCHA, MERCEDES VERDUGO-PERALES, and J. ADRIANA SAÑUDO-BARAJAS*

Centro de Investigación en Alimentación y Desarrollo A.C., Culiacán, Sinaloa, CP 80110, México

*Corresponding author. E-mail: adriana@ciad.mx

CONTENTS

ABSTRACT

The use of tomato by-products promotes the development oriented toward sustainable transformation of natural resources that complement the production chain, creating a new industrial cluster in which this waste might be used for processing into products of high value-added technologies. This creates an opportunity of technological diversification and creates a favorable environment for the communities surrounding the processing plants of fruits and vegetables. Moreover, the products generated can meet the market demand for ingredients whose nutritional and functional characteristics classify them in the market for nutraceuticals that currently have an apparent acceptance by society and its consumption is increasing. It is necessary to develop low-cost and high-impact alternatives to meet business needs and reduce environmental problems. Here, we discuss the feasibility and potential utilization of by-products derived from tomato industrialization, with emphasis in the use of the whole residual biomass as a dietary fiber, a functional ingredient in global demand. Addressing the problems of environmental impact of waste processing, a low-cost operation would be explored to add value to the processing chain and contribute to human health since might represent an alternative of fiber source diversification as well as improved biological functionality.

10.1 BOTANICAL DESCRIPTION OF TOMATO

The cultivated tomato (*Solanum lycopersicum* L.), is a member of the diverse family Solanaceae where 12 wild relatives are grouped as a section *Lycopersicon*. The family Solanaceae comprises more than 3000 species widespread in variated habitats, highlighting some crops considered to be of economic and nutritional importance like potatoes, peppers, and eggplants.

10.1.1 ORIGIN OF TOMATO

Tomato is an important crop worldwide grown with origins associated to different regions of Central and South America; Ecuador, Peru, and northern Chile have been proposed as centers of origin and Mexico is likely the country of domestication (Jenkins, 1948; Peralta et al., 2006).

Botanically, tomato belongs to the *Solanaceae* family as well as potatoes, eggplant, peppers and tobacco, all with economic importance. Tomato was previously named as *Lycopersicum esculentum* but Peralta and Spooner (2007) proposed *S. lycopersicum* based on profiles obtained from phylogenetic and genomic studies.

10.1.2 MORPHOLOGY OF TOMATO

The tomato is a berry with a variety of sizes, shapes, and colors, and this will depend on the variety of provenance. Tomatoes can shape as sphere, oblate, pear, plum, and bell pepper (Solanke and Kumar, 2013) in addition to wide range of colors (red, yellow, orange, and purple) and even striped tomatoes (Wellbaum, 2015). Tomato can contain a variable number of carpels, which, depending on the species of tomato, can be very prominent or barely be noticed. Each locule contains the seeds attached to the placenta and is surrounded by the locular gel (Fig. 10.1). The septum separates the locules and connects in the center of the fruit with the columella. In the outer layer is found the epicarp or skin, which forms a protective barrier with the surrounding environment. The area between the epicarp and locules is called outer pericarp (Yahia and Brecht, 2012). The tomato is a climacteric fruit, meaning that ethylene exposition produces a high response at the level of respiration and autocatalytic ethylene production, in consequence, impacting the physical appearance and compositional attributes (Ezura and Hiwasa-Tanase, 2010). Ethylene triggers a series of signals in the hormonal responsive chain producing irreversible changes in color, firmness, size, aroma, and flavor (Alexander and Grierson, 2002).

10.1.3 TOMATO PLANT

The tomato comes from an herbaceous perennial plant that grows mainly in warm and tropical climate at optimal temperature between 20 and 30°C, variations above 35°C might reduce yield and affect optimal fruit quality but temperature below 10°C delays seed germination and produces inadequate vegetative growth (Jones, 2007). The water requirements for cultivation are high, as the growth and development require a turgid condition, but flooding might be avoided to prevent oxygen starvation. Maintenance of adequate soil moisture is very important for seed germination (Picken et al., 1986). In relation to carbon fixation, the tomato plant performs a C3 metabolism, in consequence forming the first photosynthate as a three-carbon molecule (Jones, 2007). According to their growth habits, tomato plants are divided into determinate and indeterminate; determinate cultivars are grown mainly in areas with shorter warm seasons, and usually they have a tendency to faster and homogeneously fruit ripening, whereas indeterminate cultivars are more common in places with proper temperature ranges throughout the year, plants are taller and fruit ripening is less homogeneous (Wellbaum, 2015; Jones, 2007). An important feature is that determinate plants contains a main stem ending in flower while indeterminate plant follows its growth because they do not contain a flower in the main shoot (Coen and Nugent, 1994). Developmentally, tomato has a sympodial organization and the cymose inflorescences are formed after three leaves. The bright yellow flowers are in clusters at the end of the side branches; are hermaphroditic or bisexual; therefore, being capable for self-pollination (Pnueli et al., 1998; Solanke and Kumar, 2013).

10.2   FRUIT CHEMICAL COMPOSITION

Tomato has been studied from the different factors implied in its development, fructification, chemical composition, and bioactive constituents, including general composition, sugars, and organic acids, carotenoids, phenolics, minerals, and others (Alam et al., 1980; Acosta-Quezada et al., 2015).

In general, the content of chemical compounds in the fruit changes in function of different intrinsic and extrinsic factors, that is, variety, ripening stage, and preharvest crop management (Bruyn et al., 1971; Khan et al., 2013). In addition, chemical composition varies depending on the part analyzed of the fruit, such as epicarp, pericarp, locular gel, locule, and placenta (Fig. 10.1) (Moretti et al., 1998). Some of the main chemical components of tomato are mentioned here, for example, major composition of the fruit are sugars, the content of D-glucose, D-fructose, sucrose, ketoheptose, and raffinose represents more than 60% of the solids (Moretti et al., 1998). The starch content of tomato varies from 1% to 1.22%, and it depends upon maturity, cultivar, and ripening conditions. Pectins are associated to fruit texture and its solubilization during ripening concurred with a progressive softening (Bruyn et al., 1971). The content of ascorbic acid or vitamin C varies with the cultivars and increase with the maturation (Khan et al., 2013). The acidity in general increase to rise a maximum value and then reduce mainly as an effect of the acid consumption, that is, citric and malic acid, however, other acids such as acetic, formic, *trans*-aconitic, lactic, fumaric, and galacturonic have been detected. Fatty acids (linoleic, linolenic, oleics, stearic, palmitic, and myristic acids) and triglycerides, diglycerides, esterified, and nonesterified conform the lipid fraction (Salunkhe et al., 1974). During the ripening, different compounds are responsible of color, such as chlorophyll *a* and *b* cause the green color to immature tomato, while carotenoids as well as the ratio of the dominant pigments, lycopene (red color) and β-carotene (yellow color) are responsible for the pink-red color of ripening stages. The mineral uptake and availability increases during tomato growth and maturation, increasing cell organization and permeability, acid–base balance, and enzymatic systems are activated. Regarding aroma components, overall 65 carbonyls, 34 hydroxyl, 19 esters, 18 acids, 14 hydrocarbons, 6 nitrogenated, 5 lactones, 4 acetals and ketals, 4 sulfurs, 3 ethers and 3 chlorides compounds have been reported (Salunkhe et al., 1974).

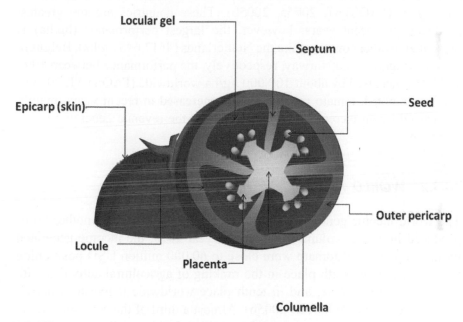

FIGURE 10.1   Tomato structure and spatial organization.

The relative concentration of chemical constituents of tomato is important for quality evaluation with respect to color, texture, appearance, nutrition value, taste, and aroma. However, there are many factors such as cultivar, cultivation method, region, and date of sampling that influence the chemical composition of tomatoes. The ripening stage decisively influences the chemical composition of the tomato (Watada and Aulenbachm, 1979; Hernández Suárez et al., 2008).

10.3   ECONOMIC IMPORTANCE OF TOMATO

10.3.1   WORLDWIDE TOMATO PRODUCTION

Tomato production is a very important activity in the world. In the year 2013, around 164 million ton worldwide were produced; with a performance of 346,983 hg/ha, tomato is placed in the position number 10 in the ranking of products at the global level. The total of the tomato production globally in a period of 20 years (1993–2013) has increased approximately 100%, this mainly due to the contribution of the Asian with around 50% of the world total production on that 20 years period. Only in the year 2013, China produced around 50.5 million tons, followed by India, United States, Turkey, and Egypt (FAOSTAT, 2005a, 2005b). These countries are the greatest producers in recent years; however, the largest performance (hg/ha) is presented in other countries like the Netherlands (4617,645 hg/ha), Belgium, United Kingdom, and Norway, respectively, the performance between 1993 and 2013 increased by about 100,000 hg/ha worldwide (FAOSTAT, 2005a). As can be noted, tomato production has increased in recent years, mainly influenced by an increase in the demand and the revenue generated by its marketing.

10.3.2   WORLD MARKET OF TOMATO

The world income generated by the tomato market exceeds another crops produced in greater volume. Only in the year 2013, the income generated by the cultivation of tomato were close to 60,000 million USD positioning this activity in the sixth place in the ranking of agricultural activities with highest economic value, and in tenth place worldwide if livestock activities are included (FAOSTAT, 2005b). Almost a third of the economic value ($18,682 million USD) is due to the market of China, followed by others such as India ($6736 million USD), United States ($4647 million USD), Turkey ($3712 million USD), and Egypt ($3,153 million USD), the values of the market of these last four countries coupled is equal to one-third of global value, leaving the rest of the income from the market to the rest of countries producers of tomato (FAOSTAT, 2005c). The increase of production and high income generated by the cultivation of tomato is due to domestic consumption and trade.

10.3.2.1   IMPORTS

In several countries of the world, the amount of tomato locally produced does not supply the domestic demand, demanding the importation from international markets. In the year 2012, about 7 million tons of tomatoes worldwide were imported; this market had a value of $8.483 million, with an estimated price of $1205.38 per hectare (FAOSTAT, 2005d). During the period of 20 years (1992–2012), there has been an increase in the imports of approximately 4.2 million tons of tomato, from 2.8 million ton in 1992 to 7 million tons in the year 2012. During that period, the increase in imports was mainly due to industrialized countries like United States which besides being the main global producers is also the largest importer of this crop (898,143 t), followed by Germany (626,637 t), France (419,153 t), Russia (347,767 t), and the United Kingdom (339,046 t). The high demand, caused by the increase in the import of the industrialized countries, has caused the increase in production values and pours out global economic during the last two decades (FAOSTAT, 2005e). In 2014, the trade value of imports of tomato was $7898 million USD with a trade of 5941,623 t worldwide (UN, 2005).

10.3.2.2   EXPORTS

International providers must supply the fresh tomatoes growing needs in some industrialized countries. World exports of tomato in 2012 were 7.25 million tons, valuated approximately around $8.180 million ($1126.54 per ton (FAOSTAT, 2005f), but in 2014, it was 7.23 million tons and the global exportation value reached $8602 million worldwide (UN, 2005). Between the years 1992 (2.47 million ton) and 2012 (7.26 million ton), the export of tomato rose steadily near to 5 million tons, the largest amount provided by Mexico (886,000 t), Spain (848,000 t), and the Netherlands (738,000 t) (UN, 2005). For example, the global consumption of tomato increased in the last 20 years. Since the year 1991, the per capita consumption of tomato has been increased from 1247 to 2023 kg/person/year in 2011 (FAOSTAT, 2005g). In conclusion by reviewing the information mentioned above, the volume of tomato exportation versus importation is needed to satisfy the consumption demand of the world's population.

10.4   TOMATO INDUSTRY

As reviewed above, the tomato production at the global level is of great relevance to international markets, both exports and imports, which generate a great amount of revenue to the economies of the countries. The volume of production responds to an increase in demand, not only due to the raw consumption, but also to the wide varietal offer, improved shelf life, and quality characteristics impacting favorable the final processed product. The processing tomato industry has broad applications for the production of paste, ketchup, concentrates, juices, and purees.

10.4.1   TOMATO PASTE

The tomato paste is obtained from the pulp but the skin and seeds are removed; this paste serves for the elaboration of other products through different processes. One of the most important parameters required for tomato paste production is the sweetness expressed in Brix (around 5–6.5°). The overall process has three general steps (1) *removal*, which consists of fruit grinding, subsequent thermal treatment, seeds and skin removal; (2) *concentration*, consisting of water evaporation to concentrate solids; and (3) *packing*, where finally the product is packed according to marketing requirements (FAO, 2009). During 2013, the main producers of tomato paste were China (1066,000 t), Italy (490,100 t), the United States (358,000 t), Turkey (333,343 t), and Iran (194,487 t) (FAOSTAT, 2005a).

10.4.2   KETCHUP

The production of ketchup is derived from the thermic concentrated pulp of tomato, usually flavored with other ingredients to confer the characteristic taste. Tomatoes cultivated for this purpose, has strict requirement of content of solids, in this case around 28–36°Brix. The process for ketchup production is divided into three stages (1) the concentrated tomato paste is flavored by the addition of various ingredients, (2) the mix is heated and sieved, and (3) the mix is packed for sale in the market (FAO, 2009). In 2014, the top five exporters of ketchup were the United States (320,230 t), the Netherlands (246,983 t), Italy (97,816 t), Germany (65,187 t), and Spain (64,719 t) (UN, 2005), while the main importers were France (164,973 t), Canada (164,465 t), United Kingdom (154,150 t), Germany (109,257 t), and Mexico (36,234 t) (UN, 2005).

10.4.3   TOMATO PUREE

The tomato puree is defined as a juice that contains pulp in its composition. The raw material for the preparation of puree is very varied and it is generally used the material that is not marketable due to the damage. The preparation of puree continues to be a process similar to that of the production of pasta, with a grinding process, warming after separation of the material that is not necessary, and packaging (FAO, 2009).

10.4.4   TOMATO JUICE

The juice is the product of maceration with filtration step to eliminate pulp residues, it might contain or not additives to enhance the flavor. The juice can be obtained by direct extraction from the raw fruit or as a secondary product from puree processing. When puree is used, the juice production passes through three stages: (1) the puree is mixed and filtered out; (2) the material is heated and (3) the product is filtered and packed (FAO, 2009). In 2014, the main exporters of tomato juice were Egypt (12,921 t), Ukraine (11,749 t), Spain (11,348 t), Germany (10,769 t), and Italy (10,087 t) (UN, 2005); on the other hand, the main importers were United Kingdom (26,654 t), Belarus (8978 t), Germany (8054 ton), France (4538 t), and Netherlands (4389 t) (UN, 2005). An important remark is that the industry of tomato processing is very important worldwide because of the large volumes of products that are imported and exported at the international market level.

10.5   BY-PRODUCTS OF TOMATO INDUSTRY

As we discussed previously, tomato is one of the most important marketable fruit worldwide, not only for its generalized fresh consumption, but also for its uses for the industrialization as a raw material to transform it into a great variety of products like paste, juice, and ketchup. This industry generates a representative source of products with a wide range of applications. In terms of yield of by-product, the large volume of tomato employed for industrialization to supply the demand of processed products generates a considerable amount of waste (Fig. 10.2), and currently, this represents an emerging market due to the compounds contained and the varied applications, from biofuel production (Rossini et al., 2013), to the exploitation of bioactive phytonutrients (Kalogeropoulos et al., 2012), or food additives (García Herrera et al., 2010; Grassino et al., 2016).

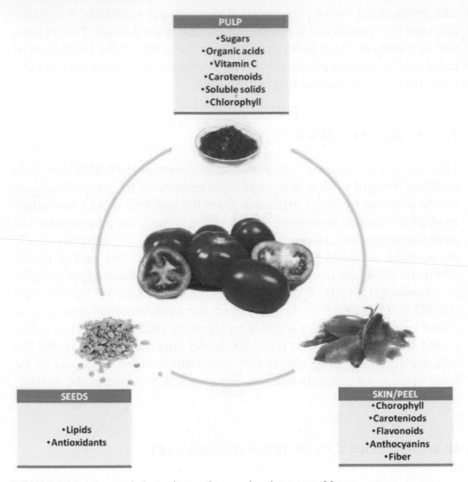

FIGURE 10.2   Tomato industry by-products and major composition.

10.5.1   TOMATO PEEL AND SEED

In the tomato industry, one of the main by-products generated are peel and seeds, with a proportion around 26.2% and 73.8%, respectively, this by-products possess high potential for the extraction of different high-value products (FAO, 2009; Sarkar and Kaul, 2014), for example, cellulose (around 62%) from tomato peel (Jiang and Hsieh et al., 2015). It is important to recognize that the success of the utilization of by-products is dependent of the process subjected, that is, when the tomato waste derives from a process where blending was used as a first operation and then was heated, the amount of vitamin C, polar isoprenoids, sucrose, alanine, gamma-aminobutyric acid and trigonelline are negatively affected in contrast to the amount of β-carotene and adenosine (Lopez-Sanchez et al., 2015). The main compounds found in tomato waste are fiber, fat, antioxidant, phenolic compounds, lycopene, β-carotene, and amino acid; those products alone are significant in the health industry and an opportunity for a variety of studies and applications.

10.5.2 FIBER

Around 86% of the total tomato peel is composed of dietary fiber, mainly in the form of insoluble dietary fiber (72%), while a minor fraction is soluble dietary fiber (14%) (García Herrera et al., 2010; Navarro-González et al., 2011), both containing neutral sugars as mannose, xylose, arabinose, and glucose, respectively. In addition to carbohydrates, dietary fiber is constituted of proteins (13.3%), moisture (5.7%), lipids (6%), and ash (3%) (Navarro-González et al., 2011).

10.5.3 FAT

The amount of fat varies according to the material, for example, the content in seeds is around 20–21% (Sarkar and Kaul et al., 2014; Mangut et al., 2006), while the peel might be composed around 3.4% (Mangut et al., 2006). The specific profile of fatty acids in seeds is around 16 different compounds; the importance of tomato fat is that 82% are of unsaturated nature, that is, linoleic (56–57%), oleic (21–22%), and palmitic (12–14%) acid, respectively (Giannelos et al., 2005; Lavelli and Torresani, 2011).

10.5.4 ANTIOXIDANTS AND PHENOLIC COMPOUNDS

The antioxidant effect of by-products of tomato has been demonstrated; freeze-dried tomato peel presented 37.5% of radical inhibition per g by DPPH (1,1-diphenyl-2-picryl-hydrazyl), meanwhile tomato seed meal presented an antioxidant activity measured by DPPH of 21% of radical inhibition by per g (Sarkar and Kaul et al., 2014). Hydrophilic extraction of tomato waste resulted in different polyphenol content; for freeze-dried tomato peel, it was around 38.6 mg of tannic acid equivalent (TAE) 100 g^{-1}, meanwhile a minor content of 20.1 mg TAE/100 g was presented in tomato seed meal (Sarkar and Kaul et al., 2014). In another study, where both main by-products were combined, a total of 9452 mg of gallic acid equivalent (GAE)/kg were obtained (Kalogeropoulos et al., 2012). Another compounds present in tomato by-products are rutin 26 mg/kg dw (dry weight) and chlorogenic acid 121 mg/kg dw (Lavelli and Torresani, 2011). Tomato peel fiber was characterized with 158 mg GAE mg/100 g and showed antioxidant activity mainly due to polar compounds to 3.9 µmol Trolox equivalent antioxidant capacity (TEAC)/g (Navarro-González et al., 2011).

10.5.4.1   LYCOPENE AND β-CAROTENE

In tomato and tomato waste, most of the total correspond to carotenoids (Strati and Oreopoulou, 2011; Rao et al., 1998), of which around 98% is lycopene (Rao et al., 1998), in tomato waste a yield of 413 mg/kg dw of lycopene and 149 mg/kg dw of β-carotene is available to obtain. In tomato peel of fresh tomato, lycopene can be found 27.5 µg/g, and some methods of drying have influence in the content of lycopene able to be extracted, for example, for freeze-drying, a recovery of 90% of lycopene is available (Sarkar and Kaul et al., 2014). Some studies realized that the case of extraction of lycopene had values as high as 1.98 mg/100 g (Kaur et al., 2008). Another kind of extraction with a mix of solvents hexane–ethyl acetate (50:50) gave a yield of 36.5 mg/kg of carotenoids, mainly constituted of lycopene (83%), β-carotene (13%), and lutein (4%) (Strati and Oreopoulou, 2011). From raw heat-treated tomato, a yield of β-carotene 116 mg/kg dw and lycopene 6420 mg/kg dw was made (Lavelli and Torresani, 2011). On the other hand, enzymatic extraction of lycopene is able to obtain around 50% of total available (Rossini et al., 2013). The above mentioned show the potential use of tomato waste as source of lycopene and carotenoids with a large range of possibilities of use in the health and food industry.

10.5.5   AMINO ACIDS

Different studies have evaluated the content of proteins and amino acids of tomato waste for potential use. Tomato seeds are rich in the proteins globulin, albumin, and gliadin (respectively, from higher to lowest), and the waste of this seed contains mainly insoluble glutenin (Sogi et al., 2002). Analyzing the composition of amino acids also in seeds, the results showed presence of histidine (25.01 mg/g), isoleucine (49.3 mg/g), leucine (77.9 mg/g), and lysine (59.63 mg/g); as well as some sulfurated (30.58 mg/g) and aromatic amino acids (87.25 mg/g), threonine (36.49 mg/g), tryptophan (12.36 mg/g), and valine (55.196 mg/g). It is important to highlight that evaluated amino acids were present in quantities higher than those required for international health and nutrition organisms (Sarkar and Kaul et al., 2014).

Therefore, tomato is source of a broad diversity of phytonutrients, some of them retained in the residual by-products and industrially extractable to add value to the chain of primary transformation. Some of this phytonutrients are highly valued and meet with the actual trends of consumption of functional compounds and the awareness of the public healthy lifestyle; those compounds are of interest as an emergent and growing market, like lycopene and dietary fiber, being the last one the most abundant by its compositional nature.

10.6   DIETARY FIBER: USES AND HEALTH BENEFITS

10.6.1   DIETARY FIBER

The recent conception of dietary fiber is accepted for the major associations and researchers around the world. According to the scientific committee of the American Association of Cereal Chemists, dietary fiber is the edible parts of plants or analogous carbohydrates that are resistant to digestion and absorption in the human small intestine with complete or partial fermentation in the large intestine. Therefore, it includes polysaccharides, oligosaccharides, lignin, and associated plant substances promoting beneficial physiological effects in one or other way. The nature of its action is dependent of the chemical composition; among physiological benefits, laxation, reduction of blood cholesterol, and/or glucose are included (AACC, 2001). Further reconsiderations were announced in the Codex Alimentarius in 2009, consisting basically in the exclusion of oligosaccharides (shorter than 10 monomeric units) (Commission, 2009; Lattimer and Haub, 2010).

10.6.2   CHEMICAL COMPOSITION

The chemical composition of the dietary fiber is generally composed of structural polysaccharides nevertheless can be very diverse depending on its source of origin. For some types of dietary fiber, the main components are cellulose, hemicellulose, pectin, and modified-starch; for some others are β-glucans, inulin, arabinoxylans, gums, or mucilages. Oligosaccharides (larger than 9 units), considered part of dietary fiber, produce positive physiological effects, as well as modified starch, lignin, cutin, suberin, and waxes (John, 2015; Wolever, 1990). Dietary fiber is divided based in water solubility as soluble and insoluble fractions. This classification is very useful to differentiate other behaviors like fermentation rate, chemical composition, and food properties. Cellulose, lignin, and some hemicelluloses are the representative components of insoluble fiber, while pectin, gum, mucilage, and oligosaccharides are for soluble (Mudgil and Barak, 2013). Tomato dietary fiber consumption and its relation to health benefits have been studied mainly by its composition as mentioned, most of the total tomato peel is dietary fiber, with a proportion of 72% of insoluble dietary fiber and 14% of soluble dietary fiber (García-Herrera et al., 2010; Navarro-González et al., 2011).

10.6.3   HEALTH BENEFITS

Tomato fiber as well as previous documented beneficial effects of fruit-derived dietary fiber are important subject of study, some of its benefits suggested on human health are improvement of bowel function; decrement of blood cholesterol, glucose and lipids; prevention of colorectal cancer; prebiotic, anti-inflammatory and immunomodulation activity; and auxiliary in weight reduction; therefore, intake of adequate quality and quantity of dietary fiber has been associated with good health and a general human wellness (Liu et al., 2015; Kaczmarczyk et al., 2012; Viuda-Martos et al., 2010; Jenkins et al., 2004).

10.6.4   GASTRIC HEALTH

Water-binding capacity is the most important characteristic that permit the dietary fiber to be useful for the improvement of bowel and gastric functionality. The insoluble fiber increases fecal volume and weight; this improves frequency and stool consistency and promotes a retard in the time of intestinal transit, and all these changes cause easy defecation and prevent disorders in the gastric functionality. To evaluate the effects of different dietary fiber source in bowel function, it was observed that although inulin causes no increase fecal weight, as grains and polydextrose, it do increase stool frequency (Raninen et al., 2011), this relation between intake of fiber and inflammatory bowel disease has been observed also in another studies (Liu et al., 2015). In rats, consumption of dietary fiber from grape has been associated with the production of phenolic acids which are in contact during large periods of time in the digestive tract, mainly due to degradation of epicatechins (Tourino et al., 2009). And, it is important to note that in animal models the effect of protection of the consumption of dietary fiber is also associated with the diminishing of negative effects of some medicaments as nonsteroidal anti-inflammatory drugs recommended for arthritis (Satoh et al., 2010; Satoh et al., 2015), and some drugs used for chemotherapy as 5-fluorouracil (Deng et al., 1999).

10.6.5   CANCER AND DIETARY FIBER

One of the most important focus on dietary fiber is the investigation of the relation between its consumption and the risk of develop some types of cancer. The consumption of fiber from grains is related with lowest incidence of some types of such as small intestine cancer (Schatzkin et al., 2008). In the same note, in Japan, a study showed association between the ingestion of dietary fiber and a minor risk of development of colon cancer (Wakai et al., 2007). In a similar study developed in European countries showed the same inverse relation between daily fiber intake and colorectal cancer (Murphy et al., 2012). This relations have been proved in different populations with distinct diet like Denmark and Spain where the last one had less incidence of colorectal cancer but higher intake of dietary fiber mainly due vegetable, legumes, and fruits (Tabernero et al., 2007), the consumption of fiber of legumes and other sources was also associated lower incidence of colorectal cancer in both women (Lin et al., 2005) and men (Nomura et al., 2007). It is important to note that intake of dietary fiber is not only associated to a lower incidence of colon cancer but also lower incidence of other types like breast cancer on women during post menopause (Suzuki et al., 2008).

10.6.6   BENEFICIAL EFFECTS ASSOCIATED WITH FIBER CONSUMPTION

Many other positive effects on health has been associated with the intake of dietary fiber as part of regular diet, for example, the relation of ingest of dietary fiber and the reduction of cholesterol LDL (Brown et al., 1999), and the less incidence of another diseases like the cardiovascular and coronary ones, without distinction of soluble and insoluble dietary fiber (Threapleton et al., 2013; Wu et al., 2015), also the consumption of fiber its associated to a good prognosis of diabetes mellitus type II (Post et al., 2012); moreover, fiber is also associated to a proliferation of *Lactobacillus* in rats and it is suggested that a regular intake can improve the natural flora and gastric health (Pozuelo et al., 2012). The overall benefits of consumption of dietary fiber makes it an emerging product for its use in food as an additive, and it is possible to track this interest by following the trends of the market.

10.7 GLOBAL MARKET OF DIETARY FIBER

Dietary fiber can be obtained from a variety of sources (Hollmann et al., 2013; Ajila and Prasada Rao, 2013; Cheung, 2013; Wang et al., 2015; Jiménez-Aguilar et al., 2015) and as mentioned above has important current uses and other potential. Among the health benefits that have been associated with the consumption of dietary fiber in various models, there are some of great impact in modern society, for example, heart diseases (Wu et al., 2015), the decrease in LDL cholesterol (Brown et al., 1999), diabetes mellitus type II (Post et al., 2012), intestinal regulation (Liu et al., 2015; Pozuelo et al., 2012), anticarcinogenic potential (Wakai et al., 2007; Murphy et al., 2012; Lizarraga et al., 2011; Sanchez-Tena et al., 2013; Aune et al., 2011) in addition other relevant to the market as the regulation of appetite (Kristensen and Jensen, 2011). Due to the foregoing, the market for functional foods has paid special attention to these products of worldwide interest; because of that, the value of fiber is tied to supply and demand.

10.7.1 DIETARY FIBER INTAKE

As demonstrated, the benefits associated with the ingestion of dietary fiber, consumption patterns have been the matter of study. In a study conducted in the United States (King et al., 2012), the consumption of dietary fiber between the population older than 18 was evaluated during the period 1999–2008, the results indicate that the pattern of consumption of dietary fiber in this country remained constant during this period, between 15 and 16 g of dietary fiber/day; at the same time, it presents an important distinction in relation to the consumption of dietary fiber and the gender of the consumer, where the highest amount of fiber during the analyzed time was part of the diet of men, in addition most of the fiber was consumed by Mexican-Americans. The study concludes that despite the proven benefit to the health of the consumption of dietary fiber, still there is a need to improve the levels of consumption in society of the United States. On the other hand, in the United Kingdom in recent years, the intake of nonstarch polysaccharides for the population equal to or greater than 19 years was ~13.8 g of dietary fiber/day, while for the population of 4–18 years was 11.1–11.8 g of dietary fiber/day (PHE, 2013). In Germany, the assessment of the levels of consumption of dietary fiber was among the population of ~19.5 g of dietary fiber/day for women and 21.8 g of dietary fiber/day for men (Linseisen et al., 2003). Taking in consideration that the American Heart Association (AHA) (Van Horn and Committee, 1997) recommends that the optimal levels of intake of dietary fiber for a healthy consumption are 25–30 g dietary fiber/day, and that in the above-mentioned countries the diet of the population is deficient in dietary fiber to achieve these levels, the market of dietary fiber as an additive has a high potential.

10.7.2   MARKET OF DIETARY FIBER

In recent years, fiber was one of the main ingredients sold in the nutraceutical market, worldwide in 2013, its value was $16.4 billion USD and in 2014. It grew to $17.5 billion USD, due to the great awareness of the health benefits of fiber consumption this market, it is predicted to have a growth rate of 7.7% annually till 2019 year where its estimated global market value will be $25.3 billion USD (Pandal, 2014). In the global market of nutraceuticals, fiber is a main ingredient for production of food and beverages, only in 2014 the revenue generated for the use of fiber as ingredient of beverages was $2.7 billion USD, similar values to 2013, in the next 5 year its projected a steady growth of 7.6% yearly to a market value of $3.9 billion USD. On the other hand, fiber leads the market as the main ingredient used in the global market of nutraceutical food with a value of $14.6 billion USD in 2014, an increment of $0.9 billion from 2013, in 2019 its estimated this market value will be around $20.9 billion USD; the main use of fiber as food ingredient was in order of market value from higher to lowest, bakery, snacks, grain and flour, confectionery, meat, cheese and frozen fruits, and vegetables (Pandal, 2014).

In 2014, the major use of this ingredient was in North America ($6.4 billion USD), followed by Europe ($5.2 billion USD) and Asia ($4.7

billion USD) (Pandal, 2014). There are many presentation of fiber and different producing companies of fiber-based products; some companies that produce major whole grain products are ConAgra, NutriTech, Cargill, meanwhile companies that produce fiber-based products (i.e., high fiber food and β-glucans) are Cargill, Alberta, Nexira, ADM, GraceLinc (Pandal, 2014).

10.8 METHODS OF LYCOPENE EXTRACTION

Tomatoes are an important part of the human diet and are commonly consumed as fresh or in processed form such as tomato juice, soup, puree, ketchup, and paste (Singh and Goyal, 2008). Tomatoes contain carotenoids that are important compounds because being historically confined to human nutrition and health. In these sense, lycopene (Fig. 10.3) represent more than 85% of total carotenoids present in this vegetable (Vasapollo et al., 2004). Commercial processing of tomato produces a large amount of waste and a major waste is skin which is particularly rich in lycopene (Topal et al., 2006). Therefore, skin can be preserved by drying and then used for lycopene extraction (Kaur et al., 2008). Various extraction techniques are available to improve the isolation and production of lycopene, and the most used extraction techniques are present in the following section.

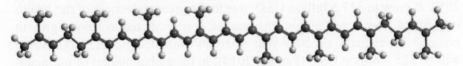

FIGURE 10.3   Model of chemical structure of lycopene.

10.8.1 SOLVENT EXTRACTION

In the past, lycopene of high purity was obtained from chemical synthesis, but today these processes are not acceptable since there are extraction processes that use natural sources for obtaining these compounds (Meyer, 1992; Ermanno et al., 2001). Within the so-called natural way, lycopene is generally extracted from plant material. The industrial process basically consists of generating a tomato paste by blending or mechanical grinding to increase efficiency of the extraction process. If the plant material is dehydrated, it requires moistening with water miscible solvent prior to extraction. Next, lycopene is extracted with organic solvent and finally the solvent is removed by evaporation (Fig. 10.4) (Ben-Yehuda et al., 2004; Levy et al., 2008).

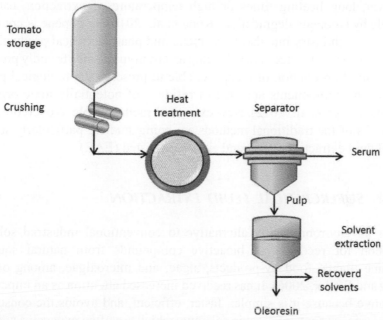

FIGURE 10.4 Simplified diagram of solvent extraction proposed for tomato lycopene.

Lycopene is fast soluble and therefore it is commonly extracted with nonpolar solvent (Kouchi et al., 2012). Thus, several organic solvents have been permitted for use in food industries such as hexane, ethyl acetate, methanol, ethanol, and acetone, but solvent mixture included hexane, acetone, and methanol or ethanol is use often (Lin and Chen, 2003; and others). Currently, ethyl acetate is most commonly used for extracting the all-trans isomers of lycopene to be used in food products. However, ethyl acetate is more efficient in extracting others carotenoids like lutein and β-carotene. Recently, ethyl lactate has been suggested as an alternative potent solvent to extract carotenoids because it has advantage in extracting all-trans lycopene isomers (Ishida and Chapman, 2009). Other normally used solvents like chloroform and tetrahydrofuran have high extraction efficiencies but have adverse effects on human health, because of its toxicity. In this sense, methods of extracting lycopene from plant material use large amounts of toxic solvents that cannot be easily removed by the process. For this reason, the extraction procedure is usually followed by a complex purification step

(Naviglio et al., 2008). Lycopene extraction solvents is a well-established method in the food industry and is advantageous compared to other methods due to low processing cost and ease of operation (Wang and Liu, 2009). However, long heating times or high temperature of extraction, can be possible by lycopene degradation (Kong et al., 2010). Lycopene is not only used in food industry but also in cosmetic and pharmaceutical productions. For these reason, it is necessary develop an environmentally friendly process for industrial production of lycopene, able to preserve the biological properties of the components and free of residues of potentially toxic organic solvents (Longo et al., 2012). New extraction methods able to overcome the drawbacks of the traditional methods are being studied, particularly supercritical fluid extraction (SFE) and enzyme-assisted (EAE).

10.8.2 SUPERCRITICAL FLUID EXTRACTION

SFE is an environmentally alternative to conventional industrial solvent extraction for recovery of bioactive compounds from natural sources including plants, food by-products, algae, and microalgae, among others (Wang and Weller, 2006). It has received increased attention as an important alternative because it is simpler, faster, efficient, and avoids the consumption of large amounts of organic solvents, which are often expensive and are completely free from residues potentially toxic (Gouveia et al., 2007). SFE is based on selective properties of the fluids, such as density, viscosity, diffusivity and dielectric constant, and usually modifies pressure and temperature to reach a supercritical fluid. Under this condition, the power of solvation increases (Machmudah et al., 2012; Strati and Oreopoulou, 2014). Basically, the fluid is delivered (solvent pump and heat exchange) into the extractor, where the solvent flows through the solid substrate. Then, the solute–solvent mixture is separate in one or more separators by rapidly reducing the pressure, increasing the temperature, or both (Separators 1 and 2). Finally, the solvent is cooled (cooler) and recompressed (compressor) and returns to the extractor (Fig. 10.5) (Pereira and Meireles, 2010).

Several hydrocarbons compounds such as hexane, pentane and butane, nitrous oxide, sulfur hexafluoride, and fluorinated hydrocarbons have been examined as SFE solvents (Reverchon and De Marco, 2006). Carbon dioxide (CO_2) is a solvent frequently used in the SFE of lycopene from a wide variety of tomato-derived substrates (especially tomato-processing by-products) due to its low critical temperature, it being neither toxic nor flammable, and its availability at low cost and high purity (Lenucci et al., 2010).

However, its low polarity reduces the effect of extraction of highly polar compounds. For this reason, the addition of a solubility enhancer, called cosolvent or modifier, can alter the characteristics of the process, and thus favor the extraction of polar compounds; cosolvents or modifiers include hexane, methanol, ethanol, isopropanol, acetonitrile, and dichloromethane, among others (Joana Gil-Chávez et al., 2013). Other solvents like ethane or propane have also been successfully applied. For example, ethane is a nonpolar solvent with a higher polarity than CO_2, which makes it presumably a better solvent for the type of carotenoids found in tomato. However, it is more expensive than CO_2, but its low critical temperature and a low critical pressure which will allow reducing the energy costs associated to the process (Nobre et al., 2012).

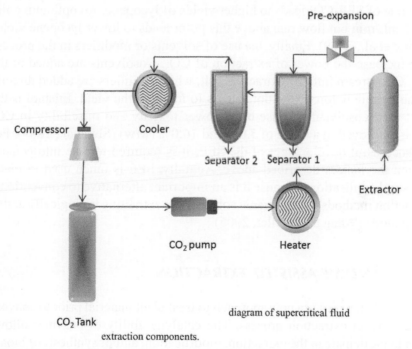

diagram of supercritical fluid extraction components.

FIGURE 10.5 Simplified diagram of supercritical fluid system components used for lycopene extraction.

Extraction parameters are directly responsible for the extract composition and component functionalities (Akanda et al., 2012). The effect of pressure and temperature are the main factors that influence on the quality of the extracts of lycopene from tomato. Extraction temperature with CO_2

varies within a wide range of 55–100°C. Notably, higher yields of lycopene can be obtained increasing the temperature of the extraction; however, it is important to consider that the isomerization and degradation of lycopene occurs at high temperatures. The effects of pressure on the extraction of lycopene from tomato by supercritical fluid extraction with CO_2 (SFE-CO_2), seem to be similar to the effects of temperature but it has been shown that 400 bar is optimum process pressure for recovering lycopene (Topal et al., 2006). Other factors may influence the quality of the lycopene extracts. Particle size of the sample may also affect the amount of lycopene obtained by SFE-CO_2. The greater extraction of lycopene results, when the particle size is smaller. This is due to higher surface to volume ratio, resulting in more solute being available for contact and interaction with SFE-CO_2 (Baysal et al., 2000). Moreover, the effect of increasing the flow rate of SFE-CO_2 leads to higher yields of lycopene. An optimum point is 2.5 mL/min but flow rate above this point leads to lower lycopene yields (Nobre et al., 2009). Finally, the use of solvents or modifiers in the process is to increase the power of extraction of CO_2. Cosolvents are added to the SFE-CO_2 stream into the extraction cell, while modifiers are added directly to the sample before extraction occurs to increase the yield. Ethanol is the most used cosolvent because of its lower toxicity and miscibility in CO_2 and is employed in a range of 5.0% and 16.0% (w/w) (Shi et al., 2009). For the successful development of the SFE, it is required to take into consideration the factors described above. Actually, SFE is much used in many industrial applications because it is an important alternative to conventional extraction methods using organic solvents for extracting biologically active compounds (Wang and Weller, 2006).

10.8.3 ENZYME-ASSISTED EXTRACTION

The use of enzymes is a new approach to treat plant material prior to conventional solvent extraction process. The catalytic ability of enzymes allows them to participate in the extraction, modification, and/or synthesis of bioactive compounds from natural source. This biotechnological method also offers a "green chemistry" concept for food industry and pharmaceutical companies (Strati and Oreopoulou, 2014).

EAE is widely used for extraction of a variety of components, including phenolic compounds, nonvolatile grape aroma precursors and carotenoids, from a variety of plant materials (Papaioannou and Karabelas, 2012). The cell wall is degraded by hydrolytic enzymes such as cellulases, hemicellulases, and pectinases. These enzymes hydrolyze complex tissue matrix of polysaccharides, thus increasing the permeability of the cell wall, resulting in higher yields of bioactive compounds extraction (Fernandes, 2010).

Enzymes have been employed for the extraction of lycopene from tomato tissues or tomato paste (Table 10.1) and several factors such as enzyme concentration, temperature, hydrolysis time, particle size or enzyme combination are recognized as key factors for extraction (Mai et al., 2013). In these sense, Ranveer et al. (2013) reported that peel of tomato had highest amount of the lycopene and the best yields are obtained with a combination of cellulase and pectinase enzymes with 1.5% and 2% concentration at 4 h incubation time. In addition, reduction in particle size improves the extraction of lycopene, whereas degradation occurs at cooking temperature. This study reports that other factors such as the type of solvent used can influence the extraction yields and suggest that the best results are obtained with a mixture of hexane:acetone:ethanol (50:25:25). Recently, "green" techniques were employed for the extraction of carotenoids and oils. In this approach, enzyme-assisted aqueous extraction is recognized as eco-friendly technology for extraction of bioactive compounds and oil because it uses water as solvent instead of organic chemicals (Puri et al., 2012).

TABLE 10.1 Summary of Cell Wall Degrading Enzymes Used for the Extraction of Lycopene from Tomato By-products.

Source	Enzyme type	Lycopene yield versus nonenzymatic treatment (folds)	Reference
Tomato peels	Pectinase	18–20×	Lavecchia and Zuorro (2008)
	Cellulase + sonication	10×	Papaioannou and Karabelas (2012)
Tomato processing wastes	Pectinase:cellulase, (50:50)	8–18×	Zuorro et al. (2011)
	Pectinase	7×	Ranveer et al. (2013)

Source: Adapted from Strati and Oreopoulou (2014).

The enzymatic-assisted extraction has several advantages such as low investment cost, environmental friendly, handling, and others. However, the enzymatic assisted-extraction obtained lowest yields and consumes more time like others method (SFE and conventional extraction) (Mai et al., 2013). The production of compounds by enzymatic extraction represents an interesting approach because when the above limitations are exceeded, the method will increase the extraction yields and improve product quality (Strati and Oreopoulou, 2014).

10.9 GLOBAL MARKET OF LYCOPENE

Carotenoids are biologically important compounds because of its anti-oxidant properties are associated with reduced risk of developing chronic degenerative diseases (Franco et al., 2013). Among these compounds is lutein, β-carotene, lycopene and others, and its demand has increased in recent years due to providing you benefit to health (Fiedor and Burda, 2014). Carotenoids are consumed around the world, mainly in Europe, Asia, North and Latin America. It is estimated that by 2018, the market for carotenoids continue the same trend in regions of major consumption and its market value reach up $ 1400 billion, being Asia the representing growing territory. The global market by carotenoid type, positions lycopene, β-carotene, and β-apo-8-carotenol ester as excellent opportunities of growing demand (BBC, nd). It should be noted that lycopene is considered one of the most popular and claimed antioxidant among carotenoids components and due to its revealed properties, this compound is considered a potential anti-cancer auxiliary (Illic et al., 2011; Trejo-Solis et al., 2013). Likewise, 85% of lycopene intake comes from tomatoes that has been used as a main source because of its content at the same time of its important cultivation extended around the world (Bramley, 2002; FAO, nd).

Lycopene is marketed as crystals and due to the extraction and purification cost; the product has always maintained a high marketing price. The price variation is also according to its manufacture method, that is, the lycopene synthetized by chemical process has a lower price than lycopene obtained via fermentation or chemical extraction from natural sources. Synthetic lycopene was priced on average $1700/kg crystalline in 2004, later in 2007 decreased its price to $1500/kg cristalline and increasing to $2000/kg crystalline in 2010. On the other hand, tomato extracts offered in average on $4500/kg in 2004, increasing until 2007 to $2500/kg and increasing again to over $3000/kg in 2010. Finally, lycopene from fermentation were offered to 2004–2010 at $2700/kg crystalline (BBC, nd).

In the past years, lycopene produced by chemical synthesis dominated the market, the development of biotechnology and fermentation technologies caused a change in the actual marketing. Despite marketing reduction in 2007, potentially caused by lower prices and lower production volumes, the world tendency over the years indicates an increase in demand of lycopene. In 2010, sales reached $ 66 million, 68% corresponding to lycopene extracted from plant material, reaching rates of 45 million. After that, lycopene obtained by chemical synthesis dominated 30% of the market with estimation of values closer to 20 million. The rest of the market belongs to fermentation technology with an estimated value of 1 million. The projected market of lycopene for 2018 is estimated above 80 million (BBC, nd).

LycoRed is having a greater presence among lycopene companies worldwide, holding around 52% of global sales. It is estimated that sales in 2010 reached 34 million. LycoRed divided its market into two segments, natural and synthetic lycopene. Then follow DSM and BASF, companies produce 38% of synthetic lycopene with estimated sales of 19 and 6 million, respectively. The rest of the market belongs to small companies, mostly of Asian origin. They added the amount of 7 million dollars in profits (BBC, nd).

The application of lycopene in the industry is divided into three major sectors, food and beverages, dietary supplements, and cosmetics. Lycopene consumption by industry has increased over the years. In this respect, the dietary supplement industry is the largest consumer of lycopene crystallized. In 2000, this sector used 6 t of lycopene produced. For 2004, the amount of lycopene produced was doubled and dietary supplement industry was the most important consumer. In 2010, over 80% of lycopene produced is used in this industrial sector, the remainder was used by the cosmetics industry and food and beverages (BBC, nd).

Lycopene consumed by industry 61% comes from the extraction of plants, especially tomatoes, the rest comes from chemical synthesis and the production of fermented lycopene is not representative in the market (BBC, nd). Lycopene is consumed around the world; however, North America (the United States and Canada) and Asia has been the most important region in the consumption of this compound. In 2010, North America and Asia consumed more than 56% and 21% of lycopene produced worldwide, respectively. On the other hand, Latin America consumed less than 1 t in the same year. This trend will continue till 2018, North America and Asia will be the most important regions in the consumption of lycopene with 52% and 24%, respectively, and Latin America will consume only 3% of this compound (BBC, nd). The crystalline lycopene must meet specific physicochemical properties like appearance, color, and range of maximum absorption for use in various regions of the world. In this sense, the lycopene must be red dust crystalline with a range of maximum absorption 471–504 nm (BBC, nd). In the United States, lycopene is considered generally recognized as safe by the Food and Drug Administration; however, its use is only permitted as a food supplement and not as a food coloring agent (BBC, nd).

10.10 CONCLUDING REMARKS

By-products constitute a serious environmental concern around the world due to two main factors: (1) an increase in the volumes of waste by the globalization of the industry and (2) emergence of stricter environmental laws. In addition to this, industrialization is promoted as an opportunity to add value to the primary horticultural production chain and there is a trend toward the consumption of functional foods, thus, emerges the need and tendency to exploite the agro-industrial wastes into useful products in demand.

Here, we discuss the feasibility and potential utilization of by-products derived from tomato industrialization, with emphasis in the use of the whole residual biomass as a dietary fiber, a functional ingredient in global demand. Addressing the problems of environmental impact of waste processing, a low-cost operation would be explored to add value to the processing chain and contribute to human health since might represent an alternative of fiber source diversification as well as improved biological functionality.

The use of tomato by-products promotes the development oriented toward sustainable transformation of natural resources that complement the production chain, creating a new industrial cluster in which this waste might be used for processing into products of high value-added technologies. This creates an opportunity of technological diversification and creates a favorable environment for the communities surrounding the processing plants of fruits and vegetables. Moreover, the products generated can meet the market demand for ingredients whose nutritional and functional characteristics classify them in the market for nutraceuticals that currently have an apparent acceptance by society and its consumption is increasing. It is necessary to develop low-cost and high-impact alternatives to meet business needs and reduce environmental problems.

Although a study of competitive intelligence (state of the art, market research, and supplier analysis) is not available for tomato fiber, enough information has been generated elsewhere providing information of competitiveness and opportunity to produce a high valuable product. We highlight the dietary fiber exploitation considering that technology to recover and process the tomato fibers according to the sanitary requirements, might be raised with low environmental impact and to meet food needs. Currently, the wheat bran fiber is the main ingredient option for food fortification; however, this precedes from different plant organs, differing to the parenchymatous cells that constitute the tomato. Therefore, the source defines the proportion of soluble and insoluble dietary fiber based in plant botanical differences and in consequence, giving the opportunity to explore source and functionality diversification. The dietary fiber linked compounds have a crucial role in functionality, emphasizing the occurrence of polymeric polyphenols or esterified phenolic acids that being present in association with cell wall components, increases the antioxidant property. Minerals are also key components of dietary fiber, and as far as we know, less information about useful benefits of silicon, boron, and calcium linked to the matrix have been documented.

ACKNOWLEDGMENTS

This document derives from critical review of information during the development of two projects supported by Fondo Mixto CONACyT-Gobierno del Estado de Sinaloa. The authors would like to thank to Jorge Humberto Siller Cepeda, Luz Gabriela Escoto González, María Julieta Acosta Barrantes, Analily Flores Álvarez and José Luis Valenzuela Lagarda as part of the team.

KEYWORDS

- functional properties
- extraction technologies
- tomato processing
- marketing

REFERENCES

AACC, The Definition of Dietary Fiber. *Cereal Foods World* **2001**, *46* (3), 112–126.

Acosta-Quezada, P. G.; Raigon, M. D.; Riofrio-Cuenca, T.; Garcia-Martinez, M. D.; Plazas, M.; Burneo, J. I.; Figueroa, J. G.; Vilanova, S.; Prohens, J. Diversity for Chemical Composition in a Collection of Different Varietal Types of Tree Tomato (*Solanum betaceum* Cav.), an Andean Exotic Fruit. *Food Chem.* **2015**, *169*, 327–335.

Ajila, C. M.; Prasada Rao, U. J. S. Mango Peel Dietary Fibre: Composition and Associated Bound Phenolics. *J. Funct. Foods* **2013**, *5* (1), 444–450.

Akanda, M. J.; Sarker, M. Z.; Ferdosh, S.; Manap, M. Y.; Ab Rahman, N. N.; Ab Kadir, M. O. Applications of Supercritical Fluid Extraction (SFE) of Palm Oil and Oil from Natural Sources. *Molecules* **2012**, *17* (2), 1764–1794.

Alam, M.; Ahmad, M.; Abrar, K. Quality and Nutritional Composition of Tomato Fruit as Influenced by Certain Biochemical and Physiological Changes. *Plant Soil* **1980**, *57*, 231–236.

Alexander, L.; Grierson, D. Ethylene Biosynthesis and Action in Tomato: A Model for Climacteric Fruit Ripening. *J. Exp. Bot.* **2002**, *53* (377), 2039–2055.

Aune, D.; Chan, D. S. M.; Lau, R.; Vieira, R.; Greenwood, D. C.; Kampman, E.; Norat, T. Dietary Fibre, Whole Grains, and Risk of Colorectal Cancer: Systematic Review and Dose-Response Meta-analysis of Prospective Studies. *BMJ* **2011**, *343* (10), d6617–d6617.

Baysal, T.; Ersus, S.; Starmans, D. A. J. Supercritical CO_2 Extraction of β-Carotene and Lycopene from Tomato Paste Waste. *J. Agric. Food Chem.* **2000**, *48* (11), 5507–5511.

BBC. *The Global Market for Carotenoids*, nd. http://www.bccresearch.com/market-research/food-and-beverage/carotenoids-global-market-fod025d.html.

Ben-Yehuda, M.; Garti, N.; Hartal, D.; Raveh, Y.; Zelkha, M. *Industrial Processing of Tomatoes and Lycopene Extraction*. Google Patents, 2004.

Bramley, P. Regulation of Carotenoid Formation during Tomato Fruit Ripening and Development. *J. Exp. Bot.* **2002**, *53* (377), 2107–2113.

Brown, L.; Rosner, B.; Willet, W.; Sacks, F. Cholesterol-Lowering Effects of Dietary Fiber: A Meta-analysis. *Am. J. Clin. Nutr.* **1999**, *69*, 30–42.

Brown, L.; Rosner, B.; Willett, W. W.; Sacks, F. M. Cholesterol-Lowering Effects of Dietary Fiber: A Meta-analysis. *Am. J. Clin. Nutr.* **1999**, *69* (1), 30–42.

Bruyn, J.; Garretsen, F.; Kooistra, E. Variation in Taste and Chemical Composition of the Tomato (*Lycopersicum esculentum* Mill.). *Euphytica* **1971**, *20* (2), 214–227.

Cheung, P. C. K. Mini-review on Edible Mushrooms as Source of Dietary Fiber: Preparation and Health Benefits. *Food Sci. Hum. Wellness* **2013**, *2* (3–4), 162–166.

Coen, E. S.; Nugent, J. M. Evolution of Flowers and Inflorescences. *Development* **1994**, *1994* (Supplement), 107–116.

Commission, C. A. *Report of the 30th session of the Codex Committee on Nutrition and Foods for Special Dietary Uses*, Cape Town, South Africa, 2009.

Deng, G.-Y.; Liu, Y.-W.; He, G.-Z.; Jiang, Z.-M. Effect of Dietary Fiber on Intestinal Barrier Function of 5-FU Stressed Rats. *Res. Exp. Med.* **1999**, *199* (2), 111–119.

Ermanno, B.; De La Vieja Alfonso J, C.; Manuel, E. M.; De Castro Antonio, E.; De Prado Emiliano, G. *Procedimiento para la obtencion de licopeno*. Google Patents, 2001.

Ezura, H.; Hiwasa-Tanase, K. *Fruit Development*; Springer: Berlin-Heidelberg, 2010.

FAO. *Agribusiness Handbook: Fruit and Vegetable Processing*. FAO: Rome, Italy, 2009.

FAO. *Crop Water Information: Tomato*, nd. http://www.fao.org/nr/water/cropinfo_tomato.html.

FAOSTAT. *Crops and Livestock Products*, 2005a. http://faostat3.fao.org/browse/T/TP/E (accessed Jul. 23, 2015).

FAOSTAT. *Crops*, 2005b. http://faostat3.fao.org/browse/Q/QC/E (accessed Jul. 23, 2015).

FAOSTAT. *Exports/Commodities by Regions*, 2005c. http://faostat3.fao.org/browse/rankings/commodities_by_regions_exports/E (accessed Jul. 23, 2015).

FAOSTAT. *Food and Agricultural Commodities Production/Commodities by Regions*, 2005d. http://faostat3.fao.org/browse/rankings/commodities_by_regions/E (accessed Jul. 22, 2015).

FAOSTAT. *Food and Agricultural Commodities Production/Countries by Commodity*, 2005e. http://faostat3.fao.org/browse/rankings/countries_by_commodity/E (accessed Jul. 23, 2015).

FAOSTAT. *Food Supply—Crops Primary Equivalent*, 2005f. http://faostat3.fao.org/browse/FB/CC/E (accessed Jul. 23, 2015).

FAOSTAT. *Imports/Commodities by Regions*, 2005g. http://faostat3.fao.org/browse/rankings/commodities_by_regions_imports/E (accessed Jul. 23, 2015).

Fernandes, P. Enzymes in Food Processing: A Condensed Overview on Strategies for Better Biocatalysts. *Enzym. Res.* **2010,** *2010,* 862537.

Fiedor, J.; Burda, K. Potential Role of Carotenoids as Antioxidants in Human Health and Disease. *Nutrients* **2014,** *6* (2), 466–488.

Franco, G.; Cartagena, J.; Correa, G. Estimating Fruit Pulp Carotenoid Content from Shell Color in Gulupa (*Passiflora edulis* Sims). *Corp. Cien. Tecnol. Agropec.* **2013,** *14* (2), 199–206.

García Herrera, P.; Sánchez-Mata, M. C.; Cámara, M. Nutritional Characterization of Tomato Fiber as a Useful Ingredient for Food Industry. *Innov. Food Sci. Emerg. Technol.* **2010,** *11* (4), 707–711.

Giannelos, P. N.; Sxizas, S.; Lois, E.; Zannikos, F.; Anastopoulos, G. Physical, Chemical and Fuel Related Properties of Tomato Seed Oil for Evaluating its Direct Use in Diesel Engines. *Ind. Crops Prod.* **2005,** *22* (3), 193–199.

Gouveia, L.; Nobre, B. P.; Marcelo, F. M.; Mrejen, S.; Cardoso, M. T.; Palavra, A. F.; Mendes, R. L. Functional Food Oil Coloured by Pigments Extracted from Microalgae with Supercritical CO_2, *Food Chem.* **2007,** *101* (2), 717–723.

Grassino, A. N.; Halambek, J.; Djaković, S.; Rimac Brnčić, S.; Dent, M.; Grabarić, Z. Utilization of Tomato Peel Waste from Canning Factory as a Potential Source for Pectin Production and Application as Tin Corrosion Inhibitor. *Food Hydrocolloids* **2016,** *52,* 265–274.

Hernández Suárez, M.; Rodríguez Rodríguez, E. M.; Díaz Romero, C. Chemical Composition of Tomato (*Lycopersicum esculentum*) from Tenerife, the Canary Islands. *Food Chem.* **2008,** *106* (3), 1046–1056.

Hollmann, J.; Themeier, H.; Neese, U.; Lindhauer, M. G. Dietary Fibre Fractions in Cereal Foods Measured by a New Integrated AOAC Method. *Food Chem.* **2013,** *140* (3), 586–589.

Illic, D.; Forbes, K.; Hassed, C. Lycopene for the Prevention of Prostate Cancer. *Cochrane Database Syst Rev* **2011,** *2011* (11), 1–27.

Ishida, B. K.; Chapman, M. H. Carotenoid Extraction from Plants Using a Novel, Environmentally Friendly Solvent. *J. Agric. Food Chem.* **2009,** *57* (3), 1051–1059.

Jenkins, D. J. A.; Marchie, A.; Augustin, L. S. A.; Ros, E.; Kendall, C. W. C. Viscous Dietary Fibre and Metabolic Effects. *Clin. Nutr. Suppl.* **2004,** *1* (2), 39–49.

Jenkins, J. A. The origin of the Cultivated Tomato. *Econ. Bot.* **1948,** *2* (4), 379–392.

Jiang, F.; Hsieh, Y. L. Cellulose Nanocrystal Isolation from Tomato Peels and Assembled Nanofibers. *Carbohydr. Polym.* **2015,** *122,* 60–68.

Jiménez-Aguilar, D. M.; López-Martínez, J. M.; Hernández-Brenes, C.; Gutiérrez-Uribe, J. A.; Welti-Chanes, J. Dietary Fiber, Phytochemical Composition and Antioxidant Activity of Mexican Commercial Varieties of Cactus Pear. *J. Food Compos. Anal.* **2015,** *41,* 66–73.

Joana Gil-Chávez, G.; Villa, J. A.; Fernando Ayala-Zavala, J.; Basilio Heredia, J.; Sepulveda, D.; Yahia, E. M.; González-Aguilar, G. A. Technologies for Extraction and Production of Bioactive Compounds to be Used as Nutraceuticals and Food Ingredients: An Overview. *Comprehens. Rev. Food Sci. Food Saf.* **2013,** *12* (1), 5–23.

John, A. M. Dietary Fiber. *Handbook of Food Analysis,* 3rd ed.—*Two Volume Set*; CRC Press: Boca Raton, FL, 2015; pp 755–767.

Jones, B. *Tomato Plant Culture: In the Field, Greenhouse, and Home Garden.* CRC Press: Boca Raton, FL, 2007.

Kaczmarczyk, M. M.; Miller, M. J.; Freund, G. G. The health Benefits of Dietary Fiber: Beyond the Usual Suspects of Type 2 Diabetes Mellitus, Cardiovascular Disease and Colon Cancer. *Metab.: Clin. Exp.* **2012,** *61* (8), 1058–1066.

Kalogeropoulos, N.; Chiou, A.; Pyriochou, V.; Peristeraki, A.; Karathanos, V. T. Bioactive Phytochemicals in Industrial Tomatoes and their Processing Byproducts. *LWT—Food Sci. Technol.* **2012,** *49* (2), 213–216.

Kaur, D.; Wani, A. A.; Oberoi, D. P.; Sogi, D. S. Effect of Extraction Conditions on Lycopene Extractions from Tomato Processing Waste Skin Using Response Surface Methodology. *Food Chem.* **2008,** *108* (2), 711–718.

Khan, I.; Azam, A.; Mahmood, A. The Impact of Enhanced Atmospheric Carbon Dioxide on Yield, Proximate Composition, Elemental Concentration, Fatty Acid and Vitamin C Contents of Tomato (*Lycopersicum esculentum*). *Environ. Monit. Assess.* **2013,** *185* (1), 205–214.

King, D. E.; Mainous, A. G., 3rd; Lambourne, C. A. Trends in Dietary Fiber Intake in the United States, 1999–2008. *J. Acad. Nutr. Diet.* **2012,** *112* (5), 642–648.

Kong, K. W.; Khoo, H. E.; Prasad, K. N.; Ismail, A.; Tan, C. P.; Rajab, N. F. Revealing the Power of the Natural Red Pigment Lycopene. *Molecules* **2010,** *15* (2), 959–987.

Kouchi, H.; Moosavi-Nasab, M.; Shabanpour, B. *Extraction of Carotenoids from Crustacean Waste Using Organic Solvent.* In The 1st International and the 4th National Congress on Recycling of Organic Waste in Agriculture, Isfahan, Iran, 2012.

Kristensen, M.; Jensen, M. G. Dietary Fibres in the Regulation of Appetite and Food Intake. Importance of Viscosity. *Appetite* **2011,** *56* (1), 65–70.

Lattimer, J. M.; Haub, M. D. Effects of Dietary Fiber and Its Components on Metabolic Health. *Nutrients* **2010,** *2* (12), 1266–1289.

Lavecchia, R.; Zuorro, A. Improved Lycopene Extraction from Tomato Peels Using Cell-Wall Degrading Enzymes. *Eur. Food Res. Technol.* **2008,** *228* (1), 153–158.

Lavelli, V.; Torresani, M. C. Modelling the Stability of Lycopene-Rich By-products of Tomato Processing. *Food Chem.* **2011,** *125* (2), 529–535.

Lenucci, M. S.; Caccioppola, A.; Durante, M.; Serrone, L.; Leonardo, R.; Piro, G.; Dalessandro, G. Optimisation of Biological and Physical Parameters for Lycopene Supercritical CO_2 Extraction from Ordinary and High-Pigment Tomato Cultivars. *J. Sci. Food Agric.* **2010,** *90* (10), 1709–1718.

Levy, J.; Walfisch, S.; Walfisch, Y.; Nahum, A.; Hirsch, K.; Danilenko, M.; Sharoni, Y. Tomato Carotenoids and the IGF System in Cancer. In *Tomatoes and Tomato Products: Nutritional, Medicinal and Therapeutic Properties*; Preedy, V., Watson, R., Eds.; Science Publishers: Enfield, New Hampshire, USA, 2008; pp 395–410.

Lin, C. H.; Chen, B. H. Determination of Carotenoids in Tomato Juice by Liquid Chromatography. *J. Chromatogr. A* **2003,** *1012* (1), 103–109.

Lin, J.; Zhang, S. M.; Cook, N. R.; Rexrode, K. M.; Liu, S.; Manson, J. E.; Lee, I. M.; Buring, J. E. Dietary Intakes of Fruit, Vegetables, and Fiber, and Risk of Colorectal Cancer in a Prospective Cohort of Women (United States). *Cancer Causes Control* **2005,** *16* (3), 225–233.

Linseisen, J.; Schulze, M. B.; Saadatian-Elahi, M.; Kroke, A.; Miller, A. B.; Boeing, H. Quantity and Quality of Dietary Fat, Carbohydrate, and Fiber Intake in the German EPIC Cohorts. *Ann. Nutr. Metab.* **2003,** *47* (1), 37–46.

Liu, J.; Willför, S.; Xu, C. A Review of Bioactive Plant Polysaccharides: Biological Activities, Functionalization, and Biomedical Applications. *Bioact. Carbohydr. Diet. Fibre* **2015,** *5* (1), 31–61.

Liu, X.; Wu, Y.; Li, F.; Zhang, D. Dietary Fiber Intake Reduces Risk of Inflammatory Bowel Disease: Result from a Meta-Analysis. *Nutr. Res.* **2015,** *35* (9), 753–758.

Lizarraga, D.; Vinardell, M. P.; Noe, V.; van Delft, J. H.; Alcarraz-Vizan, G.; van Breda, S. G.; Staal, Y.; Gunther, U. L.; Carrigan, J. B.; Reed, M. A.; Ciudad, C. J.; Torres, J. L.; Cascante, M. A Lyophilized Red Grape Pomace Containing Proanthocyanidin-Rich Dietary Fiber Induces Genetic and Metabolic Alterations in Colon Mucosa of Female C57BL/6J Mice. *J. Nutr.* **2011,** *141* (9), 1597–1604.

Longo, C.; Leo, L.; Leone, A. Carotenoids, Fatty Acid Composition and Heat Stability of Supercritical Carbon Dioxide-Extracted-Oleoresins. *Int. J. Mol. Sci.* **2012,** *13* (4), 4233–4254.

Lopez-Sanchez, P.; de Vos, R. C.; Jonker, H. H.; Mumm, R.; Hall, R. D.; Bialek, L.; Leenman, R.; Strassburg, K.; Vreeken, R.; Hankemeier, T.; Schumm, S.; van Duynhoven, J. Comprehensive Metabolomics to Evaluate the Impact of Industrial Processing on the Phytochemical Composition of Vegetable Purees. *Food Chem.* **2015,** *168,* 348–355.

Machmudah, S.; Zakaria; Winardi, S.; Sasaki, M.; Goto, M.; Kusumoto, N.; Hayakawa, K. Lycopene Extraction from Tomato Peel By-product Containing Tomato Seed Using Supercritical Carbon Dioxide. *J. Food Eng.* **2012,** *108* (2), 290–296.

Mai, H.; Truong, V.; Debaste, F. Optimisation of Enzyme-Assisted Extraction of Oil Rich in Carotenoids from Gac Fruit (*Momordica cochinchinensis* Spreng.). *Food Technol. Biotechnol.* **2013,** *51* (4), 488–499.

Mangut, V.; Sabio, E.; Gañán, J.; González, J. F.; Ramiro, A.; González, C. M.; Román, S.; Al-Kassir, A. Thermogravimetric Study of the Pyrolysis of Biomass Residues from Tomato Processing Industry. *Fuel Process. Technol.* **2006,** *87* (2), 109–115.

Meyer, K. *Method for the Manufacture of Carotinoids and the Novel Intermediates.* Google Patents, 1992.

Moretti, C. L.; Sargent, S. A.; Huber, D. J.; Calbo, A. G.; Puschmann, R. Chemical Composition and Physical Properties of Pericarp, Locule, and Placental Tissues of Tomatoes with Internal Bruising. *J. Am. Soc. Hort. Sci.* **1998,** *123* (4), 656–660.

Mudgil, D.; Barak, S. Composition, Properties and Health Benefits of Indigestible Carbohydrate Polymers as Dietary Fiber: A Review. *Int. J. Biol. Macromol.* **2013,** *61,* 1–6.

Murphy, N.; Norat, T.; Ferrari, P.; Jenab, M.; Bueno-de-Mesquita, B.; Skeie, G.; Dahm, C. C.; Overvad, K.; Olsen, A.; Tjønneland, A.; Clavel-Chapelon, F.; Boutron-Ruault, M. C.; Racine, A.; Kaaks, R.; Teucher, B.; Boeing, H.; Bergmann, M. M.; Trichopoulou, A.; Trichopoulos, D.; Lagiou, P.; Palli, D.; Pala, V.; Panico, S.; Tumino, R.; Vineis, P.; Siersema, P.; van Duijnhoven, F.; Peeters, P. H.; Hjartaker, A.; Engeset, D.; Gonzalez, C. A.; Sanchez, M. J.; Dorronsoro, M.; Navarro, C.; Ardanaz, E.; Quiros, J. R.; Sonestedt, E.; Ericson, U.; Nilsson, L.; Palmqvist, R.; Khaw, K. T.; Wareham, N.; Key, T. J.; Crowe, F. L.; Fedirko, V.; Wark, P. A.; Chuang, S. C.; Riboli, E. Dietary Fibre Intake and Risks of Cancers of the Colon and Rectum in the European Prospective Investigation into Cancer and Nutrition (EPIC). *PLoS ONE* **2012,** *7* (6), e39361.

Navarro-González, I.; García-Valverde, V.; García-Alonso, J.; Periago, M. J. Chemical Profile, Functional and Antioxidant Properties of Tomato Peel Fiber. *Food Res. Int.* **2011,** *44* (5), 1528–1535.

Naviglio, D.; Pizzolongo, F.; Ferrara, L.; Naviglio, B.; Aragón, A.; Santini, A. Extraction of Pure Lycopene from Industrial Tomato Waste in Water Using the Extractor Naviglio®. *Afr. J. Food Sci.* **2008,** *2* (2), 37–44.

Nobre, B. P.; Gouveia, L.; Matos, P. G.; Cristino, A. F.; Palavra, A. F.; Mendes, R. L. Supercritical Extraction of Lycopene from Tomato Industrial Wastes with Ethane. *Molecules* **2012,** *17* (7), 8397–8407.

Nobre, B. P.; Palavra, A. F.; Pessoa, F. L. P.; Mendes, R. L. Supercritical CO_2 Extraction of Trans-Lycopene from Portuguese Tomato Industrial Waste. *Food Chem.* **2009,** *116* (3), 680–685.

Nomura, A. M.; Hankin, J. H.; Henderson, B. E.; Wilkens, L. R.; Murphy, S. P.; Pike, M. C.; Le Marchand, L.; Stram, D. O.; Monroe, K. R.; Kolonel, L. N. Dietary Fiber and Colorectal Cancer Risk: The Multiethnic Cohort Study. *Cancer Causes Control* **2007,** *18* (7), 753–764.

Pandal, N. Nutraceutical Market by Ingredients. In *Nutraceuticals: Global Markets*; BCC Research: Wellesley, MA, **2014;** pp 45–69.

Papaioannou, E.; Karabelas, A. Lycopene Recovery from Tomato Peel under Mild Conditions Assisted by Enzymatic Pre-treatment and Non-ionic Surfactants. *Acta Biochim. Pol.* **2012,** *59* (1), 71–74.

Peralta, I.; Knapp, S.; Spooner, D. Nomenclature for Wild and Cultivated Tomatoes. *Tomato Genetics Cooperative Report*, 2006; pp 6–12.

Peralta, I.; Spooner, D. History, Origin and Early Cultivation of Tomato (Solanaceae). In *Genetic Improvement of Solanaceous Crops*; Razdan, M. K., Mattoo, A. K., Eds.; Science Publishers: Enfield, New Hampshire, USA, 2007; Vol. 2 (Tomato), pp 1–24.

Pereira, C.; Meireles, M. A. Supercritical Fluid Extraction of Bioactive Compounds: Fundamentals, Applications and Economic Perspectives. *Food Bioprocess Technol.* **2010,** *3* (3), 340–372.

PHE. *National Diet and Nutrition Survey. Results from Years 1–4 (combined) of the Rolling Programme (2008/2009–2011/12)*, 2013; p 83.

Picken, A. J. F.; Stewart, K.; Klapwijk, D. Germination and Vegetative Development. In *The Tomato Crop*; Atherton, J., Rudich, J., Eds.; Springer: Amsterdam, Netherlands, 1986; pp 111–166.

Pnueli, L.; Carmel-Goren, L.; Hareven, D.; Gutfinger, T.; Alvarez, J.; Ganal, M.; Zamir, D.; Lifschitz, E. The SELF-PRUNING gene of Tomato Regulates Vegetative to Reproductive Switching of Sympodial Meristems and Is the Ortholog of CEN and TFL1. *Development* **1998,** *125* (11), 1979–1989.

Post, R. E.; Mainous, A. G., 3rd; King, D. E.; Simpson, K. N. Dietary Fiber for the Treatment of Type 2 Diabetes Mellitus: A Meta-analysis. *J. Am. Board Fam. Med.* **2012,** *25* (1), 16–23.

Pozuelo, M. J.; Agis-Torres, A.; Hervert-Hernandez, D.; Elvira Lopez-Oliva, M.; Munoz-Martinez, E.; Rotger, R.; Goni, I. Grape Antioxidant Dietary Fiber Stimulates *Lactobacillus* Growth in Rat Cecum. *J. Food Sci.* **2012,** *77* (2), H59–H62.

Puri, M.; Sharma, D.; Barrow, C. J. Enzyme-Assisted Extraction of Bioactives from Plants. *Trends Biotechnol.* **2012,** *30* (1), 37–44.

Raninen, K.; Lappi, J.; Mykkanen, H.; Poutanen, K. Dietary Fiber Type Reflects Physiological Functionality: Comparison of Grain Fiber, Inulin, and Polydextrose. *Nutr. Rev.* **2011,** *69* (1), 9–21.

Ranveer, R. C.; Patil, S. N.; Sahoo, A. K. Effect of Different Parameters on Enzyme-Assisted Extraction of Lycopene from Tomato Processing Waste. *Food Bioprod. Process.* **2013,** *91* (4), 370–375.

Rao, A. V.; Waseem, Z.; Agarwal, S. Lycopene Content of Tomatoes and Tomato Products and their Contribution to Dietary Lycopene. *Food Res. Int.* **1998,** *31* (10), 737–741.

Reverchon, E.; De Marco, I. Supercritical Fluid Extraction and Fractionation of Natural Matter. *J. Supercrit. Fluids* **2006,** *38* (2), 146–166.

Rossini, G.; Toscano, G.; Duca, D.; Corinaldesi, F.; Foppa Pedretti, E.; Riva, G. Analysis of the Characteristics of the Tomato Manufacturing Residues Finalized to the Energy Recovery. *Biomass Bioenergy* **2013,** *51*, 177–182.

Salunkhe, D.; Jadhav, S.; Yu, M. Changes Quality and Nutritional Composition of Tomato Fruit as Influenced by Certain Biochemical and Physiological Changes. *Qual. Plant.* **1974,** *24* (1), 85–113.

Sanchez-Tena, S.; Lizarraga, D.; Miranda, A.; Vinardell, M. P.; Garcia-Garcia, F.; Dopazo, J.; Torres, J. L.; Saura-Calixto, F.; Capella, G.; Cascante, M. Grape Antioxidant Dietary Fiber Inhibits Intestinal Polyposis in ApcMin/+ Mice: Relation to Cell Cycle and Immune Response. *Carcinogenesis* **2013,** *34* (8), 1881–1888.

Sarkar, A.; Kaul, P. Evaluation of Tomato Processing By-Products: A Comparative Study in a Pilot Scale Setup. *J. Food Process Eng.* **2014,** *37* (3), 299–307.

Satoh, H.; Hara, T.; Murakawa, D.; Matsuura, M.; Takata, K. Soluble Dietary Fiber Protects against Nonsteroidal Anti-inflammatory Drug-Induced Damage to the Small Intestine in Cats. *Digest. Dis. Sci.* **2010,** *55* (5), 1264–1271.

Satoh, H.; Matsumoto, H.; Hirakawa, T.; Wada, N. Soluble Dietary Fibers Can Protect the Small Intestinal Mucosa without Affecting the Anti-inflammatory Effect of Indomethacin in Adjuvant-Induced Arthritis Rats. *Digest. Dis. Sci.* **2015,** *61* (1), 91–98.

Schatzkin, A.; Park, Y.; Leitzmann, M. F.; Hollenbeck, A. R.; Cross, A. J. Prospective Study of Dietary Fiber, Whole Grain Foods, and Small Intestinal Cancer. *Gastroenterology* **2008,** *135* (4), 1163–1167.

Shi, J.; Yi, C.; Xue, S. J.; Jiang, Y.; Ma, Y.; Li, D. Effects of Modifiers on the Profile of Lycopene Extracted from Tomato Skins by Supercritical CO_2. *J. Food Eng.* **2009,** *93* (4), 431–436.

Singh, P.; Goyal, G. Dietary Lycopene: Its Properties and Anticarcinogenic Effects. *Comprehens. Rev. Food Sci. Food Saf.* **2008,** *52* (19), 5796–5802.

Sogi, D. S.; Arora, M. S.; Garg, S. K.; Bawa, A. S. Fractionation and Electrophoresis of Tomato Waste Seed Proteins. *Food Chem.* **2002,** *76* (4), 449–454.

Solanke, A.; Kumar, P. A. Phenotyping of Tomatoes. In *Phenotyping for Plant Breeding*; Punguluri, S. K., Kumar, A. A., Eds.; Springer: New York, USA, 2013, pp 169–204.

Strati, I. F.; Oreopoulou, V. Process Optimisation for Recovery of Carotenoids from Tomato Waste. *Food Chem.* **2011,** *129* (3), 747–752.

Strati, I. F.; Oreopoulou, V. Recovery of Carotenoids from Tomato Processing By-products—A Review. *Food Res. Int.* **2014,** *65,* 311–321.

Suzuki, R.; Rylander-Rudqvist, T.; Ye, W.; Saji, S.; Adlercreutz, H.; Wolk, A. Dietary Fiber Intake and Risk of Postmenopausal Breast Cancer Defined by Estrogen and Progesterone Receptor Status—A Prospective Cohort Study among Swedish Women. *Int. J. Cancer* **2008,** *122* (2), 403–412.

Tabernero, M.; Serrano, J.; Saura-Calixto, F. Dietary Fiber Intake in Two European Diets with High (Copenhagen, Denmark) and Low (Murcia, Spain) Colorectal Cancer Incidence. *J. Agric. Food Chem.* **2007,** *55* (23), 9443–9449.

Threapleton, D. E.; Greenwood, D. C.; Evans, C. E.; Cleghorn, C. L.; Nykjaer, C.; Woodhead, C.; Cade, J. E.; Gale, C. P.; Burley, V. J. Dietary Fibre Intake and Risk of Cardiovascular Disease: Systematic Review and Meta-analysis. *BMJ* **2013,** *347,* f6879.

Topal, U.; Sasaki, M.; Goto, M.; Hayakawa, K. Extraction of Lycopene from Tomato Skin with Supercritical Carbon Dioxide: Effect of Operating Conditions and Solubility Analysis. *J. Agric. Food Chem.* **2006,** *54* (15), 5604–5610.

Tourino, S.; Fuguet, E.; Vinardell, M. P.; Cascante, M.; Torres, J. L. Phenolic Metabolites of Grape Antioxidant Dietary Fiber in Rat Urine. *J. Agric. Food Chem.* **2009,** *57* (23), 11418–11426.

Trejo-Solis, C.; Pedraza-Chaverri, J.; Torres-Ramos, M.; Jimenez-Farfan, D.; Cruz Salgado, A.; Serrano-Garcia, N.; Osorio-Rico, L.; Sotelo, J. Multiple Molecular and Cellular Mechanisms of Action of Lycopene in Cancer Inhibition. *Evid.-Based Complement. Alternat. Med.: eCAM* **2013,** *2013*, 705121.

UN. *UN Comtrade Database*, 2005. http://comtrade.un.org/data/ (accessed Aug. 25, 2015).

Van Horn, L.; Committee, F. T. N. Fiber, Lipids, and Coronary Heart Disease: A Statement for Healthcare Professionals From the Nutrition Committee, American Heart Association, *Circulation* **1997,** *95* (12), 2701–2704.

Vasapollo, G.; Longo, L.; Rescio, L.; Ciurlia, L. Innovative Supercritical CO_2 Extraction of Lycopene from Tomato in the Presence of Vegetable Oil as Co-solvent. *J. Supercrit. Fluids* **2004,** *29* (1–2), 87–96.

Viuda-Martos, M.; López-Marcos, M. C.; Fernández-López, J.; Sendra, E.; López-Vargas, J. H.; Pérez-Álvarez, J. A. Role of Fiber in Cardiovascular Diseases: A Review. *Comprehens. Rev. Food Sci. Food Saf.* **2010,** *9* (2), 240–258.

Wakai, K.; Date, C.; Fukui, M.; Tamakoshi, K.; Watanabe, Y.; Hayakawa, N.; Kojima, M.; Kawado, M.; Suzuki, K.; Hashimoto, S.; Tokudome, S.; Ozasa, K.; Suzuki, S.; Toyoshima, H.; Ito, Y.; Tamakoshi, A.; Group, J. S. Dietary Fiber and Risk of Colorectal Cancer in the Japan Collaborative Cohort Study. *Cancer Epidemiol., Biomark. Prevent.* **2007,** *16* (4), 668–675.

Wang, L.; Liu, Y. Optimization of Solvent Extraction Conditions for Total Carotenoids in Rapeseed Using Response Surface Methodology. *Nat. Sci.* **2009,** *01* (01), 23–29.

Wang, L.; Weller, C. L. Recent Advances in Extraction of Nutraceuticals from Plants. *Trends Food Sci. Technol.* **2006,** *17* (6), 300–312.

Wang, L.; Xu, H.; Yuan, F.; Fan, R.; Gao, Y. Preparation and Physicochemical Properties of Soluble Dietary Fiber from Orange Peel Assisted by Steam Explosion and Dilute Acid Soaking. *Food Chem.* **2015,** *185*, 90–98.

Watada, A.; Aulenbach, B. Chemical and Sensory Qualities of Fresh Market Tomatoes. *J. Food Sci.* **1979,** *44*, 1013–1016.

Wellbaum, G. *Vegetable Production and Practices*; CAB International: Wallingford, 2015.

Wolever, T. M. Relationship between Dietary Fiber Content and Composition in Foods and the Glycemic Index. *Am. J. Clin. Nutr.* **1990,** *51* (1), 72–75.

Wu, Y.; Qian, Y.; Pan, Y.; Li, P.; Yang, J.; Ye, X.; Xu, G. Association between Dietary Fiber Intake and Risk of Coronary Heart Disease: A Meta-analysis. *Clin. Nutr.* **2015,** *34* (4), 603–611.

Yahia, E. M.; Brecht, J. K. Tomatoes. *Crop Post-Harvest: Science and Technology*; Wiley-Blackwell: Hoboken, NJ, 2012; pp 5–23.

Zuorro, A.; Fidaleo, M.; Lavecchia, R. Enzyme-Assisted Extraction of Lycopene from Tomato Processing Waste. *Enzym. Microb. Technol.* **2011,** *49* (6–7), 567–573.

CHAPTER 11

BY-PRODUCTS FROM ESSENTIAL OIL EXTRACTION

M. M. GUTIERREZ-PACHECO[1], C. A. MAZZUCOTELLI[2],
G. A. GONZÁLEZ-AGUILAR[1], J. F. AYALA-ZAVALA[1], and
B. A. SILVA-ESPINOZA[1*]

[1]Centro de Investigación en Alimentación y Desarrollo, A.C. (CIAD, AC), Carretera a la Victoria Km 0.6, La Victoria, Hermosillo, Sonora 83000, México

[2]Grupo de Investigación en Ingeniería en Alimentos, Universidad Nacional de Mar del Plata, Buenos Aires, Argentina

*Corresponding author. E-mail: bsilva@ciad.mx

CONTENTS

ABSTRACT

Today, essential oils production has rapidly increased as a consequence of the growing number of consumer's preferences for healthier foods due their negative perception of synthetic additives in health. For this reason, trends have targeted the use of natural plant extracts as source of food additives. As consequence, enormous quantities of aromatic plants and fruit peels are used in the essential oils production around the world, thereby creating a huge amount of by-products. In some cases, these are disposed to produce compost or as part of livestock feed; however, others are not properly managed that lead to pollution and environmental problems worldwide. In this sense, a more efficient use of these by-products is its use as food additives because of their bioactive properties. This chapter provides an overview of what has been investigated about the by-products of essential oils extraction, from their characterization to its potential use as food additives.

11.1   CHARACTERIZATION OF THE DERIVED BY-PRODUCTS

Essential oils are complex mixtures of volatile substances produced by aromatic plants as secondary metabolites, characterized by their powerful aroma, and antiseptic and medicinal properties, making them very useful in the perfume industries, pharmaceutical sector, and in the food and human nutrition field (De Castro et al., 1999; Bakkali et al., 2008). Essential oils comprise mixtures of more than 200 compounds, which can be grouped basically into two fractions, a volatile fraction, that constitutes 90–95% of the whole oil that contains terpenes as their main components, sesquiterpenes, hydrocarbons, aliphatic aldehydes, alcohols, and esters (Lucchesi et al., 2004); and a nonvolatile residue, that constitutes from 5% to 10% of the whole oil and generally contains hydrocarbons, fatty acids, sterols, carotenoids, waxes, coumarins, and flavonoids (De Castro et al., 1999).

Essential oils can be synthesized by all plant organs, that is, buds, flowers, leaves, stems, twigs, seeds, fruits, roots, wood, or bark and are stored in secretory cells, cavities, canals, epidermic cells, or glandular trichomes (Bakkali et al., 2008; Sanchez-Vioque et al., 2013). There are several methods for extracting essential oils, which are mainly based on solvent extraction, hydrodistillation, steam distillation, or from the epicarp of citrus fruits by mechanical process (Okoh et al., 2010). These processes cost is expensive but can induce thermal degradation, losses of some volatile compounds, hydrolysis, and water solubilization of some fragrance constituents, long extraction times, and degradation of unsaturated or ester compounds. These disadvantages can be avoided by using new alternative extraction processes with supercritical fluids, ultrasound, and microwave (De Castro et al., 1999; Lucchesi et al., 2004, 2007; Bakkali et al., 2008).

The average yield of essential oil production from aromatic plants is below 5% (w/w) depending on the applied extraction method and on the class of plant utilized (Santana-Méridas et al., 2012). Lucchesi et al. (2004) compared the yields on essential oil production of three aromatic herbs by hydrodistillation and with supercritical fluids, obtaining by the two different methods approximately the same yields: 0.028–0.029% for basil; 0.095% for crispate mint, and 0.160% for thyme. Golmakani and Rezaei (2008) realized the comparison of microwave-assisted hydrodistillation with the traditional hydrodistillation method for *Thymus vulgaris*, obtaining yields of 2.52% and 2.39%, respectively. In the same way, there have been mentioned yields of 0.44–1.2% in the extraction of rosemary essential oil (Boutekedjiret et al., 2003); 0.13–0.24% from garlic (Kimbaris et al., 2006); 1% from lavender flowers (Santana-Méridas et al., 2012); and 2–5% from cardamoms seeds (Lucchesi et al., 2007), among others.

In all cases, a considerable amount of both liquid and solid residues is generated during the essential oils extraction, which causes a growing concern. In some industries, the biomass is used for generating energy or for preparing compost. However, recycling the by-product to energy requires a huge investment, and composting is not always satisfactory due to the antigerminative properties of some aromatic plants (Santana-Méridas et al., 2012; Méndez-Tovar et al., 2015). Therefore, these residues may result in environmental concerns if they are not properly managed. These plant origin wastes are also a significant source of bioactive compounds, so it is possible to increase the overall profitability of the aromatic plants improving the rational utilization of these natural resources (Sanchez-Vioque et al., 2013).

The search for natural additives in wastes of plant origin is being explored as an alternative to synthetic ones in food and pharmaceutical industries or as anti-aging ingredients in cosmetic products (Parejo et al., 2002). By-products from essential oil production are a potential pool of compounds with strong bioactivity, especially rich in polyphenols that rarely form part of essential oils (Jordan et al., 2010; Farhat et al., 2014). This is due to the water-soluble properties of phenolic compounds and because much of phenolic compounds are not volatile and are not degraded with thermal treatments, they remain in the aromatic plants wastes after distillation of raw plant materials (Torras-Claveria et al., 2007).

Among plant residues, citric wastes that remain after juice and essential oil extraction also have a high potential for use. These residues are an important source for dietary fiber production. Dietary fibers have specific technological functions and functional properties that allow its use as an ingredient in foods. Moreover, citrus fruits have better quality than other sources of dietary fiber due to the presence of associated bioactive compounds, as flavonoids and ascorbic acid, which possess antioxidant properties and may provide additional health-promoting effects (Marin et al., 2007; Rezzadori et al., 2012). Also, pectin present in the citrus peel wastes can be utilized as an important additive in the food industry because of their gelling capacity (Mandalari et al., 2006). However, in all the cases, the literature concerning chemical and biological characterization of these residues is scarce (Sanchez-Vioque et al., 2013). In Table 11.1, a resume of the different compounds identified in essential oil by-products is shown.

11.2   CONSIDERATION FOR PRACTICAL USES OF THE DERIVED BY-PRODUCTS

Actually, by-products of essential oil industry have been used as a source of many bioactive and functional compounds that could be exploited as new food additive alternatives. However, this requires, among others necessary considerations, the knowledge of by-products chemical composition, the security of being free of toxic compounds derived from solvents used for extraction, economic feasibility, sensorial impact, and bioavailability.

Various bioactive compounds found in plants appear to have a wide spectrum of bioactivities, as chemoprotective, anti-inflammatory, antioxidant, antimicrobial, among others (Santos, 2004; Rios and Recio, 2005; Vanamala et al., 2006). Some of these are phenols and polyphenols, flavonoids, isoflavones, terpenes, and glucosinolates (Drewnowski and Gomez-Carneros, 2000; Viuda-Martos et al., 2008). Besides, the potential of these compounds to be included in food formulations to improve food quality and provide health benefits, their bitter, acrid, and astringent characteristics could limit their use as food additives. Some, but not all of the compounds found in essential oil by-products, have shown some of these characteristics. This could be a dilemma for the food technologists, because the trends are focused on providing added value to by-products; however, their addition to food could be wholly incompatible with consumer acceptance (Drewnowski and Gomez-Carneros, 2000).

TABLE 11.1 Bioactive Compounds Identified in By-products from Essential Oil Extraction.

Source/residue plant material	Bioactive compounds	References
Lavandin (*Lavandula × intermedia*)	Lactones (coumarin[1], herniarin[1]), sesquiterpenes (cadinol[1], α-bisabolol[1]), alcanes (pentacosane[1], hexacosane[1], heptacosane[1]), phenolic acids and derivates (protocatechuic acid[2], rosmarinic acid[1,3], chlorogenic acid[1,2], caffeic acid[2], glycosylated coumaric acid[1,2,3], glycosylated caffeic acid[2], glycosylated ferulic acid[1,3]), flavonoids (lutein-*O*-hexoside[1], apigenin[1,2], luteolin[1,2])	[1]Lesage-Meessen et al. (2015) [2]Torras-Claveria et al. (2007) [3]Sanchez-Vioque et al. (2013)
Rosemary (*Rosmarinus officinalis*)	Carnosol[1,2,3], phenolic acids (rosmarinic[1,2,3], carnosic[1,2,3], caffeic[1], and chlorogenic acid[1]); hispidulin-7-*O*-glucoside[2]; triterpenic acids (ursolic[4], oleanolic[4], and micromeric acids[4])	[1]Navarrete et al. (2011) [2]Zibetti et al. (2013) [3]Nieto et al. (2011) [4]Altinier et al. (2007)
Sage (*Salvia officinalis*, *Salvia glutinosa*)	Coumarins (2,3-dihydrobenzofuran)[1,2], monoterpenes (thujone, 1,8-cineole, camphor, carane-3-on, borneol)[1], hydrocarbons (*n*-undecane, *n*-dodecane, *n*-tridecane)[1], sesquiterpenes (α-humulene, viridiflorol)[1], diterpenes (*trans*-totarol), triterpenes (13-epi-manool)[1], hydroxycinnamic acid derivatives[2], flavonoids (luteolin[1,2,3], apigenin-7-glucoside[3]); rosmarinic acid[3]; phenolic diterpenes[3] (carnosic acid, carnosol, methyl carnosate)	[1]Veličković et al. (2008) [2]Guinot et al. (2007) [3]Farhat et al. (2014)
Thyme (*Thymus vulgaris*)	Hydroxycinnamic acids[1,2] (ferulic[1], caffeic[1], and rosmarinic[1,3] acids); flavonoids and derivates (apigenin[1]; luteolin[2,3,4], kaempferol[3], quercetin derivatives[2,3], eriodictyol[4], thymonin[4], luteolin-7-*O*-glucoside[4]); phenolic diterpenes (carnosic acid[1]); triterpenic acids (ursolic[4], oleanolic[4])	[1]Nieto et al. (2011) [2]Guinot et al. (2007) [3]Sanchez-Vioque et al. (2013) [4]Ismaili et al. (2002)
Cistus ladanifer	Kaempferol flavonoid and derivates (kaempferol methyl ether, kaempferol dimethyl ether, kaempferol diglycoside); ellagic acid; gallic acid	[1]Sanchez-Vioque et al. (2013)
Santolina rosmarinifolia	Hydroxy-cinnamoylquinic acid; isoquercetin; flavonoids (quercetin, luteolin, and apigenin)	[1]Sanchez-Vioque et al. (2013)

TABLE 11.1 *(Continued)*

Source/residue plant material	Bioactive compounds	References
Mentha (*Mentha piperita*)	Menthol, sesquiterpenes, and derivates (cadinene, α-cadinol, caryophyllene oxide); terpenes (limonene, menthone); cuminol	[1]Sapra et al. (2010)
Bergamot (*Citrus bergamia*)	Flavonoids (neoeriocitrin[1], naringin[1], neohesperidin[1], hesperetin[1,3], narirutin[1], apigenin[1,3], eriodictyol[1], naringenin[1,3], brutieridin[2], melitidin[2], luteolin[3]); pectins[3]	[1]Mandalari et al. (2007) [2]Di Donna et al. (2011) [3]Mandalari et al. (2006)
Lemon (*Citrus limon*)	Flavanone glycosides (eriocitrin[1], hesperidin[1]); dietary fiber[2] (pectin[3], lignin[3], cellulose[3], hemicellulose[3])	[1]Coll et al. (1998) [2]Lario et al. (2004) [3]Marin et al. (2007)
Orange (*Citrus sinensis*)	Flavonoids (hesperidin[1,2], naringin[2], didymin[2], eriocitrin[2], narirutin[2]); dietary fiber (cellulose[1,3]; hemicellulose[1,3]; pectin[1,3], lignin[3])	[1]Bicu and Mustata (2011) [2]Anagnostopoulou et al. (2005) [3]Marin et al. (2007)

Phenolic compounds have been reported as responsible for the bitterness and astringency of many foods and beverages (Lesschaeve and Noble, 2005). Naringin (flavanone) and quercetin (flavonol), phenolic compounds found in by-products of bergamot (*Citrus bergamia*) and *Santolina rosmarinifolia* essential oil extraction, showed a bitter taste (Drewnowski and Gomez-Carneros, 2000; Mandalari et al., 2007). As an alternative, it has been reported that the encapsulation of these type of compounds could delay their negative taste when applied in food systems. Fathi et al. (2013) reported that the flavonoid hesperetin was encapsulated in solid–lipid nanoparticles and proved as fortified agent in milk. These authors reported that the fortified milk samples did not show any significant difference with blank milk sample and could well mask the bitter taste and obviate the poor solubility of hesperetin.

On the other hand, other consideration for the use of essential oil by-products as a source of food additives is the economic feasibility, taking into account the supplies needed for extraction and feasibility with respect to the amount of compound obtained. It has been reported that the amount of essential oil obtained from aromatic plants is below 5% (w/w) and is dependent of the applied extraction method and on the class of plant utilized (Santana-Méridas et al., 2012). In this sense, it would be important to consider what amount of compounds in these by-products would remain? These products are known to have a variety of bioactive compounds; however, the cost benefit must take into consideration. Some of the methods used to obtain these compounds comprise a second hydrodistillation after essential oil extraction. This technique requires around 4 h for extraction, which requires an additional cost of energy (Golmakani and Rezaei, 2008). For this reason, it is important take into account all considerations needed to exploit the use of by-products from essential oil extraction in a better way due its high potential to be used as food additives.

11.2.1 FOOD ADDITIVES

Many consumers are demanding foods without artificial and harmful chemicals, including many used as antimicrobials and preservatives. Consequently, interest in more natural and nonsynthesized bioactive compounds as potential alternatives to conventional food additives has heightened (Ayala-Zavala et al., 2011). Essential oils from different sources have been widely promoted for their functional properties capabilities (Calo et al., 2015). However, it is highlighting that the essential oil extraction generates different by-products such as hydrolats and plant tissues which are valuable sources for the recovery of many bioactive compounds, like terpenes, polyphenols, pectins, dietary fiber, and proteins (Fig. 11.1), that could be used as natural antioxidant, antimicrobial, and functional food ingredients (Kammerer et al., 2014).

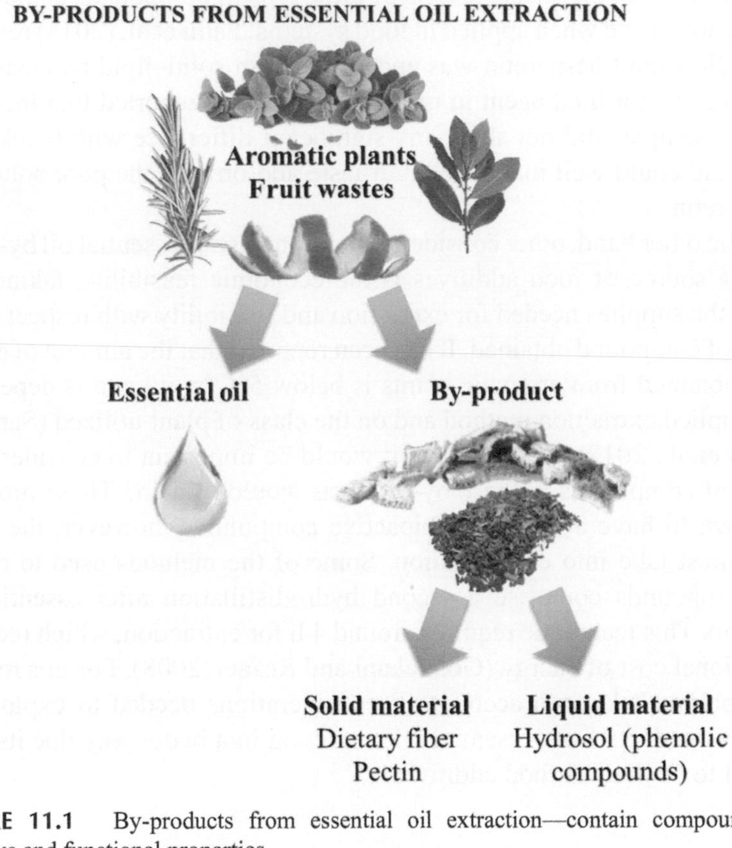

BY-PRODUCTS FROM ESSENTIAL OIL EXTRACTION

Aromatic plants
Fruit wastes

Essential oil   **By-product**

Solid material   **Liquid material**
Dietary fiber   Hydrosol (phenolic
Pectin   compounds)

FIGURE 11.1   By-products from essential oil extraction—contain compounds with bioactive and functional properties.

11.2.2   ANTIMICROBIAL AND ANTIOXIDANT USES

Many natural compounds, including plant phenolics and terpenoids, have been widely researched because of their strong antimicrobial and antioxidant properties (Friedman et al., 2002). Despite that, there are numerous reports of the bioactive activity of crude plant extracts like essential oils (Elgayyar et al., 2001; Benkeblia, 2004), neither the antimicrobial and antioxidant potential of the by-products nor their way to be obtained as food additives are scarce. However, some authors have reported the content of bioactive compounds and their potential to inhibit food-related pathogens and to inactivate free radicals (Torras-Claveria et al., 2007; Ulusoy et al., 2009; Méndez-Tovar et al., 2015). Taking into account the antioxidant and antimicrobial properties of the compounds found in these by-products, it is valid mentioning that some of these have been tested not only in in vitro studies but also in food matrices.

By-products of *Satureja hortensis* L. (hydrosol and ground material) have been shown antifungal activity against pathogenic fungi *Alternaria mali* Roberts and *Botrytis cinerea* Pers. A 100% of mycelial growth inhibition of both fungi was observed after their exposition to 15% and 1% of hydrosol and ground material, respectively (Boyraz and Özcan, 2006). This is important from the point of view of food preservation because these are the most common pathogenic fungi that are causing major decay in fruits and vegetables. On the other hand, Ranasinghe et al. (2003) evaluated the composition and the in vitro and in vivo antimicrobial activity of the by-product extract from cinnamon bark oil extraction against fungal pathogens causing anthracnose and crown rot of Embul banana. These authors reported that the main component on the by-product was cinnamaldehyde (66.2%), which showed a fungistatic and fungicidal activity in the range of 0.64–1.00 mg/mL. Additionally, the extract application on fruit surface controlled crown rot enabling banana to be stored for up to 14 days at ambient temperature (28°C) and 21 days at 14°C without observing any detrimental effect on physicochemical properties.

Cinnamaldehyde also has been incorporated into gliadin films to control fungal growth of *Penicillium expansum* and *Aspergillus niger* on sliced bread and cheese spread. Mold growth was observed on sliced bread after 27 days of storage at 23°C with cinnamaldehyde incorporated films, whereas in the noncovered bread fungal growth appeared around the fourth day. In cheese spread, no fungi were observed after 26 days of storage at 4°C; however, fungi growth was observed in noncovered cheese after 16 days of storage (Balaguer et al., 2013). The antimicrobial activity of purified flavonoids has also been described (Taguri et al., 2006). Various flavanones (hesperetin and naringenin) have been shown to be active against *Helicobacter pylori* (Bae et al., 1999). The present study has demonstrated that bergamot peel, a by-product of the citrus fruit processing and essential oil industries, is a potential source of natural antimicrobials.

Some phenolic compounds founded in Lavandin (*Lavandula intermedia*) by-products include rosmarinic acid, caffeic acid, chlorogenic acid, apigenin, and luteolin (Torras-Claveria et al., 2007), which have shown antimicrobial properties. Bowles and Miller (1994) reported that caffeic acid at 0.78 and 3.25 mM inhibited the germination of *Clostridium botulinum* spores for 6 and 24 h, respectively; and >100 mM was necessary to render the spores nonviable. Moreover, these authors reported a reduction of thermal resistance at 50 mM and retention of sporostatic activity when tested in commercial meat broths. In addition, these compounds also have shown antioxidant properties; Farhat et al. (2013) reported a content of 941.81, 1377.03, and 432.00 µg/g of rosmarinic acid (phenolic acid), methyl carnosate (phenolic diterpene), and naringenin (flavonoid) in *Salvia officinalis* L. by-products, respectively. A methanolic extract of this by-product showed an antioxidant capacity of 69.35 µg/mL (expressed as EC_{50} = concentration of antioxidant that causes a 50% decrease in the DPPH· radical absorbance).

On the other hand, Méndez-Tovar et al. (2015) evaluated the antioxidant properties of *Lavandula latifolia* essential oil by-product extraction. This by-product presented total phenol content in a range of 1.89–3.54 mg of gallic acid equivalents/g of dry weight and an EC_{50} of 5–14.30 mg/mL. These authors showed that although the by-product showed less antioxidant activity than the original plant, it could be possible to recover appreciable amounts of antioxidants from the hydrodestilled by-product. Ulusoy et al. (2009) evaluated the composition, phenol content, and antibacterial activity of *Rosa damascene* Mill. hydrosol. These authors reported that citrenellol, nerol, phenylethyl alcohol, and geraniol were the major components identified in this by-product, which could be contributed to the total phenol content of 5.2 mg GAE/L. However, no antibacterial activity was observed against *Pseudomonas aeruginosa*, *Chromobacterium violaceum*, and *Escherichia coli*. This behavior could be attributed to the concentration and type of phenolic compounds of hydrosol which are lowest than that found in rose essential oil.

11.2.3 THICKENER USES

During the processing of some fruits like oranges, lemons, limes, etc., for essential oils extraction, peels represent between 50% and 65% of the total weight of the fruits and remain as the primary by-product (Mandalari et al., 2006). One of the major compounds present in citrus fruit peels is pectin, one of the main components of dietary fiber, which is an important additive in the food industry for their gelling, texturizer, emulsifier, stabilizer, and thickener capacity (Thakur et al., 1997; Mandalari et al., 2006). The use of this and other derived by-products has brought with it a growing interest from ingredient companies because they are continually looking for cheaper but value-added ingredients (O'Shea et al., 2012).

Some authors reported that lemon peel, a by-product of essential oil extraction, showed higher levels of dietary fiber (Gorinstein et al., 2001). Chau and Huang (2003) investigated the dietary fiber content of the peel of orange cv. "Liucheng." They found that peel contain 57% (dry weight) of total dietary fiber. It was also determined that pectic polysaccharides and cellulose were the main constituents of the fiber. The gelling and thickening abilities of orange fiber were explored in an enriched yogurt. Sendra et al. (2010) found that orange fiber increased the viscosity of the yogurt. At low concentrations, the authors proved that the presence of fiber has a disruptive effect on the structure of the yogurt. Consequently, after pasteurization at higher levels of fiber (greater than 6%), it was found to strengthen the gel.

Pectin plays an important role in food quality, particularly in dairy products, due its capacity to stabilize acidified milk drinks. It has been mentioned that this property is related to their capacity to bind electrostatically to the surface of casein particles, preventing the aggregation and sedimentation (Willats et al., 2006). Although pectins are added to foods for the gelling and stabilizing effects as described above, they can, in sufficient concentrations, also exert physiological effects.

11.2.4 NUTRACEUTICAL USES

In recent years, a greater emphasis has been placed on the link between the prevention of chronic diseases and the human diet. Consumers began to view food in a new way since several clinical studies demonstrated a close relationship between dietary habits and disease risk showing that food has a direct impact on health (Espin et al., 2007; Moreno et al., 2012).

The capacity of some plant-derived food to reduce the risk of chronic diseases has been associated to the presence of a wide variety of bioactive compounds, being phytochemicals and dietary fiber the substances with greater implication on health (Beecher, 1999; Espin et al., 2007; Bernal et al., 2011). Phytochemicals have been linked to health benefits principally because their antioxidant properties since oxidative stress induced by free radicals is involved in the etiology of a wide range of chronic degenerative diseases as atheroma plaque development, cancer, aging, etc. (Lee et al., 2004; Liu, 2004; Moreno et al., 2012).

Between major groups of phytochemicals, carotenoids, phenolic compounds, alkaloids, and glucosinolates are found. On the other way, dietary fiber consumption was related with the prevention of cancer, coronary heart disease, hypercholesterolemia, between others affections. Dietary fibers mostly include nonstarch polysaccharides (NSP) such as celluloses, hemicelluloses, gums, pectins, and lignin (Larrauri, 1999; Marin et al., 2007; Das et al., 2012). All these bioactive compounds found in fruits and vegetables have low potency when compared to pharmaceutical drugs, but since they are ingested regularly and in significant amounts as part of the diet, they may have a noticeable long-term physiological effect (Espin et al., 2007).

A new food concept was incorporated for consumers, where food constituents go beyond their role as dietary essentials for sustaining life and growth, to one of preventing, managing, or delaying the premature onset of chronic disease later in life (Wrick, 1995; Tsao and Akhtar, 2005). Zeisel (1999) defined the term nutraceuticals as dietary supplements that deliver a concentrated form of a presumed bioactive agent from a food, presented in a nonfood matrix (pills, capsules, powders, granulates, etc.) and used to enhance health in dosages that exceed those that could be obtained from normal foods. This type of health-promoting products is getting more popular amongst health-conscious consumers and, thus, a large list of nutraceuticals containing phytochemicals from foods is now available in the market.

11.2.4.1 ANTIOXIDANTS

Benefits of antioxidative nutraceuticals in the prevention of diseases and promotion of health have been extensively reported in recent years. The interest in antioxidants has been increasing because of their high capacity in scavenging free radicals (Lee et al., 2004; Liu, 2004). Free radicals and reactive oxygen species (ROS) are considered the most important among the causes of the major health problems since these species have been implicated as mediators of degenerative and chronic deteriorative, inflammatory and autoimmune diseases such as atherosclerosis, myocardial injury, brain dysfunction, diabetes, hypertension, cancer, arthritis, aging process, and diseases of the nervous system (Villeponteau et al., 2000; Cui et al., 2004; Tsao and Akhtar, 2005; Prakash and Gupta, 2009).

Phytochemicals with antioxidant capacity naturally present in food are of great interest due to their beneficial effects on human health as they offer protection against oxidative deterioration. The reducing properties of these chemicals (as hydrogen or electron-donating agents) predicts their potential for action as free-radical scavengers (antioxidants) (Rice-Evans et al., 1997; Prakash and Gupta, 2009). Carotenoids, tocopherols, ascorbates, lipoic acids, and polyphenols are strong natural antioxidants with free-radical scavenging activity (Lee et al., 2004; Prakash and Gupta, 2009; Bernal et al., 2011). Antioxidants are known to defuse free radicals leading to limited risk of oxidative stress and associated disorders. At cellular and molecular levels, they inactivate ROS and under specific low concentration inhibit or delay oxidative processes by interrupting the radical's chain reaction (Prakash and Gupta, 2009).

Therefore, to prevent or slow the oxidative stress induced by free radicals, sufficient amounts of antioxidants need to be consumed, since these bioactive compounds could provide novel and safe therapeutic options for these disorders (Liu, 2004; Kelsey et al., 2010). As mentioned above, by-products from essential oil production are a potential source of a variety of compounds with strong antioxidant activity. The most important groups of phytochemical antioxidants are the carotenoids (as lutein, β-carotene, and lycopene), glucosinolates, and polyphenols. Polyphenols is a highly inclusive term that covers many different subgroups of phenols and phenolic acids (hydroxybenzoic acids, hydroxycinnamic acids, flavonoids, catechins, anthocyanidins, procyanidins, isoflavones, tocopherols, and tocotrienols) (Tsao and Akhtar, 2005).

Molecular studies have revealed that in addition to their antioxidant capacity, phenolics can exert modulatory actions in cell by interacting with a wide spectrum of molecular targets central to the cell-signaling machinery. These include activation of mitogen-activated protein kinase (MAPK), protein kinase C, serine/threonine protein kinase Akt/PKB, phase II antioxidant detoxifying enzymes, downregulation of pro-inflammatory enzymes (COX-2 and iNOS) through the activation of peroxisome proliferator-activated receptor gamma (PPAR), regulation of calcium homeostasis, inhibition of phosphoinositide 3-kinase (PI3-kinase), tyrosine kinases, among others (Soobrattee et al., 2005). Phytochemicals with antioxidant capacity founded in different by-products from essential oil production are presented in Table 11.2, and their beneficial effects on human health associated with their consumption are also indicated.

11.2.4.2   FIBERS

Dietary fiber is the plant material that is not hydrolyzed by human digestive enzymes, but digested by microflora in the gut. Dietary fibers mostly

TABLE 11.2 Phytochemicals with Antioxidant Capacity and their Beneficial Effects on Human Health.

Nutraceutical	Source/essential oil extraction by-producta	Possible health effect	References
AP	Lavandin, Sage, Thyme, Santolina rosmarinifolia, Bergamot	*Anti-inflamatory.* AP inhibit HIF-1α and VEGF expression by blocking PI3K/Akt signaling or LPS-induced pro-inflammatory cytokines expression by inactivating NF-κB through the suppression of p65 phosphorylation[2] *Neuropotective* AP ameliorate AD through relieving amyloid-β peptides burden, which play a critical role in the onset and progression of AD, suppressing amyloidogenic process, inhibiting oxidative stress, and restoring ERK/CREB/BDNF pathway[3] AP modulate microglial activation via inhibition of STAT1-induced CD40 expression[4] *Anticarcinogen.* Proteasome-inhibitory and apoptosis-inducing abilities in human tumor cells[1]	[1]Chen et al. (2005) [2]Pan et al. (2009) [3]Zhao and Zhao (2013) [4]Rezai-Zadeh et al. (2008)
CFA	Lavandin, Thyme, Rosemary	*Antioxidant* *Anticarcinogen.* CFA show strong inhibitory effect of MMP-9 activity, which is known to be involved in tumor cell invasion and metastasis[1]	[1]Park et al. (2005)
CNA	Sage, Thyme, Rosemary	*Antioxidant*[1,2] *Neuroprotective* CNA promote neuronal survival signals and is effective against neuronal injury and neurodegenerative disease[1] CNA protect to the brain against injury induced by middle cerebral artery ischemia[1] *Anti-inflamatory.* CNA inhibit the formation of proinflammatory leukotrienes and downregulate pro-inflammatory activity of leukocytes and platelets, causing a diminished of inflammatory reactions[2]	[1]Kelsey et al. (2010) [2]Poeckel et al. (2008)

TABLE 11.2 *(Continued)*

Nutraceutical	Source/essential oil extraction by-product[a]	Possible health effect	References
CS	Sage	*Antioxidant*[1,2]	[1]Dörrie et al. (2001)
	Rosemary	*Anticarcinogen*[1,3]	[2]Poeckel et al. (2008)
		CS inhibits the anti-apoptotic protein Bcl-2 in B-lineage leukemia cells, so apoptosis can be induced these cancer cells[1]	[3]Johnson (2011)
		CS is involved in multiple deregulated pathways associated with inflammation and cancer that include inhibition of NF-B, apoptotic related proteins, PI3K/Akt, androgen and estrogen receptors[3]	
		CS has been evaluated for anticancer property in prostate, breast, skin, leukemia, and colon cancer[3]	
		Anti-inflamatory. CS inhibit the formation of proinflammatory leukotrienes and downregulate pro-inflammatory activity of leukocytes and platelets, causing a diminished of inflammatory reactions[2]	
CGA	Lavandin	*Antioxidant*	[1]Lee et al. (2012)
	Rosemary	*Neuroprotective.* CGA reduces brain damage, sensory-motor functional deficits, BBB, damage and brain edema through a free radical scavenging effect, and the inhibition of matrix metalloproteinase MMP-2 and MMP-9[1]	
KF	Thyme	*Antioxidant*[1]	[1]Prakash and Gupta (2009)
	Cistus ladanifer	*Cardioprotective.* LT inhibit oxidation of low-density lipoprotein, a major factor in the promotion of atherosclerosis[1]	[2]Aggarwal et al. (2009)
		Anticarcinogen	[3]Hämäläinen et al. (2007)
		KF inhibits the anti-apoptotic protein Bcl-2, so apoptosis can be induced in cancer cells[2]	[4]Soobrattee et al. (2005)
		Anti-inflamatory	

TABLE 11.2 (Continued)

Nutraceutical	Source/essential oil extraction by-product[a]	Possible health effect	References
		KF iNOS protein, decreasing NO production which is associated with inflammatory effects. Inhibited the activation of NF-κB, which is a significant transcription factor for iNOS[3]	
		KF suppresses COX-2 expression by inhibiting tyrosine kinases important for induction of COX-2 gene expression in colon cancer[4]	
LM	Mentha	*Anti-inflammatory*	[1]Aggarwal et al. (2009)
		Anticarcinogen. LM Inhibits the anti-apoptotic protein Bcl-2, so apoptosis can be induced in cancer cells	
LT	Lavandin	*Antioxidant*[1,2]	[1]Prakash and Gupta (2009)
	Sage	*Cardioprotective*	[2]Qi et al. (2011)
	Thyme	LT inhibit oxidation of low-density lipoprotein, a major factor in the promotion of atherosclerosis[1]	[3]Fang et al. (2011)
	Santolina rosmarinifolia	LT prevents ischemia–reperfusion injury by reducing necrosis and apoptosis in rat cardiomyocytes[2,3]	[4]Fang et al. (2006)
	Bergamot	*Anticarcinogen*	[5]Xavier et al. (2009)
		LT inhibits insulin-like growth factor 1 receptor signaling. Suppressed proliferation and induced apoptosis of prostate cancer cells[4]	[6]Rezai-Zadeh et al. (2008)
		LT inhibit proliferation and induce apoptosis in both KRAS and BRAF mutated human colorectal cancer cells through regulation of both MAPK/ERK and PI3K pathways[5]	
		Neuroprotective. LT modulate microglial activation via inhibition of STAT1-induced CD40 expression[6]	

TABLE 11.2 *(Continued)*

Nutraceutical	Source/essential oil extraction by-product[a]	Possible health effect	References
MT	Mentha	*Antioxidant*	[1]Pan et al. (2009)
		Anti-inflamatory. MT inhibits LPS-induced cytokine production through blocking phosphorylation of IkBR and translocation of NF-κB[1]	
NR	Bergamot	*Antioxidant*[1]	[1]Gao et al. (2006)
		Anticarcinogen.[1] NR stimulates DNA repair in prostate cancer cells	[2]Hämäläinen et al. (2007)
		Anti-inflammatory	
		NR iNOS protein, decreasing NO production which is associated with inflammatory effects. Inhibited the activation of NF-κB, which is a significant transcription factor for iNOS[2]	
Oleanolic acid	Rosemary	*Antioxidant*	[1]Altinier et al. (2007)
	Thyme	*Anti-inflamatory.* Reduction of edematous response	[2]Ismaili et al. (2002)
QC	Thyme	*Antioxidant*[1,5]	[1]Prakash and Gupta (2009)
	Santolina rosmarinifolia	*Cardioprotective*[1,5]	[2]Xavier et al. (2009)
		QC inhibit oxidation of low-density lipoprotein, a major factor in the promotion of atherosclerosis[1]	[3]Chen et al. (2005)
		QC quenches the cytokine-induced expression of human CRP, a sensitive inflammation marker and risk marker for future CVD[5]	[4]Hämäläinen et al. (2007)
		QC affects lesional SMC proliferation and specific inflammatory factors implicated in atherogenesis[5]	[5]Kleemann et al. (2011)
		Anticarcinogen	
		QC shows proteasome-inhibitory and apoptosis-inducing abilities in human tumor cells[3]	

TABLE 11.2 *(Continued)*

Nutraceutical	Source/essential oil extraction by-product[a]	Possible health effect	References
		QC inhibit proliferation and induce apoptosis in both KRAS and BRAF mutated human colorectal cancer cells through regulation of both MAPK/ERK and PI3K pathways[2]	
		Anti-inflamatory[4,5]. QC iNOS protein, decreasing NO production which is associated with inflammatory effects. Inhibited the activation of NF-κB, which is a significant transcription factor for iNOS[4]	
RA	Lavandin Sage Thyme Rosemary	*Antioxidant*[1,2]	[1]Kelsey et al. (2010)
		Neuroprotective. RA promote neuronal survival signals and is effective against neuronal injury and neurodegenerative disease. Alleviates memory impairment associated with neurotoxicity[1]	[2]Kolettas et al. (2006)
		Anti-inflamatory[2,3]. RA increased the secretion of anti-inflammatory cytokine IL-10[3]	[3]Mueller et al. (2010)
		Anticarcinogen. RA Inhibits the anti-apoptotic protein Bcl-2, so apoptosis can be induced in cancer cells[2]	
UA	Thyme Rosemary	*Anti-inflamatory.* Reduction of edematous response[1,2]	[1]Altinier et al. (2007)
		Anticarcinogen	[2]Ismaili et al. (2002)
		UA inhibits the anti-apoptotic protein Bcl-2, so apoptosis can be induced in cancer cells[3]	[3]Aggarwal et al. (2009)
		UA inhibit proliferation and induce apoptosis in both KRAS-and BRAF-mutated human colorectal cancer cells through regulation of both MAPK/ERK and PI3K pathways[4]	[4]Xavier et al. (2009)

AD, Alzheimer's disease; *AP,* apigenin; *CFA,* caffeic acid; *MMP,* matrix metalloproteinase; *CNA,* carnosic acid; *CS,* carnosol; *CGA,* chlorogenic acid; *KF,* kaempferol; *LM,* limonene; *LT,* luteolin; *MT,* menthone; *NR,* naringenin; *QC,* quercetin; *RA,* rosmarinic acid; *UA,* ursolic acid; *NF-κB,* nuclear factor kappa B; *PI3K,* phosphatidylinositol-3-kinase; *BBB,* blood brain barrier; *SMC,* smooth muscle cell; *iNOS,* inhibited nitric oxide synthase; *IL-10,* interleukin-10. [a]References in Table 11.1.

include NSP such as celluloses, hemicelluloses, gums, pectins and lignin, resistant dextrins and starches. These NSP may be divided in soluble dietary fiber (SDF), also called viscous fibers, and insoluble dietary fiber (IDF) (nonviscous fibers). SDF includes β-glucans, pectins, gums, mucilages, and hemicelluloses and could be fermented in the colon. On the other hand, IDF, which includes celluloses, some hemicelluloses, and lignins, only could be fermented to a limited extent in the colon (Das et al., 2012).

Among the many sources of dietary fiber, citrus by-products have a high potential for use. There has been much interest in the use of fibers from the peel and dehydrated membranes. During citrus juice production and essential oil extraction, only around the half of the fresh fruit weight is transformed into products, generating great amounts of residue which can negatively affect the soil and the ground and superficial waters (Lario et al., 2004; Rezzadori et al., 2012). Therefore, waste of citrus industry are a good alternative for obtaining dietary fibers, which are a product of a high added value because they can be used for the enrichment of usually consumed foods or for the production of dietary fiber tablets (Lario et al., 2004; Rezzadori et al., 2012).

Lemon peel, a by-product of essential oil extraction, showed higher levels of dietary fiber (14 g/100 g of dry material) than in the peeled fruit itself (7.34 g/100 g of dry material). From the total dietary fiber, 9.04 g/100 g was IDF and 4.93 g/100 g was SDF (Gorinstein et al., 2001). Lario et al. (2004) studied the obtaining of dietary fiber from lemon juice by-products and found that crude fiber was about 20% of raw material and showed good functional and microbial quality, as well as favorable physicochemical characteristics to be used in food formulations. On the other hand, Marin et al. (2007) evaluated by-products from different citrus processes as a source of functional fibers (pectin, lignin, cellulose, and hemicellulose) and found that orange by-products showed a total amount of chemical fiber of approximately 60% (dry weight), lemon 64%, and grapefruit by-products 52% of dietary fiber.

In addition, Bortoluzzi and Marangoni (2006) described the obtaining of dietary fiber from orange dry residue after juice extraction and obtained yields of 47.9% and 20.7% of IDF and SDF, respectively. Also, Chau and Huang (2003) investigated the dietary fiber content of the peel of orange cv. "Liucheng." They found that peel contain 57% (dry weight) of total dietary fiber; being 47.6% the insoluble fraction and 9.41% the soluble fraction. Russo et al. (2015) reported that orange by-products have a total dietary fiber of 68.2%, of which 56.8 and 11.4% are IDF and SDF, respectively. On the other hand, Ubando-Rivera et al. (2005) reported that Persian and Mexican lime peels contain high levels of dietary fiber of 70.4% and 66.7%, respectively. Spent residue from cumin after oil extraction was found to contain 62.1%, 51.7% and 10.4% of total, IDF and SDF, respectively (Sowbhagya et al., 2007). The insoluble fraction is the dominant fraction thus providing health benefits such as intestinal regulation and increased stool volume (Russo et al., 2015).

The consumption of a high-fiber diet provides many health benefits. Numerous studies linking fiber consumption to lower blood pressure and hypertension treatment (Pereira and Pins, 2000; Burke et al., 2001), to the improves of glycemic control, hyperinsulinemia and to the low of plasma lipid concentrations in diabetes treatment (Chandalia et al., 2000; Montonen et al., 2003). Moreover, dietary fiber consumption is associated with hypocholesterolemic effect in human, and the significant decrease in cardiovascular diseases (CVD) observed in patients with a high intake of fiber in their diet is a result of the conjunction of the above mentioned effects, which may prevent or delay the development of atherosclerosis (Pereira and Pins, 2000; Liu et al., 2002; King, 2005; Eshak et al., 2010).

Other effects associated with fiber consumption were the decrease of diverticular diseases (Frieri et al., 2006), constipation (Schaefer and Cheskin, 1998), and hemorrhoids (Ho et al., 2000). The hypothesis that dietary fiber may prevent or delay the progression of colon cancer has been debated for years. But in most cases, inconclusive results without a significant difference in the prevalence of colorectal cancer or adenomatous polyps in groups with higher fiber intakes were obtained (Terry et al., 2001; Park et al., 2005; Anderson et al., 2009).

11.2.4.3   PREBIOTICS

A prebiotic has been defined as a nondigestible food ingredient that beneficially affects the host by selectively stimulating the growth and/or activity of one or a limited number of bacteria in the colon, resulting in improved host health (Gomez et al., 2013). A wide variety of dietary carbohydrates have such prebiotic characteristics, among which inulin, oligofructose, and lactulose are commercially the most known (Roberfroid, 2002; Mamma and Christakopoulos, 2014). The interest of consumers in healthy foods has enhanced the importance in producing new prebiotic oligosaccharides with improved properties from readily available, renewable carbohydrate sources (Gomez et al., 2013).

In this sense, citrus wastes have become a new source of prebiotics. Many studies have reported that nondigestible pectic-oligosaccharides (POS), components derived from pectin, have a prebiotic effect, exerting a number of health-promoting effects (Hotchkiss et al., 2003; Gomez et al., 2013). POS could be produced from pectins of different sources, as orange, lemons, and citrus wastes in general, where the albedo offers a suitable pectin source (Gomez et al., 2013; Mamma and Christakopoulos, 2014).

Among the beneficial effects attributed to the consumption of POS that can be named are as follows: protection of colonic cells blocking the binding of food pathogen toxins, prevention of the adhesion of uropathogenic micro-organisms, stimulation of apoptosis of human colonic adenocarcinoma cells, regulation of lipid and glucose metabolism with decreased glycemic response, and blood cholesterol levels and potential for cardiovascular protection in vivo, antiobesity effects, antitoxic, anti-infection, antibacterial, and antioxidant properties (Hotchkiss et al., 2003; Gomez et al., 2013; Mamma and Christakopoulos, 2014).

The prebiotic effect of essential oil by-products is scarce; however, some authors have reported the activity of fruits peels commonly used to extract essential oils. Mandalari et al. (2007) reported a prebiotic activity of POS-rich extract from bergamot peel. These authors observed that individual species of bifidobacteria and lactobacilli responded positively to the addition of the bergamot peel extract, showing higher prebiotic index (6.90) than fructo-oligosacharides (6.12). On the other hand, Li et al. (2016) studied the obtaining of POS from orange peel by hydrolysis with fungal multi enzyme complexes and evaluated their prebiotic potentials. The hydrolyzate was fractionated via membrane separation into three fractions: POS1 < 1 kDa; 1 kDa < POS2 < 3 kDa; POS3 > 3 kDa, with yields of 6.6%, 28%, and 4.3% (w/w), respectively. The results showed that POS2 had higher prebiotic property than POS1 and 3, and it was comparable to fructo-oligosaccharide. Hotchkiss et al. (2003) evaluated the orange peel albedo as a source of POS with prebiotic properties. They tested POS production by microwave and autoclave extraction and with an ultrafiltration dead-end membrane enzyme reactor and found that these POS may have greater persistence through the colon, making them excellent candidate for second generation prebiotic product development.

11.3 CONCLUSIONS AND FUTURE TRENDS

By-products from essential oils extraction have a varied content of useful bioactive compounds for use in the food formulations as well as in nutraceutical products due to its broad spectrum of bioactivities. Therefore, it is important to find new and effective techniques for bioactives extraction that allow the exploitation of these poorly managed resources and allow a reduction in disposal costs.

KEYWORDS

- solid residues
- oil extraction
- essential oils
- volatile substances
- bioactive compounds

REFERENCES

Aggarwal, B. B.; Van Kuiken, M. E.; et al. Molecular Targets of Nutraceuticals Derived from Dietary Spices: Potential Role in Suppression of Inflammation and Tumorigenesis. *Exp. Biol. Med.* **2009**, *234* (8), 825–849.

Altinier, G.; Sosa, S.; et al. Characterization of Topical Antiinflammatory Compounds in *Rosmarinus officinalis* L. *J. Agric. Food Chem.* **2007**, *55* (5), 1718–1723.

Anagnostopoulou, M. A.; Kefalas, P.; et al. Analysis of Antioxidant Compounds in Sweet Orange Peel by HPLC–Diode Array Detection–Electrospray Ionization Mass Spectrometry. *Biomed. Chromatogr.* **2005**, *19* (2), 138–148.

Anderson, J. W.; Baird, P.; et al. Health Benefits of Dietary Fiber. *Nutr. Rev.* **2009**, *67* (4), 188–205.

Ayala-Zavala, J.; Vega-Vega, V.; et al. Agro-industrial Potential of Exotic Fruit Byproducts as a Source of Food Additives. *Food Res. Int.* **2011**, *44* (7), 1866–1874.

Bakkali, F.; Averbeck, S.; et al. Biological Effects of Essential Oils—A Review. *Food Chem. Toxicol.* **2008**, *46* (2), 446–475.

Balaguer, M. P.; Lopez-Carballo, G.; et al. Antifungal Properties of Gliadin Films Incorporating Cinnamaldehyde and Application in Active Food Packaging of Bread and Cheese Spread Foodstuffs. *Int. J. Food Microbiol.* **2013**, *166* (3), 369–377.

Beecher, G. R. Phytonutrients' Role in Metabolism: Effects on Resistance to Degenerative Processes. *Nutr. Rev.* **1999**, *57* (9), 3–6.

Benkeblia, N. Antimicrobial Activity of Essential Oil Extracts of Various Onions (*Allium cepa*) and Garlic (*Allium sativum*). *LWT—Food Sci. Technol.* **2004**, *37* (2), 263–268.

Bernal, J.; Mendiola, J.; et al. Advanced Analysis of Nutraceuticals. *J. Pharm. Biomed. Anal.* **2011**, *55* (4), 758–774.

Bicu, I.; Mustata, F. Cellulose Extraction from Orange Peel Using Sulfite Digestion Reagents. *Bioresour. Technol.* **2011**, *102* (21), 10013–10019.

Bortoluzzi, R. C.; Marangoni, C. Caracterizac¸ão da fibra dietética obtida da extrac¸ão do suco de laranja. *Rev. Bras. Prod. Agroind.* **2006,** *8* (1), 61–66.

Boutekedjiret, C.; Bentahar, F.; et al. Extraction of Rosemary Essential Oil by Steam Distillation and Hydrodistillation. *Flav. Fragr. J.* **2003,** *18* (6), 481–484.

Bowles, B. L.; Miller, A. J. Caffeic Acid Activity against *Clostridium botulinum* Spores. *J. Food Sci.* **1994,** *59* (4), 905–908.

Boyraz, N.; Özcan, M. Inhibition of Phytopathogenic Fungi by Essential Oil, Hydrosol, Ground Material and Extract of Summer Savory (*Satureja hortensis* L.) Growing Wild in Turkey. *Int. J. Food Microbiol.* **2006,** *107* (3), 238–242.

Burke, V.; Hodgson, J. M.; et al. Dietary Protein and Soluble Fiber Reduce Ambulatory Blood Pressure in Treated Hypertensives. *Hypertension* **2001,** *38* (4), 821–826.

Calo, J. R.; Crandall, P. G.; et al. Essential Oils as Antimicrobials in Food Systems—A Review. *Food Control* **2015,** *54,* 111–119.

Coll, M.; Coll, L.; et al. Recovery of Flavanones from Wastes of Industrially Processed Lemons. *Zeitschr. Lebensmitteluntersuch. Forsch. A* **1998,** *206* (6), 404–407.

Cui, K.; Luo, X.; et al. Role of Oxidative Stress in Neurodegeneration: Recent Developments in Assay Methods for Oxidative Stress and Nutraceutical Antioxidants. *Progr. Neuro-Psychopharmacol. Biol. Psychiatry* **2004,** *28* (5), 771–799.

Chandalia, M.; Garg, A.; et al. Beneficial Effects of High Dietary Fiber Intake in Patients with Type 2 Diabetes Mellitus. *N. Engl. J. Med.* **2000,** *342* (19), 1392–1398.

Chau, C.-F.; Huang, Y.-L. Comparison of the Chemical Composition and Physicochemical Properties of Different Fibers Prepared from the Peel of *Citrus sinensis* L. Cv. Liucheng. *J. Agric. Food Chem.* **2003,** *51* (9), 2615–2618.

Chen, D.; Daniel, K. G.; et al. Dietary Flavonoids as Proteasome Inhibitors and Apoptosis Inducers in Human Leukemia Cells. *Biochem. Pharmacol.* **2005,** *69* (10), 1421–1432.

Das, L.; Bhaumik, E.; et al. Role of Nutraceuticals in Human Health. *J. Food Sci. Technol.* **2012,** *49* (2), 173–183.

De Castro, M. L.; Jimenez-Carmona, M.; et al. Towards More Rational Techniques for the Isolation of Valuable Essential Oils from Plants. *TrAC: Trends Anal. Chem.* **1999,** *18* (11), 708–716.

Di Donna, L.; Gallucci, G.; et al. Recycling of Industrial Essential Oil Waste: Brutieridin and Melitidin, Two Anticholesterolaemic Active Principles from Bergamot Albedo. *Food Chem.* **2011,** *125* (2), 438–441.

Dörrie, J.; Sapala, K.; et al. Carnosol-Induced Apoptosis and Downregulation of Bcl-2 in B-Lineage Leukemia Cells. *Cancer Lett.* **2001,** *170* (1), 33–39.

Drewnowski, A.; Gomez-Carneros, C. Bitter Taste, Phytonutrients, and the Consumer: A Review. *Am. J. Clin. Nutr.* **2000,** *72* (6), 1424–1435.

Elgayyar, M.; Draughon, F. A.; et al. Antimicrobial Activity of Essential Oils from Plants against Selected Pathogenic and Saprophytic Microorganisms. *J. Food Protect.* **2001,** *64* (7), 1019–1024.

Eshak, E. S.; Iso, H.; et al. Dietary Fiber Intake is Associated with Reduced Risk of Mortality from Cardiovascular Disease among Japanese Men and Women. *J. Nutr.* **2010,** *140* (8), 1445–1453.

Espin, J. C.; García-Conesa, M. T.; et al. Nutraceuticals: Facts and Fiction. *Phytochemistry* **2007,** *68* (22), 2986–3008.

Fang, F.; Li, D.; et al. Luteolin Inhibits Apoptosis and Improves Cardiomyocyte Contractile Function through the PI3K/Akt Pathway in Simulated Ischemia/Reperfusion. *Pharmacology* **2011,** *88* (3–4), 149–158.

Fang, J.; Zhou, Q.; et al. Luteolin Inhibits Insulin-Like Growth Factor 1 Receptor Signaling in Prostate Cancer Cells. *Carcinogenesis* **2006,** *28* (3), 713–723.

Farhat, M. B.; Chaouch-Hamada, R.; et al. Antioxidant Potential of *Salvia officinalis* L. Residues as Affected by the Harvesting Time. *Ind. Crops Prod.* **2014,** *54,* 78–85.

Farhat, M. B.; Landoulsi, A.; et al. Profiling of Essential Oils and Polyphenolics of *Salvia argentea* and Evaluation of its By-products Antioxidant Activity. *Ind. Crops Prod.* **2013,** *47,* 106–112.

Fathi, M.; Varshosaz, J.; et al. Hesperetin-Loaded Solid Lipid Nanoparticles and Nanostructure Lipid Carriers for Food Fortification: Preparation, Characterization, and Modeling. *Food Bioprocess Technol.* **2013,** *6* (6), 1464–1475.

Friedman, M.; Henika, P. R.; et al. Bactericidal Activities of Plant Essential Oils and Some of their Isolated Constituents against *Campylobacter jejuni*, *Escherichia coli*, *Listeria monocytogenes*, and *Salmonella enterica*. *J. Food Protect.* **2002,** *65* (10), 1545–1560.

Frieri, G.; Pimpo, M. T.; et al. Management of Colonic Diverticular Disease. *Digestion* **2006,** *73* (Suppl. 1), 58–66.

Gao, K.; Henning, S. M.; et al. The Citrus Flavonoid Naringenin Stimulates DNA Repair in Prostate Cancer Cells. *J. Nutr. Biochem.* **2006,** *17* (2), 89–95.

Golmakani, M.-T.; Rezaei, K. Comparison of Microwave-Assisted Hydrodistillation with the Traditional Hydrodistillation Method in the Extraction of Essential Oils from *Thymus vulgaris* L. *Food Chem.* **2008,** *109* (4), 925–930.

Gomez, B.; Gullon, B.; et al. Pectic Oligosaccharides from Lemon Peel Wastes: Production, Purification, and Chemical Characterization. *J. Agric. Food Chem.* **2013,** *61* (42), 10043–10053.

Gorinstein, S.; Martín-Belloso, O.; et al. Comparison of Some Biochemical Characteristics of Different Citrus Fruits. *Food Chem.* **2001,** *74* (3), 309–315.

Guinot, P.; Benonge, I.; et al. Combined Dyeing and Antioxidative Properties of Some Plant By-products. *Acta Bot. Gall.* **2007,** *154* (1), 43–52.

Hämäläinen, M.; Nieminen, R.; et al. Anti-inflammatory Effects of Flavonoids: Genistein, Kaempferol, Quercetin, and Daidzein Inhibit STAT-1 and NF-κB Activations, Whereas Flavone, Isorhamnetin, Naringenin, and Pelargonidin Inhibit Only NF-κB Activation along with their Inhibitory Effect on iNOS Expression and NO Production in Activated Macrophages. *Mediat. Inflamm.* **2007,** 1-10.

Ho, Y.-H.; Cheong, W.-K.; et al. Stapled Hemorrhoidectomy—Cost and Effectiveness. Randomized, Controlled Trial Including Incontinence Scoring, Anorectal Manometry, and Endoanal Ultrasound Assessments at up to Three Months. *Dis. Colon Rect.* **2000,** *43* (12), 1666–1675.

Hotchkiss, A.; Olano-Martin, E.; et al. Pectic Oligosaccharides as Prebiotics. *ACS Symposium Series*; ACS Publications, 2003.

Ismaili, H.; Sosa, S.; et al. Topical Anti-inflammatory Activity of Extracts and Compounds from *Thymus broussonettii*. *J. Pharmacy Pharmacol.* **2002,** *54* (8), 1137–1140.

Johnson, J. J. Carnosol: A Promising Anti-cancer and Anti-inflammatory Agent. *Cancer Lett.* **2011,** *305* (1), 1–7.

Jordan, M. J.; Moñino, M. I.; et al. Introduction of Distillate Rosemary Leaves into the Diet of the Murciano-Granadina goat: Transfer of Polyphenolic Compounds to Goats' Milk and the Plasma of Suckling Goat Kids. *J. Agric. Food Chem.* **2010,** *58* (14), 8265–8270.

Kammerer, D. R.; Kammerer, J.; et al. Recovery of Polyphenols from the By-products of Plant Food Processing and Application as Valuable Food Ingredients. *Food Res. Int.* **2014,** *65,* 2–12.

Kelsey, N. A.; Wilkins, H. M.; et al. Nutraceutical Antioxidants as Novel Neuroprotective Agents. *Molecules* **2010,** *15* (11), 7792–7814.

Kimbaris, A. C.; Siatis, N. G.; et al. Comparison of Distillation and Ultrasound-Assisted Extraction Methods for the Isolation of Sensitive Aroma Compounds from Garlic (*Allium sativum*). *Ultrasonics Sonochem.* **2006,** *13* (1), 54–60.

King, D. E. Dietary Fiber, Inflammation, and Cardiovascular Disease. *Mol. Nutr. Food Res.* **2005,** *49* (6), 594–600.

Kleemann, R.; Verschuren, L.; et al. Anti-inflammatory, Anti-proliferative and Anti-atherosclerotic Effects of Quercetin in Human *In Vitro* and *In Vivo* Models. *Atherosclerosis* **2011,** *218* (1), 44–52.

Kolettas, E.; Thomas, C.; et al. Rosmarinic Acid Failed to Suppress Hydrogen Peroxide-Mediated Apoptosis but Induced Apoptosis of Jurkat Cells which was Suppressed by Bcl-2. *Mol. Cell. Biochem.* **2006,** *285* (1–2), 111–120.

Lario, Y.; Sendra, E.; et al. Preparation of High Dietary Fiber Powder from Lemon Juice By-products. *Innov. Food Sci. Emerg. Technol.* **2004,** *5* (1), 113–117.

Larrauri, J. New Approaches in the Preparation of High Dietary Fibre Powders from Fruit By-products. *Trends Food Sci. Technol.* **1999,** *10* (1), 3–8.

Lee, J.; Koo, N.; et al. Reactive Oxygen Species, Aging, and Antioxidative Nutraceuticals. *Comprehens. Rev. Food Sci. Food Saf.* **2004,** *3* (1), 21–33.

Lee, K.; Lee, J.-S.; et al. Chlorogenic Acid Ameliorates Brain Damage and Edema by Inhibiting Matrix Metalloproteinase-2 and 9 in a Rat Model of Focal Cerebral Ischemia. *Eur. J. Pharmacol.* **2012,** *689* (1), 89–95.

Lesage-Meessen, L.; Bou, M.; et al. Essential Oils and Distilled Straws of Lavender and Lavandin: A Review of Current Use and Potential Application in White Biotechnology. *Appl. Microbiol. Biotechnol.* **2015,** *99* (8), 3375–3385.

Lesschaeve, I.; Noble, A. C. Polyphenols: Factors Influencing their Sensory Properties and their Effects on Food and Beverage Preferences. *Am. J. Clin. Nutr.* **2005,** *81* (1), 330S–335S.

Li, P. J.; Xia, J. L.; et al. Pectic Oligosaccharides Hydrolyzed from Orange Peel by Fungal Multi-enzyme Complexes and their Prebiotic and Antibacterial Potentials. *LWT—Food Sci. Technol.* **2016,** *69,* 203–210.

Liu, R. H. Potential Synergy of Phytochemicals in Cancer Prevention: Mechanism of Action. *J. Nutr.* **2004,** *134* (12), 3479S–3485S.

Liu, S.; Buring, J. E.; et al. A Prospective Study of Dietary Fiber Intake and Risk of Cardiovascular Disease among Women. *J. Am. Coll. Cardiol.* **2002,** *39* (1), 49–56.

Lucchesi, M. E.; Chemat, F.; et al. Solvent-Free Microwave Extraction of Essential Oil from Aromatic Herbs: Comparison with Conventional Hydro-distillation. *J. Chromatogr. A* **2004,** *1043* (2), 323–327.

Lucchesi, M. E.; Smadja, J.; et al. Solvent-Free Microwave Extraction of *Elletaria cardamomum* L.: A Multivariate Study of a New Technique for the Extraction of Essential Oil. *J. Food Eng.* **2007,** *79* (3), 1079–1086.

Mamma, D.; Christakopoulos, P. Biotransformation of Citrus By-products into Value Added Products. *Waste Biomass Valor.* **2014,** *5* (4), 529–549.

Mandalari, G.; Bennett, R.; et al. Antimicrobial Activity of Flavonoids Extracted from Bergamot (*Citrus bergamia* Risso) Peel, a Byproduct of the Essential Oil Industry. *J. Appl. Microbiol.* **2007,** *103* (6), 2056–2064.

Mandalari, G.; Bennett, R. N.; et al. Characterization of Flavonoids and Pectins from Bergamot (*Citrus bergamia* Risso) Peel, a Major Byproduct of Essential Oil Extraction. *J. Agric. Food Chem.* **2006,** *54* (1), 197–203.

Marin, F. R.; Soler-Rivas, C.; et al. By-products from Different Citrus Processes as a Source of Customized Functional Fibres. *Food Chem.* **2007,** *100* (2), 736–741.

Méndez-Tovar, I.; Herrero, B.; et al. By-product of *Lavandula latifolia* Essential Oil Distillation as Source of Antioxidants. *J. Food Drug Anal.* **2015,** *23* (2), 225–233.

Montonen, J.; Knekt, P.; et al. Whole-Grain and Fiber Intake and the Incidence of Type 2 Diabetes. *Am. J. Clin. Nutr.* **2003,** *77* (3), 622–629.

Moreno, S.; Sana, A. M. O.; et al. *Rosemary Compounds as Nutraceutical Health Products.* InTech Open Access Publisher: Rijeka, Croatia, 2012.

Mueller, M.; Hobiger, S.; et al. Anti-inflammatory Activity of Extracts from Fruits, Herbs and Spices. *Food Chem.* **2010,** *122* (4), 987–996.

Navarrete, A.; Herrero, M.; et al. Valorization of Solid Wastes from Essential Oil Industry. *J. Food Eng.* **2011,** *104* (2), 196–201.

Nieto, G.; Huvaere, K.; et al. Antioxidant Activity of Rosemary and Thyme By-products and Synergism with Added Antioxidant in a Liposome System. *Eur. Food Res. Technol.* **2011,** *233* (1), 11–18.

O'Shea, N.; Arendt, E. K.; et al. Dietary Fibre and Phytochemical Characteristics of Fruit and Vegetable By-products and their Recent Applications as Novel Ingredients in Food Products. *Innov. Food Sci. Emerg. Technol.* **2012,** *16,* 1–10.

Okoh, O. O.; Sadimenko, A. P.; et al. Comparative Evaluation of the Antibacterial Activities of the Essential Oils of *Rosmarinus officinalis* L. Obtained by Hydrodistillation and Solvent Free Microwave Extraction Methods. *Food Chem.* **2010,** *120* (1), 308–312.

Pan, M.-H.; Lai, C.-S.; et al. Modulation of Inflammatory Genes by Natural Dietary Bioactive Compounds. *J. Agric. Food Chem.* **2009,** *57* (11), 4467–4477.

Parejo, I.; Viladomat, F.; et al. Comparison between the Radical Scavenging Activity and Antioxidant Activity of Six Distilled and Nondistilled Mediterranean Herbs and Aromatic Plants. *J. Agric. Food Chem.* **2002,** *50* (23), 6882–6890.

Park, W.-H.; Kim, S.-H.; et al. A New Matrix Metalloproteinase-9 Inhibitor 3,4-Dihydroxycinnamic Acid (Caffeic Acid) from Methanol Extract of *Euonymus alatus*: Isolation and Structure Determination. *Toxicology* **2005,** *207* (3), 383–390.

Pereira, M. A.; Pins, J. J. Dietary Fiber and Cardiovascular Disease: Experimental and Epidemiologic Advances. *Curr. Atherosclerosis Rep.* **2000,** *2* (6), 494–502.

Poeckel, D.; Greiner, C.; et al. Carnosic Acid and Carnosol Potently Inhibit Human 5-Lipoxygenase and Suppress Pro-inflammatory Responses of Stimulated Human Polymorphonuclear Leukocytes. *Biochem. Pharmacol.* **2008,** *76* (1), 91–97.

Prakash, D.; Gupta, K. The Antioxidant Phytochemicals of Nutraceutical Importance. *Open Nutraceut. J.* **2009,** *2,* 20–35.

Qi, L.; Pan, H.; et al. Luteolin Improves Contractile Function and Attenuates Apoptosis Following Ischemia–Reperfusion in Adult Rat Cardiomyocytes. *Eur. J. Pharmacol.* **2011,** *668* (1), 201–207.

Ranasinghe, L.; Jayawardena, B.; et al. Use of Waste Generated from Cinnamon Bark Oil (*Cinnamomum zeylanicum* Blume) Extraction as a Post Harvest Treatment for Embul Banana. *J. Food Agric. Environ.* **2003,** *1,* 340–344.

Rezai-Zadeh, K.; Ehrhart, J.; et al. Apigenin and Luteolin Modulate Microglial Activation via Inhibition of STAT1-Induced CD40 Expression. *J. Neuroinflamm.* **2008,** *5,* 41.

Rezzadori, K.; Benedetti, S.; et al. Proposals for the Residues Recovery: Orange Waste as Raw Material for New Products. *Food Bioprod. Process.* **2012,** *90* (4), 606–614.

Rice-Evans, C.; Miller, N.; et al. Antioxidant Properties of Phenolic Compounds. *Trends Plant Sci.* **1997,** *2* (4), 152–159.

Rios, J.; Recio, M. Medicinal Plants and Antimicrobial Activity. *J. Ethnopharmacol.* **2005,** *100* (1), 80–84.

Roberfroid, M. B. Functional Foods: Concepts and Application to Inulin and Oligofructose. *Br. J. Nutr.* **2002,** *87* (S2), S139–S143.

Russo, M.; Bonaccorsi, I.; et al. Underestimated Sources of Flavonoids, Limonoids and Dietary Fiber: Availability in Orange's By-products. *J. Funct. Foods* **2015,** *12*, 150–157.

Sanchez-Vioque, R.; Polissiou, M.; et al. Polyphenol Composition and Antioxidant and Metal Chelating Activities of the Solid Residues from the Essential Oil Industry. *Ind. Crops Prod.* **2013,** *49*, 150–159.

Santana-Méridas, O.; González-Coloma, A.; et al. Agricultural Residues as a Source of Bioactive Natural Products. *Phytochem. Rev.* **2012,** *11* (4), 447–466.

Santos, A. Anti-inflammatory Compounds of Plant Origin. Part II. Modulation of Pro-inflammatory Cytokines, Chemokines and Adhesion Molecules. *Planta Med.* **2004,** *70*, 93–103.

Sapra, S.; Nepali, K.; et al. Analysis of Mentha Waste Products Using GC–MS. *Int. J. Pharm. Sci. Res.* **2010,** *1* (4), 53–55.

Schaefer, D. C.; Cheskin, L. J. Constipation in the Elderly. *Am. Fam. Phys.* **1998,** *58* (4), 907–914.

Sendra, E.; Kuri, V.; et al. Viscoelastic Properties of Orange Fiber Enriched Yogurt as a Function of Fiber Dose, Size and Thermal Treatment. *LWT—Food Sci. Technol.* **2010,** *43* (4), 708–714.

Soobrattee, M. A.; Neergheen, V. S.; et al. Phenolics as Potential Antioxidant Therapeutic Agents: Mechanism and Actions. *Mut. Res./Fundam. Mol. Mech. Mutagenesis* **2005,** *579* (1), 200–213.

Sowbhagya, H.; Suma, P. F.; et al. Spent Residue from Cumin—A Potential Source of Dietary Fiber. *Food Chem.* **2007,** *104* (3), 1220–1225.

Taguri, T.; Tanaka, T.; et al. Antibacterial Spectrum of Plant Polyphenols and Extracts Depending upon Hydroxyphenyl Structure. *Biol. Pharm. Bull.* **2006,** *29* (11), 2226–2235.

Terry, P.; Giovannucci, E.; et al. Fruit, Vegetables, Dietary Fiber, and Risk of Colorectal Cancer. *J. Natl. Cancer Inst.* **2001,** *93* (7), 525–533.

Thakur, B. R.; Singh, R. K.; et al. Chemistry and Uses of Pectin—A Review. *Crit. Rev. Food Sci. Nutr.* **1997,** *37* (1), 47–73.

Torras-Claveria, L.; Jauregui, O.; et al. Antioxidant Activity and Phenolic Composition of Lavandin (*Lavandula* × *intermedia* Emeric ex Loiseleur) Waste. *J. Agric. Food Chem.* **2007,** *55* (21), 8436–8443.

Tsao, R.; Akhtar, M. H. Nutraceuticals and Functional Foods. I. Current Trend in Phytochemical Antioxidant Research. *J. Food Agric. Environ.* **2005,** *3* (1), 10–17.

Ubando-Rivera, J.; Navarro-Ocaña, A.; et al. Mexican Lime Peel: Comparative Study on Contents of Dietary Fibre and Associated Antioxidant Activity. *Food Chem.* **2005,** *89* (1), 57–61.

Ulusoy, S.; Boşgelmez-Tınaz, G.; et al. Tocopherol, Carotene, Phenolic Contents and Antibacterial Properties of Rose Essential Oil, Hydrosol and Absolute. *Curr. Microbiol.* **2009,** *59* (5), 554–558.

Vanamala, J.; Leonardi, T.; et al. Suppression of Colon Carcinogenesis by Bioactive Compounds in Grapefruit. *Carcinogenesis* **2006,** *27* (6), 1257–1265.

Veličković, D. T.; Milenović, D. M.; et al. Ultrasonic Extraction of Waste Solid Residues from the *Salvia* sp. Essential Oil Hydrodistillation. *Biochem. Eng. J.* **2008,** *42* (1), 97–104.

Villeponteau, B.; Cockrell, R.; et al. Nutraceutical Interventions May Delay Aging and the Age-Related Diseases. *Exp. Gerontol.* **2000,** *35* (9), 1405–1417.

Viuda-Martos, M.; Ruiz-Navajas, Y.; et al. Functional Properties of Honey, Propolis, and Royal Jelly. *J. Food Sci.* **2008,** *73* (9), R117–R124.

Willats, W. G.; Knox, J. P.; et al. Pectin: New Insights into an Old Polymer Are Starting to Gel. *Trends Food Sci. Technol.* **2006,** *17* (3), 97–104.

Wrick, K. L. Consumer Issues and Expectations for Functional Foods. *Crit. Rev. Food Sci. Nutr.* **1995,** *35* (1–2), 167–173.

Xavier, C. P.; Lima, C. F.; et al. Luteolin, Quercetin and Ursolic Acid Are Potent Inhibitors of Proliferation and Inducers of Apoptosis in both KRAS and BRAF Mutated Human Colorectal Cancer Cells. *Cancer Lett.* **2009,** *281* (2), 162–170.

Zeisel, S. H. Regulation of "nutraceuticals." *Science* **1999,** *285* (5435), 1853–1855.

Zhao, Y.; Zhao, B. Oxidative Stress and the Pathogenesis of Alzheimer's Disease. *Oxidat. Med. Cell. Long.* **2013,** 1–10.

Zibetti, A. W.; Aydi, A.; et al. Solvent Extraction and Purification of Rosmarinic Acid from Supercritical Fluid Extraction Fractionation Waste: Economic Evaluation and Scale-Up. *J. Supercrit. Fluids* **2013,** *83*, 133–145.

CHAPTER 12

ANTIOXIDANT ENRICHMENT OF ICE CREAM USING FRUIT BY-PRODUCTS

MOHAMMED WASIM SIDDIQUI[1*] and VASUDHA BANSAL[2]

[1]Department of Food Science and Postharvest Technology, Bihar Agricultural University, Sabour, Bhagalpur, Bihar, India

[2]Department of Civil and Environmental Engineering, Hanyang University, 222 Wangsimni-Ro, Seoul 133791, South Korea

*Corresponding author. E-mail: wasim_serene@yahoo.com

CONTENTS

ABSTRACT

Owing to health consciousness by consumers, the demand of functional foods has increased many folds in the recent few years. This demand resulted in the introduction of several nutritionally enriched foods. Ice cream is one of most popular delicacies for all age groups and several amendments have been made accordingly. The present chapter deals with the nutritionally enriched ice-cream formulation developed by incorporating plant by-products.

12.1 INTRODUCTION

Ice cream is one of those deserts which are rich in carbohydrates, fats, proteins, and some micronutrients in the form of calcium and vitamins A, D, and E. However, this delicacy has been resisted by the large group of health-conscious people owing to poor concentration of functional compounds in the form of antioxidants (Williams et al., 2004; Sun-Waterhouse et al., 2013). Due to the incrementing preferences of healthy foods, there is a great need to accommodate the replacement of fat-rich foods with the antioxidants. That's is why taking into this consideration, various food scientists all over the world are trying to transform these nutrient-deficient delicacies into enriched products (Starling, 2005). However, scanty studies have been reported but the results are very promising. The main concern while preparing these enriched products is that organoleptic characteristics should not be altered in the form of taste, color, and aroma.

There has been a growing awareness about healthy ingredients to be included in our diet (Dervisoglu and Yazici, 2006). Synthetic fat is one of the ingredients that prevents the people from relishing product. Since ice-cream has been included in high-calorie dessert, fiber is one of the ingredients that can be added in the fat-rich food products and act as fat replacers (de Moraes Crizel et al., 2013). The addition of fruit by-products in the form of seed, pulp, or peel not only adds fiber rather but also increases the ice cream quality in terms of antioxidant activity, color, odor, and texture (Fig. 12.1). Moreover, the amount of associated bioactive compounds present in by-products in the form of flavonoids, polyphenols, vitamin C, and carotenoids, render the functional and health-gaining properties (Marín et al., 2007). In addition of providing the nutritional aspects, addition of fruit or fiber by-products also ameliorate the characteristics of ice cream such as water-binding capacity, structure of gelly building, enhanced gel formation, viscosity, and water-retention ability (de Moraes Crizel et al., 2013; O'Shea et al., 2012). de Moraes Crizel et al. (2013) reported the nutritional and sensory properties of ice cream after incorporation of multi-by-products of orange fruit (peels, pulp, and seeds). It was observed that addition of orange by-products to ice cream resulted in 70% reduction of fat with no significant effect on color, odor, and texture of ice cream.

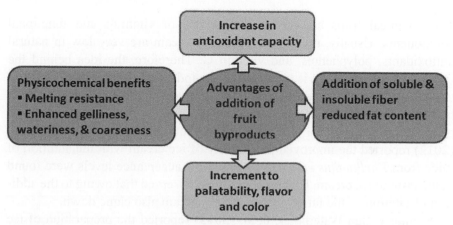

FIGURE 12.1   Advantages offered by addition of fruit by-products to ice cream.

Similarly, Çam et al. (2013) reported the enrichment of ice cream with the incorporation of pomegranate peels and pomegranate seed oil. The blunt improvements were observed in the functional quality of ice cream. It was observed the color of the ice cream got increased with the addition of pomegranate seed oil. The sensory evaluation of ice cream depicted that there was no difference in hardness, gumminess, coarseness, iciness, and wateriness of the ice cream. Rather, the addition of pomegranate peels and oil resulted in incremented palatability and astringency of the ice cream. It is attributed to the concentration of phenolics present, that is, high molecular weight tannins called punicalagins (Çam et al., 2013). The phenolic content of the ice cream was increased up to 0.4% (w/w) without degrading the organoleptic quality of ice cream. Therefore, the role of adding peels and seed oil proved to be advantageous in both terms of nutrition and functional property of ice cream. Not only are the parts of fruits added, but the protein contents are also fortified in ice cream with the addition of fish protein (Shaviklo et al., 2011).

Dervisoglu and Yazici (2006) reported the effect of adding citrus fiber (0.4–1.2%) on the sensory and physical properties of ice cream. The positive result was noticed in the melting resistance of ice cream with improved viscosity. No defects were observed in flavor, texture, and appearance of the ice cream.

12.2 EFFECT OF ORGANOLEPTIC PROPERTY OF ICE CREAM ON ADDITION OF BOTANICAL FRUITS

The botanical fruits are the ample sources of vitamins and functional components. Usually, the products like ice cream are very low in natural antioxidants, polyphenols, and vitamin C. Therefore, the idea behind the enrichment of ice cream is to lower the fat and avoid the usage of synthetic colorants, flavors, and additives (El-Nagar et al., 2002). Since, they cannot be consumed raw but their addition to the products like ice cream will not only improve their taste also adds potent nutrients to them. Yusoh et al. (2012) reported the improved palatability of ice cream with the addition of juice from *Phyllanthus emblica*. The positive acceptance levels were found in the taste of ice cream. Further, it has been observed that owing to the addition of vitamin C, the fat content of the ice cream also came down.

Similarly, Sun-Waterhouse et al. (2013) reported the preparation of ice cream by adding the substantial amount of juice of kiwi fruit. The final product of the ice cream was found to be significantly incremented with beneficial functional compounds in the form of caffeic acid, beta-carotene, lutein, and salicylic acid. The higher amount of phenolic compounds leads to the enhanced levels of antioxidant activity.

However, the addition of these compounds is not that easy to manage, as the incorporation of juices loaded with different pigments can cause interactions among milk proteins and can lead to a formation of protein complexes (Pérez-Jiménez and Saura-Calixto, 2006). It may change the pH of the milk, precipitation of milk on heating; therefore, conditions of processing are needed to be well managed and under controlled. Also, optimization of the processing methods and the ratios of the ingredients will develop a natural fortified ice cream with the appealing taste, flavor, and color.

12.3 PURPOSE OF REPLACING FAT FORMULATIONS

Since, the nature of resisting the ice cream is the presence of large amount of fat and it makes the consumer to refrain from these deserts. The desired temptation of the mouth-feel largely depends on the fat globules and their concentration (Dalgleish, 2006). Further, the absence of the fat, these droplets lose their strength during aggregation (Thivilliers et al., 2008). Consequently, rapid loss of air occurs and makes the ice cream to melt rapidly. Thus, replacing the fat with the ratios of healthier liquids is a potent challenge to encounter (Méndez-Velasco and Goff, 2011).

Here, the role of the dietary fiber arrives which comprises heterogeneous compounds in terms of cellulose, pectins, hemicellulose, lignins, gum, etc. The addition of these fibers fosters the improvement in texture, improves water-holding capacity, gel-forming capability, prevention of clumping, and sticking (Staffolo et al., 2004; Soukoulis et al., 2009). Also, the enrichment of ice cream with fiber can improve the crystal growth of fat droplets which will be restored during freezing and storage. It shows that the fiber incorporation has two major leading advantages that make the ice cream to be accepted at large front: (1) reducing the fat content to a significant level, and (2) perseverance of the freezing and crystallization property of ice cream. Dervisoglu and Yazici (2006) showed the remarkable improvement in the antimelting property of ice cream with the addition of citrus fiber. Similarly, Cody et al. (2007) reported the rice flour as fat replacer in the vanilla ice cream. Likewise, the enhanced concentration of wheat and oat fibers also showed improvement in the state of viscosity (Soukoulis et al., 2009). Moreover, there is another term as freezing point depression which is considered as the essential feature for the production of ice cream and for the growth of the ice crystals (Hartel, 2001). It was observed that the addition of 2% apple fibers increased the temperature of freezing point and improved the composition of the solutes. The addition of pectins in the form of soluble apple fiber is found to increase the binding of water characteristic.

Likewise, Erkaya et al. (2012) studied the effect of Cape gooseberry (5%, 10%, and 15%) on the fat formulations of the ice cream. It also found that the concentration of micronutrients such as iron, zinc, and manganese were also enhanced along with high scores on the sensory features of the ice cream (tested by the panelists) with the 15% addition of Cape gooseberry. However, no significant effects were observed on the flavor, color, and the texture properties.

12.4 ROLE OF ADDITION OF FUNCTIONAL COMPOUNDS ON THE SENSORY AND RHEOLOGICAL PROPERTY

It is not only the concentration of the fat that can be reduced with the addition of fruit components, rather sugar levels can also come down with the addition of natural juices. Soukoulis and Tzia (2010) added the concentrates of grape juice and sugarcane molasses which surpassed the necessity of addition of sucrose. The replacement of synthetic sugar with natural sugar defines the incremented acceptability among consumers. The study revealed the appropriateness in the sensorial characteristics. This step is a great initiation toward combatting obesity, reduction of cardiovascular risk, and relief to intestinal problems.

In spite of the functional addition of compounds, the fruit-based addition provides numerous flavoring compounds. El-Samahy et al. (2009) reported the new product of ice cream with the usage of concentrated cactus pear pulp which is a great source of polyphenols, flavonoids, betalins, and vitamin C. It was found that substituting 5% of cactus pulp achieved the flavor close to the controlled sample of ice cream. Therefore, the qualities of fruits as low acidity, natural sweetness, flavors, colors, and flavoring compounds are the excellent choices to be assimilated in ice creams. Güven and Karaca (2002) added the varying concentration of fruit (strawberry) on the vanilla and the fruit ice cream. The structural property of the ice cream was found to be hardened. Hwang et al. (2009) assessed the addition of grape wine lees on the antioxidant concentration of the ice cream. The amount of 50, 100, and 150 g/kg was found to increase the viscosity and destabilization of fat in the ice cream. Also, DPPH-scavenging activity was also improved. Grape wine lees are considered as the waste product of the grape wine and consist of grape peel, grape pomace, and grape seed. The addition of this end product delivered the product with better phenolic compounds give protection against the heart diseases in the form of atherosclerosis and also improved the textural properties of ice cream. Thus, these value-added ingredients can offer a tremendous potential to the ice cream industry.

The effect of addition of fruit by-products in terms of pomegranate peel, grape seed extract, and oil of peppermint were added to investigate the effects on nutritional and sensorial attributes of ice cream (Sagdic et al., 2012). It was found that the phenolic compounds were incremented in the final processed product of ice cream. The enhanced concentration of ellagic acid and gallic acid were found to increase the DPPH-scavenging activity of ice cream and improved taste was perceived in the end.

Similarly, natural plant fibers in the form of polysaccharides are also reported. One of the good examples was showcased in ice creams with the addition of inulin (El-Nagar et al., 2002). It was found that the meltdown characteristics of ice cream was improved, also hardness in the texture of texture of ice cream was also noticed (increased on addition of 7–9% of inulin) which will impart more firmness in taste. Therefore, these fibers are not only reducing the concentration of fats but also improving the quality of ice cream in terms of its texture and taste. Likewise, de Moraes Crizel et al. (2013) evaluated the addition of orange juice fibers on the property of fat replacer in ice cream. They have studied the fibers from the peel, pulp, and seeds of the orange fruit. The ice cream was with the increased ratio of the phenolic compounds and carotenoid pigments. A remarkable reduction of 70% was shown in fat without causing any change in the properties of color, flavor, and texture.

Guven et al. (2003) evaluated the use of locust bean gum on the quality of Kahramanmaras type ice cream. These ice creams are bit different from the regular ice creams found in Turkey. One of the major problems in the ice creams on the coagulation of the milk is the syneresis that results in the leaking of the crystals and loss in frozen consistency of the ice cream. This breaking of crystals have been tried to improve by the addition of locust beam which can provide the structured texture to the ice cream to form strong gels. Moreover, on the addition of combination of Salep and locust bean extract, the storage life of the ice cream was also found to be increased. The concentration of 0.3–0.4% locust bean gum was found to be optimum with combination of other stabilizers in the form of guar gum and sodium alginate.

The peels of Kinnow were investigated for the nutritional enrichment of the ice cream (Kaur et al., 2013). The addition of Kinnow peel showcased the improvement in appearance, flavor, and acceptability of the ice cream. The functional quality in terms of flavonoids and ascorbic acid were enhanced followed by fostering the free-radical scavenging activity. Three percent of the unblanched Kinnow peels rendered the best results in the formulation of the ice cream. Also, the extract of wild terrestrial orchid, "Salep" (*Orchis anatolica*) was tested in the ice cream (Kus et al., 2005). It has been regarded as the natural stabilizing agent which provides the gummier and stickiness to the ice cream. This signifies the improvement in the rheological characteristic of the ice cream. Similar results on the effect of rheological behavior of the ice cream were noticed by Kaya and Tekin (2001). The superior changes were observed in the viscosity of the ice cream followed by better consistency.

Among the list of functional components, herbal teas share the major part. Karaman and Kayacier (2012) evaluated the rheological features of ice cream by the addition of black and herbal tea (sage, chamomile, and linden). It was observed that the sensorial quality of the ice cream was decreased with the addition of herbal tea, whereas the rheological behavior was improved. It was suggested that the new product in the formulations of the ice cream can be produced with herbal teas in terms of increased flavor.

It is clearly perceived that the fortification of the ice cream with dietary fiber render all the advantages such as functional and nutritional improvement in ice cream, improving the freezing and crystal growth of ice crystals, strengthen the rheological property of the processed product, retaining the viscosity, and acceptable taste by lowering the fat content. Further, it is very much convincing way of encountering consumers with a low calorie ice cream which they can have without psychological taboos.

12.5 CONCLUSION

The theme of addition of natural components in the form of fruit by-products has demarcated a new direction to the desert industry. The effort will greatly benefit the health of consumers along with boosting the economic growth the ice-cream industry. The purpose of rendering functional compounds in the ice-cream not only enhances the taste and flavors but also strengthens the basic quality restoring properties of the ice cream which are very much desirable for the acceptability of the ice cream. The other major advantage comes in the field of fat replacement by allowing more firmness in the texture of the ice-cream. Furthermore, the organoleptic acceptability has been achieved as the major parameter by the score panelists that showcased the better retaining of quality in terms of flavor and taste.

KEYWORDS

- **ice cream**
- **functional compounds**
- **by-products**
- **bioactive compounds**
- **phenolic content**

REFERENCES

Çam, M.; Erdoğan, F.; Aslan, D.; Dinç, M. Enrichment of Functional Properties of Ice Cream with Pomegranate By-products. *J. Food Sci.* **2013**, *78* (10), C1543–C1550.

Cody, T. L.; Olabi, A.; Pettingell, A. G.; Tong, P. S.; Walker, J. H. Evaluation of Rice Flour for Use in Vanilla Ice Cream. *J. Dairy Sci.* **2007**, *90* (10), 4575–4585.

Dalgleish, D. G. Food Emulsions—Their Structures and Structure-Forming Properties. *Food Hydrocolloids* **2006**, *20* (4), 415–422.

de Moraes Crizel, T.; Jablonski, A.; de Oliveira Rios, A.; Rech, R.; Flôres, S. H. Dietary Fiber from Orange Byproducts as a Potential Fat Replacer. *LWT—Food Sci. Technol.* **2013**, *53* (1), 9–14.

Dervisoglu, M.; Yazici, F. Note. The Effect of Citrus Fibre on the Physical, Chemical and Sensory Properties of Ice Cream. *Food Sci. Technol. Int.* **2006**, *12* (2), 159–164.

El-Nagar, G.; Clowes, G.; Tudorică, C. M.; Kuri, V.; Brennan, C. S. Rheological Quality and Stability of Yog-Ice Cream with Added Inulin. *Int. J. Dairy Technol.* **2002**, *55* (2), 89–93.

El-Samahy, S. K.; Youssef, K. M.; Moussa-Ayoub, T. E. Producing Ice Cream with Concentrated Cactus Pear Pulp: A Preliminary Study. *J. Profess. Assoc. Cactus Dev.* **2009**, *11*, 1–12.

Erkaya, T.; Dağdemir, E.; Şengül, M. Influence of Cape Gooseberry (*Physalis peruviana* L.) Addition on the Chemical and Sensory Characteristics and Mineral Concentrations of Ice Cream. *Food Res. Int.* **2012**, *45* (1), 331–335.

Guven, M.; Karaca, O. B.; Kacar, A. The Effects of the Combined Use of Stabilizers Containing Locust Bean Gum and of the Storage Time on Kahramanmaraş-Type Ice Creams. *Int. J. Dairy Technol.* **2003**, *56* (4), 223–228.

Güven, M.; Karaca, O. B. The Effects of Varying Sugar Content and Fruit Concentration on the Physical Properties of Vanilla and Fruit Ice-Cream-Type Frozen Yogurts. *Int. J. Dairy Technol.* **2002**, *55* (1), 27–31.

Hartel, R. W. *Crystallization in Foods*, 1st ed.; Aspen Publishers Inc.: Gaithersburg, MD, 2001.

Hwang, J. Y.; Shyu, Y. S.; Hsu, C. K. Grape Wine Lees Improves the Rheological and Adds Antioxidant Properties to Ice Cream. *LWT—Food Sci. Technol.* **2009**, *42* (1), 312–318.

Karaman, S.; Kayacier, A. Rheology of Ice Cream Mix Flavored with Black Tea or Herbal Teas and Effect of Flavoring on the Sensory Properties of Ice Cream. *Food Bioprocess Technol.* **2012**, *5* (8), 3159–3169.

Kaur, S.; Minhas, K. S.; Aggarwal, P. Development of Phytochemical Rich Ice Cream Incorporating Kinnow Peel. *Glob. J. Sci. Front. Res.* **2013**, *13* (4). https://www.journalofscience.org/index.php/GJSFR/article/view/829 (accessed 09 Sept. 2017).

Kaya, S.; Tekin, A. R. The Effect of Salep Content on the Rheological Characteristics of a Typical Ice-Cream Mix. *J. Food Eng.* **2001**, *47* (1), 59–62.

Kus, S.; Altan, A.; Kaya, A. Rheological Behavior and Time-Dependent Characterization of Ice Cream Mix with Different Salep Content. *J. Text. Stud.* **2005**, *36* (3), 273–288.

Marín, F. R.; Soler-Rivas, C.; Benavente-García, O.; Castillo, J.; Pérez-Alvarez, J. A. By-products from Different Citrus Processes as a Source of Customized Functional Fibres. *Food Chem.* **2007**, *100* (2), 736–741.

Méndez-Velasco, C.; Goff, H. D. Enhancement of Fat Colloidal Interactions for the Preparation of Ice Cream High in Unsaturated Fat. *Int. Dairy J.* **2011**, *21* (8), 540–547.

O'Shea, N.; Arendt, E. K.; Gallagher, E. Dietary Fibre and Phytochemical Characteristics of Fruit and Vegetable By-products and their Recent Applications as Novel Ingredients in Food Products. *Innov. Food Sci. Emerg. Technol.* **2012**, *16*, 1–10.

Pérez-Jiménez, J.; Saura-Calixto, F. Effect of Solvent and Certain Food Constituents on Different Antioxidant Capacity Assays. *Food Res. Int.* **2006**, *39* (7), 791–800.

Sagdic, O.; Ozturk, I.; Cankurt, H.; Tornuk, F. Interaction between Some Phenolic Compounds and Probiotic Bacterium in Functional Ice Cream Production. *Food Bioprocess Technol.* **2012**, *5* (8), 2964–2971.

Shaviklo, G. R.; Thorkelsson, G.; Sveinsdottir, K.; Rafipour, F. Chemical Properties and Sensory Quality of Ice Cream Fortified with Fish Protein. *J. Sci. Food Agric.* **2011,** *91* (7), 1199–1204.

Soukoulis, C.; Lebesi, D.; Tzia, C. Enrichment of Ice Cream with Dietary Fibre: Effects on Rheological Properties, Ice Crystallisation and Glass Transition Phenomena. *Food Chem.* **2009,** *115* (2), 665–671.

Soukoulis, C.; Tzia, C. Response Surface Mapping of the Sensory Characteristics and Acceptability of Chocolate Ice Cream Containing Alternate Sweetening Agents. *J. Sens. Stud.* **2010,** *25* (1), 50–75.

Staffolo, M. D.; Bertola, N.; Martino, M. Influence of Dietary Fiber Addition on Sensory and Rheological Properties of Yogurt. *Int. Dairy J.* **2004,** *14* (3), 263–268.

Starling, S. Fruit Dons 'Healthy' Halo. *Funct. Foods Nutraceut.* **2005,** *December*, 6–6.

Sun-Waterhouse, D.; Edmonds, L.; Wadhwa, S. S.; Wibisono, R. Producing Ice Cream Using a Substantial Amount of Juice from Kiwifruit with Green, Gold or Red Flesh. *Food Res. Int.* **2013,** *50* (2), 647–656.

Thivilliers, F.; Laurichesse, E.; Saadaoui, H.; Leal-Calderon, F.; Schmitt, V. Thermally Induced Gelling of Oil-in-Water Emulsions Comprising Partially Crystallized Droplets: The Impact of Interfacial Crystals. *Langmuir* **2008,** *24* (23), 13364–13375.

Williams, E.; Stewart-Knox, B.; Rowland, I. A Qualitative Analysis of Consumer Perceptions of Mood, Food and Mood-Enhancing Functional Foods. *J. Nutraceut. Funct. Med. Foods* **2004,** *4* (3–4), 61–83.

Yusoh, H. N. M.; Aznan, S. N. A.; Nor, N. J. M. Producing Ice Cream Using a Substantial Amount of Juice from *Phyllanthus emblica* (Pokok Melaka Fruit). In *Business, Engineering and Industrial Applications (ISBEIA), 2012 IEEE Symposium on.* IEEE, 2012; pp 600–603.

CHAPTER 13

BY-PRODUCTS FROM GRAINS AND CEREALS PROCESSING

ERICK PAUL GUTIÉRREZ-GRIJALVA,
LAURA GRECIA FLORES-ACOSTA, GABRIELA VAZQUEZ-OLIVO,
and JOSÉ BASILIO HEREDIA*

Functional Foods and Nutraceuticals Laboratory, Centro de Investigación en Alimentación y Desarrollo A. C., Unidad Culiacán, AP 32-A, Sinaloa 80129, México

*Corresponding author. E-mail: jbheredia@ciad.mx

CONTENTS

ABSTRACT

Cereals and grains are among the most commonly cultivated crops in the world. However, as a result of the production of cereals and grains, a vast quantity of biomass is discharged as waste. On this regard, several studies have recognized that agricultural waste can result in a negative environmental impact. Hence, strategies to obtain value-added products from the biomass are needed. Interestingly, waste from cereals and grains are a rich-source of phenolic compounds and dietary fiber; these components have been of special interest in research due to their potential health-promoting properties, since have been suggested as promising antioxidants, anti-inflammatory, and anticancer compounds. This chapter explores the antioxidant and dietary fiber constituents of cereals and grains as potential sources of nutraceuticals in the production of by-products of economic importance.

13.1   TYPES OF PROCESSING AND STATISTICS IN PRODUCTION

13.1.1   CEREALS PROCESSING

Cereals are the staple diet of most of the world's population (Moore et al., 1995). Cereal or grains are members of the grass family. The Food and Agriculture Organization of the United Nations defines cereals as the most important group of crops, mainly composed of starches and in a modest amount of protein and fat, also the moisture content is low. Briefly, the definition of cereal is "Annual plants, generally of the gramineous family, yielding grains used for food, feed, seed and industrial purposes, e.g., ethanol" (FAO, 2011). The most economically important cereals are wheat, rice, corn (maize), barley, oat, sorghum, rye, and as a minor cereal grain in terms of global economic importance, millets (calculated from statistics sourced from http://faostat3.fao.org/browse/Q/QC/E, last accessed 12 August 2015).

All grains need to be processed, milled, or cooked prior to consumption. The milling process increases the palatability of the grain, but at the same time, minimizes the composition of micronutrients, fiber, and phytochemicals of the final product. The aim of this process is different for each type of grain but the essential purpose is the same, to prepare the grain for the consequent process. Moreover, the grain processing is important in terms of food safety and acceptability by the consumers (Delcour-Jan and Hoseney, 2010; Hemery et al., 2007).

In the traditional milling process of grains, the starchy endosperm is separated from the outer layers (Hemery et al., 2007). For corn, there are two different types of corn milling operations, dry milling and wet milling.

The dry milling process consists of the following steps: (1) cleaning, (2) conditioning, (3) degerming, (4) drying, (5) cooling, (6) grading, (7) aspirating, (8) grinding, and (9) sifting and classifying (Matz, 1991).

Corn bran is a by-product derived from the production of corn-flaking grits. The corn bran mainly consists of the corn hull (De Vasconcelos et al., 2013).

Some corn by-products of dry milling are corn flour, corn bran, and hominy feed, whereas the frequent by-products of corn wet milling are liquefied corn product, starch molasses, germ meal, gluten meal, hydrolyzed corn protein, and condensed fermented corn extractives (De Vasconcelos et al., 2013; ElMekawy et al., 2013).

Similar to corn, sorghum, and wheat can be processed by dry milling. Wheat bran is a by-product of the wheat milling industry and contain dietary fiber, antioxidants, and minerals, and this product can be incorporated in cereal-based products to improve their nutritional properties (Delcour-Jan and Hoseney, 2010).

Wet milling is a maceration process and this makes the differences between dry milling. The main purpose is the dissociation of the endosperm cell contents with the release of starch granules from the protein network in which they are enclosed. In wet milling of corn, the final products and by-products are starch, crude oil, germ cake, solubles, fiber, gluten meal, and other. The germ cake, fiber, and solubles are usually combined and sold as gluten feed (Matz, 1991).

Sorghum milling produces some by-products, that is, gluten and grits. Also, several researchers have found antioxidant activity in corn and wheat by-products (De Vasconcelos et al., 2013).

Barley is milled or malted. Malting process is consists of three stages: (1) steeping, (2) germination, and (3) kilning. The malting industry generates a by-product that is the barley malt sprouts which consist of roots, sprouts, and malt hulls. Also, sorghum can be malted (Van Nierop et al., 2009). Barley is also a raw material for brewing industry, the brewers' spent grain is the main solid by-product that remains following wort production, the first step of the brewing process (McCarthy et al., 2013; Kitrytė et al., 2014). Recently, this by-product is mainly used as a raw material for animal feeds, fertilizers, as an alternative source of energy and, more recently, as substrate for microorganism cultivation in some biotechnological processes and enzyme production (Mussatto et al., 2006). However, there are some investigations that has evaluated the antioxidant content of this by-product and concluded that it could be utilized as a source of antioxidants with potential applications in food industry as well (Kitrytè et al., 2014).

On the other hand, kilning and milling are two important steps of oat processing (Lehtinen and Kaukovirta-Norja, 2011). The oat milling industry generates a by-product that is oat meal (Van Nierop et al., 2009).

Milling is also a step in postproduction of rice. The main rice milling by-product is rice polishing and rice bran, the first is a good source of thiamin and crude fat and relatively a poor source of crude fiber, the other has high oil content and is a good source of B-complex vitamins, protein, and amino acids (Moongngarm et al., 2012).

13.1.2 STATISTICS IN PRODUCTION

The main cereal producer in the world is China followed by the United States (FAO, 2013). FAO's latest forecast for 2015 world cereal production stands at 2527 million tons, 1.1% (27 million tons) below the 2014 record (FAO, 2015). In Mexico, the total cereal production in 2011 was 28,409,477 t (Table 13.1), and the principal cereal produce is maize, followed by wheat.

TABLE 13.1 Total Cereal Production in Mexico 2006–2011.

Year	Tons
2006	32,158,037
2007	34,314,527
2008	36,110,440
2009	31,286,661
2010	34,926,332
2011	28,409,477

The main producer of wheat in America is the United States with 57 millions of tons and Mexico is the sixth producer with 3 millions of tons (FAO, 2013). In Tables 13.2 and 13.3, the production of the most economically important cereals and the main countries that produce them are shown.

TABLE 13.2   Production of the Main Cereals Produced in Mexico (FAO, 2013).

Commodity	Production (t)	Yield (Hg/Ha)
Maize	22,663,953.00	31,940.72
Sorghum	6,308,146.00	37,350.24
Wheat	3,357,307.00	52,934.25
Rice, paddy	179,776.00	54,252.35
Oats	91,049.00	18,448.15
Millet	853.00	9,477.78
Rye	31.00	17,222.22

TABLE 13.3   Production of Wheat, Sorghum, Rye, Maize, Rice, and Oats of Main Countries in America (from FAO, 2013).

Commodity	Production (t)					
	Wheat	Sorghum	Rye	Maize	Rice	Oat
Country						
USA	57,966,658	9,881,788	194,793	353,699,441	8,613,094	1,016,024
Mexico	3,357,307	6,308,146	31	22,663,953	179,776	91,049
Canada	37,529,600 [a]		207,600	14,193,800	[a]	3,888,000
Argentina	9,188,339	3,635,837	52,130	32,119,211	1,563,450	444,820
Brazil	5,738,473	2,126,179	5743	80,273,172	11,782,549	520,397
Bolivia	223,572.45	408,459.56	100	1,063,696.48	426,050.6	5975
Venezuela	443.01	401,311.98	[a]	2,247,043.99	1,005,000.01	[a]
Chile	1,474,663	[a]	4147	1,518,549	130,307	680,382

[a]Not mentioned.

13.2   CHARACTERIZATION OF THE DERIVED BY-PRODUCTS

There are several studies that characterize cereal by-products; the most studied by-product is bran. Bran is composed of endosperm and germ, and it contains fiber, vitamin E, antioxidants, and phytoestrogens. Bioactive phenolic compounds in cereal grains are mainly located in this fraction and covalently bound to indigestible polysaccharides, although they have very low bioavailability because of the complex bran (Slavin et al., 2002). Cereal brans contains xylo-oligosaccharides, this polysaccharide has been characterized by its antioxidant activity (Veenashri and Muralikrishna, 2011). Ndolo and Beta (2013) reported the carotenoid composition of oat (943 µg/kg), wheat (776 µg/kg), and yellow corn (9652 µg/kg) aleurone layer and their antioxidant activity and conclude that they may be utilized as food ingredients with enhanced carotenoid content. Data in the literature often shows variation in the content of bioactive compounds in gran tissues depending on the process technology.

Another by-product is corn gluten meal, it contain a minimum of 600 g/kg proteins, consisting of alcohol-soluble zein (68%) and alkali-soluble glutelin (27%) (Homco-Ryan et al., 2004).

13.2.1 CONSIDERATION FOR PRACTICAL USES OF THE DERIVED BY-PRODUCTS

There are different uses of by-products of cereal and grain processing. Nowadays, the main use is for animal feed and in addition to functional foods. Most of the by-products are underexploited. There is a need on the study of the fate of individual tissues of the processing cereal grains and determine the tissue composition of the milling fractions to produce new flours and ingredients with optimized technofunctional and nutritional attributes (Dei-Piu et al., 2014; Hemery et al., 2007; Antoine et al., 2003).

There have been some recent attempts to use this by-product as a source of value-added products, bioactive compounds, arabinoxylans, among others. Oat meal has been evaluated for its use in bread-making industry because of the recently investigations on gluten-free products (Londono et al., 2015), rice starch has been used as a raw material for the production of proteins (Dei-Piu et al., 2014), and brewer's spent grain has been used for extraction of arabinoxylans. However, further investigations have to be done to optimize this process (Severini et al., 2015).

13.3 CHARACTERIZATION OF DERIVED BY-PRODUCTS

There are several studies that focus on the characterization of cereal by-products; the most studied by-product is bran. Bran is composed of endosperm and germ, and it is a rich source of fiber, vitamin E, antioxidants, and phytoestrogens. Phenolic compounds are among the most studied because of their potential health benefits. Bioactive phenolic compounds in cereal grains are mainly located in the bran fraction and are covalently bound to indigestible polysaccharides, although they have very low bioavailability because the complex cell-wall structure and nondigestible linkages between the carbohydrates that form them (Slavin et al., 2002). One of the most studied group of cereal brans polysaccharides are the xylo-oligosaccharides, these polysaccharides are of scientific interest because of their antioxidant activity (Veenashri and Muralikrishna, 2011). In addition, Ndolo and Beta (2013) reported the carotenoid composition of oat (943 μg/kg), wheat (776 μg/kg), and yellow corn (9652 μg/kg) aleurone layer and their antioxidant activity and conclude that they may be utilized as food ingredients with enhanced carotenoid content. Data in the literature often shows variation in the content of bioactive compounds in gran tissues depending of the process technology.

Another by-product is corn gluten meal, it contain a minimum of 600 g/kg proteins, consisting of alcohol-soluble zein (68%) and alkali-soluble glutelin (27%) (Homco-Ryan ct al., 2004).

13.3.1 CONSIDERATION FOR PRACTICAL USES OF THE DERIVED BY-PRODUCTS

There is a need for a grain dry fractionation technology that can efficiently separate desirable and nondesirable elements and increase the bioavailability of the latter, to produce new flours and ingredients with optimized technofunctional and nutritional attributes (Hemery et al., 2007).

In addition, optimum milling procedures needs to be developed for the removal of undesirable parts of the grain and obtain specifically parts that optimize the further processing (Li et al., 2014). On the other hand, oat meal has been evaluated for its use in the bread-making industry because of the recent investigations on gluten-free products, and oats are an interesting alternative for this purpose. However, some studies conclude that it is not yet a proper material for bread applications because its high beta-glucan content interferes with dough rheology (Londono et al., 2015).

13.4 FOOD ADDITIVES

Cereals are one of most harvested crops in the world, and they represent over 60% of the world food production. By-products from grains and cereals processing are a potential source of added-value components. Dietary fiber, proteins, energy, minerals, and some vitamins are among the most extracted compounds from cereal by-products. Charalampopoulos et al. (2002) stated that the following potential applications of by-products from cereals:

- as fermentable substrates for growth of probiotic microorganisms, especially lactobacilli and bifidobacteria;
- as a thickener in food;
- as a medium for compounds bioproduction;
- as dietary fiber promoting several beneficial physiological effects;
- as prebiotics due to their content of specific nondigestible carbohydrates;
- as encapsulation materials for probiotic to enhance their stability;
- as source of antioxidants;
- natural antimicrobials;
- among others

So far, grain and cereal by-products constitute an underexploited source for the recovery/production of compounds for food applications (Naziri et al., 2014). With some additional processing, these compounds have the potential to be used as raw materials for compounds with beneficial properties for health (Perretti et al., 2003). Value-added processing of cereals produces high-value fractions for food and bioprocessing application and by-products.

The FDA (2014) defined to food additive as any substance that is reasonably expected to become a component of food and that impact on any properties of the food. Taking this as a basis, in this section, the properties of the most important by-products of cereals and grains and their use as food additives (antioxidant, antimicrobial, and thickener use) is reviewed.

Lately, the study of encapsulation techniques has arisen using several parts of cereal to improve the viability of the probiotic strains in functional foods. Wang et al. (1999) investigated the use of high amylose maize (amylomaize) starch granules as a delivery system for bifidobacterium strain. These encapsulation techniques allowed probiotic strains growth to enhanced survival. Therefore, the cell had a better recovery in amylose-containing medium.

Some authors have studied ways to improve the encapsulation of probiotic bacteria. For example, incorporation of maize with calcium alginate and use of capsules of liquid starch core with calcium alginate membranes (Jankowski et al., 1997; Sultana et al., 2000). The use of this components retained sensory characteristics of the product (Charalampopoulos et al., 2002).

The rice industry generates important by-products as hulls and rice bran. Rice bran may be used in pasta because of it causes enrichment on the color, cooking, sensory quality, and shelf life (Kaur et al., 2012). Moreover, rice bran improves the content of dietary fiber, minerals and protein, cookies, and pan bread (Ajmal et al., 2006; Kong et al., 2012). Also, rice bran has effect on functional properties in pizza (de Delahaye et al., 2005) and bakery products (Lima et al., 2002; Hu et al., 2009).

Another industry that generates high amounts of by-products and wastes is the brewing industry. The barley-spent grain is the most important by-product. Because spent grain is too granular for direct addition to food, it must first be converted to flour (Gupta et al., 2010). Spent grain was successfully incorporated into a number of bakery products, including breads, muffins, cookies, mixed grain cereals, fruit and vegetable loaves, cakes, waffles, pancakes, tortillas, snacks, doughnuts, and brownies (Townsley, 1979). However, its use as a food additive is limited because it changes color and flavors of the final product (Gupta et al., 2010). Another use of barley-spent grain is its incorporation into food products as a source of dietary fiber, for example, in flour-mixed breads (Vasanthan et al., 2002), cookies (Kissell and Prentice, 1979; Öztürk et al., 2002), and animal and fish feed (Batajoo and Shaver, 1994; Dung et al., 2002).

13.4.1 ANTIOXIDANT USE

Some grain processes of cereals, like milling, polishing, and steeping, can generate by-products. These can be used to improve the technological performance and/or integrate into foods with healthy compounds as antioxidant compounds, carbohydrates, peptides and fiber–antioxidant complex. These compounds are present in almost all parts of grains; these compounds can be extracted with different techniques; the most common are acid or basic extraction, enzyme treatment, and supercritical fluids extraction.

Bran is formed by aleurone and pericarp and accounts for most of the micronutrient, phytochemicals, and fiber content of the grain (Fava et al., 2013). Cereal bran is a by-product of the milling industry. Cereal bran contains a high amount of phenolic acids, in particular, ferulic acid, that is mainly esterified to arabinoxylans; these feruoylayted arabinoxylans give wheat bran the potential to be used as a source of polymers and oligosaccharides with antioxidant properties for food use (Okarter and Liu, 2010).

Eposito et al. (2005) studied antioxidant activity in fractions of durum wheat bran. In the internal bran fraction, the antioxidant activity was higher, and it increased in fractions having reduced granulometry.

On the other hand, the rice bran applications in food have increased in nutraceutical and pharmaceutical industries, this is due to the presence of important bioactive components present in rice bran. The most important bioactive components present in bran are anthocyanins, flavonoids, polymeric carbohydrates, phenolics, cinnamic acids, and steroidal compounds. As a result, Kong et al. (2012) used pasta supplemented with rice bran. They reported high polyphenols, flavonoids, and anthocyanins and high antioxidant activity. That is why rice bran can be exploited as a source of tocopherols, tocotrienols, γ-oryzanols and phytosterols, compounds with significant health-promoting properties (Naziri et al., 2014). Also, Wanio et al. (2009) added rice bran to flake products. This resulted in higher DPPH and FRAP values (Kaur et al., 2012).

Moreover, rice bran oil is also rich in natural antioxidant as oryzanols and psitosterols (Gul et al., 2015). It is known that oryzanol have sundry biological and physiological effects, specially serving as an antioxidant (Duve and White, 1991), and it has important properties as antiblood cholesterol lowering agent, hypolipidemic effect, growth promotion, stimulation of the hypothalamus (Seetharamaiah and Chandrasekhara, 1989), induction of natural killer to inhibit tumor-bearing growth mice, inactivation of macrophages, inhibition of angiogenesis (Kim et al., 2012), and effective in reduction of serum cholesterol levels (Sugano and Tsuji, 1997).

Rice hull is a good source of lignin and phenolic compounds. The most important phenolics in rice hull are p-coumaric and ferulic acids, the latter is used as a food additive with antioxidant properties in Japan (Naziri et al., 2014). However, p-coumaric and ferulic acids are bound to cell walls polysaccharides, lignin and silicon; this makes it necessary to apply drastic conditions to liberate these compounds from the hulls (Butsat and Siriamornpun, 2010; Lee et al., 2003). In this sense, Wang et al. (2015) extracted phenolic compounds with ethyl acetate based on alkaline digestion and isolated these compounds. Among the isolated compounds, the most attractive are (1) cycloeucalenol cis-ferulate, (2) cycloeucalenol trans-ferulate, (3) trans-ferulic acid, (4) trans-ferulic acid methyl ester, (5) cis-ferulic acid, (6) cis-ferulic acid methyl ester, and (7) methyl caffeate. The compound 3, 5 and 7 exhibited higher DPPH and ABTS+ radical scavenging activities, followed by compounds 4 and 6; compound 1 and 2 showed potent DPPH and ABTS radical scavenging activities. It is clear that cereals and grains by-products are a great source of antioxidant compounds; therefore, it is important to improve the existing methods of extraction.

Perretti et al. (2003) evaluated the use of supercritical fluid extraction technology for recovery of antioxidants compounds of rice by-products and investigated the novel conversion processes to manufacture value-added food products. Their results suggest that by-products may be valuable sources of antioxidants, and it is possible to improve extraction conditions for their enrichment.

Cereals and grains have a highly consumed by humans, and they are a great source of tocols (tocopherols and tocotrienols). Barley is one of the best sources of these compounds that have well known antioxidant properties. Sheppard et al. (1993) reported that each tocol shows, to a different extent, vitamin E activity depending on chemical structure and physiological factors ($\alpha T > \beta T > \alpha T3 > \gamma T > \beta T3 > \delta T$) or no activity ($\gamma T3$ and $\delta T3$). Barley is processed through pearling; this generates by-products that contain interesting amounts of bioactive tocopherols and tocotrienols (Zheng et al., 2000). In this sense, Panfili et al. (2008) investigated the content of tocols (tocopherols and tocotrienols) and β-glucans of pearling by-products of 36 barley varieties; and they observed that the pearling by-products had from five to sevenfold content tocopherols and tocotrienols, respectively.

One of the most economically important cereals is maize. Maize bran is a by-product of the commercial maize dry-milling process in Mexico. This waste is a by-product scarcely used for human consumption, that is generally destined to fodder and animal feed in spite of having a high content of phenolic compounds, particularly ferulic acid. Phenolics can act as free-radical terminators, chelators of metal catalysts, or singlet oxygen quenchers (Shahidi and Wanasundra, 1992). Corn bran has a low cost and is a good source of antioxidant fiber; this led Holguín-Acuña et al. (2008) to develop an extruded breakfast cereal with added corn bran as a source of antioxidants and complex polysaccharides. They report that a 100-g serving of this cereal formulation provides 0.2 g of ferulic acid, and 8 g of complex polysaccharides, that includes 1.2 g of β-glucans and 6.8 g of arabinoxylans. This proves that corn bran can also be used as a food additive.

13.4.2 ANTIMICROBIAL USE

Several cereal by-products have been used as supports and/or substrates for production of metabolites (Schieber et al., 2011) with special properties like antimicrobial and antioxidant activity. The most important by-product compounds used as intermediaries to produce these metabolites are ferulic and ellagic acid.

As already mentioned, ferulic acid is an important compound in cereals by-products and can be released from agricultural residues by physicochemical and/or enzymatic treatments. Ferulic acid is precursor of a compound with strong antimicrobial and antioxidant properties: vanillin which is used as food preservative (Naziri et al., 2014). Several authors have studied the bioconversion of ferulic acid to vanillin from enzymatically hydrolyzed wheat bran by an engineered *Escherichia coli* strain (Di Gioia et al., 2007), and through an alkaline hydrolysis which corn cob was hydrolyzed to obtain ferulic acid that was used as a medium for vanillin bioproduction by the engineered *E. coli* (Torres et al., 2009).

Zheng et al. (2007) developed a fungal technique (*Aspergillus niger* and *Pycnoporus cinnabarinus*) in which vanillic acid was fermented into vanillin by *P. cinnabarinus*. Vanillin is also used as flavoring agent in the food industry, as intermediate in the herbicides industry, drugs or anti-foaming agents, and as a component of household products (Walton et al., 2003). In conclusion, cereal by-products are an excellent source ferulic acid that is imprescindible for biotechnological production of vanillin from agro-industrial wastes.

Another compound with antioxidant, anti-inflammatory, and antimicrobial properties (Ascacio-Valdés et al., 2011) is ellagic acid. Ellagic acid belongs to the ellagitannin group, which are hydrolysable polyphenolic compounds (Wilson and Hagerman, 1990).

Ellagic acid is result of acid or basic hydrolysis of hexahydroxydyphenic acid which spontaneously rearranged to form ellagic acid (Aguilera-Carbo et al., 2008; Gross, 2009); however, this method generates high costs and high volumes of chemical waste, for this reason microbial hydrolysis of ellagitannins using enzymes such ellagitanase has been investigated in solid state fermentation for biodegradation of ellagitannins and ellagic acid production.

Buenrostro-Figueroa et al. (2014) described the ellagitannase enzyme production from partial purified ellagitannins induced by *A. niger* GH1 grown in solid-state fermentation carried out on four different lignocellulosic materials (sugarcane bagasse, corn cobs, coconut husks, and candelilla stalks) as matrix support. They found that the best lignocellulosic materials as solid-state fermentation for ellagitannin production as inducers by *A. niger* GH1 were sugarcane bagasse and corn cobs. Moreover, the highest specific productivity was obtained with corn cobs that enable an increase in ellagitannin productivity up to 140 times. Corn cobs, a cereal by-product, have great potential as source for the production of fungal ellagitannase in solid-state fermentation.

As already stated in the previous section, barley by-products have been of great importance as food additives; in this case, brewer's spent grain has been investigated as a by-product with different properties, such as antimicrobial activity. Kotlar et al. (2013) investigated the characteristics of brewer's spent grain hydrolyzed by *Bacillus cereus* sp. extracellular peptidase. They demonstrated that an enzymatic hydrolysis was an efficient tool for preparing high-quality protein hydrolyzates of this by-product with potential antimicrobial activity. It is important to do more research to understand the antimicrobial properties of this hydrolases to potentially use them as natural antimicrobials against *E. coli* O157:H7.

13.4.3 THICKENER USES

Several polysaccharides are used as thickening additives, and the vast majority are derived from cell walls from a wide variety of plants. The structure of these polysaccharides is very variable; nonetheless, they all have one common structural feature: they all have water-absorbing properties (Southgate, 2001). These polysaccharides are commonly known as thickening agents in the food industry. A thickener is a substance that can increase the viscosity of a liquid without substantially affecting its other properties.

Thickening agents are widely used in the food industry; they are common additives in soup, jam, and salad dressing production, among others. One of the most widely used thickening agents in the food industry are galactomannans; they are used as additives in soups and other foods (Southgate, 2001).

Another carbohydrate suitable for this purpose is guar gum. Guar gum is a polysaccharide made of D-galacto-D-mannans isolated from ground endosperm of guar seeds. Guar gum is mainly used as food thickener in salad dressing and ice-cream processing (Southgate and Spiller, 2001). It is important to note that these polysaccharides are nondigestible to humans due to the lack of enzymes that hydrolyze β linkages.

Some other carbohydrates used as thickening agents are agar (agarose and agaropectin), carrageenan (a 4-*O*-sulfato-β-D-galactopyranosyl unit and a 3,6-anhydro-2-*O*-sulfato-α-D-galactopyranosyl unit), dextran [α-D-(1⇒6)-glucan], gellan [tetrasaccharide repeating unit of a β-D-(1⇒3)-gluconopyranosyl unit, a β-D-(1⇒4)-glucuronopyranosyl unit, β-D-(1⇒4)-gluconopyranosyl and an α-D-(1⇒4)-rhamnopyranosyl unit with acetyl and glycerol groups attached], gum ghatti [arabinogalactan with a β-D-(1⇒6)-galactopyranosyl main chain with β-D-(1⇒4)-glucuronic acid, (1⇒2)-mannose, and some L-arabinose attached] (Smith and Hong-Shum, 2003). Humans do not metabolize most of the polysaccharides previously mentioned; nonetheless, their property to modulate the growth and composition of gut microbiota has been a major topic of recent study. This topic will be further explained.

However, because of the lack of extraction technologies, some of these carbohydrates are not yet obtainable from cereal by-products; this sets an opportunity for food technologists for the establishment of new research projects regarding cereal waste potential as food thickeners and new extraction technologies.

13.5 SENSORIAL IMPACT

Most of the material that is obtained from by-product industry comes from the cell walls of grains and cereals; therefore, their main constituents are cell wall polysaccharides, and other components linked to cell walls. Lignin and tannins are among the most common of these components in cereals and grains waste (Hagerman, 1992; Vitaglione et al., 2008; Dunn et al., 2015).

13.5.1 TANNINS AND ASTRINGENCY

Tannins are flavonoids that occur in a wide variety of plants. Tannins are classified into (1) condensed tannins (proanthocyanidins), formed by the polymerization of catechin, and (2) hydrolysable tannins, derived from the esterification of gallic or ellagic acid with glucose.

Protoanthocyanidins structure makes them prone to interact with several proteins. Protoanthocyanidins interact more easily with proline-rich proteins. Salivary proteins are a group of proline-rich proteins. Around 20 salivary proteins are proline rich (Bennick et al., 2015).

During mastication, tannins can be released from starch by salivary enzymes like α-amylase. After PA release, they can interact with salivary proteins causing precipitation. This protein–tannin precipitation is responsible for astringency.

For example, sorghum has a PA content that varies from 0.13% to 7.2% and a tannic acid content varying from 0.37% to 1.57% (Rostango et al., 1973; Bennick et al., 2002).

13.6 NUTRACEUTICAL USES

DeFelice (2002) defines nutraceuticals as a wide-ranging term that includes foods, dietary supplements, and medical foods that have a health benefit, including the prevention and/or treatment of a disease. Phenolic acids and dietary fiber are the major representatives of plant based-foods nutraceuticals. These nutraceutical components can be extracted from several plant sources and it has been of recent interest the study of the potential of agricultural waste to extract these components to increase the value of these by-products.

Several compounds can be extracted from cereal and grain by-products to promote human health. Phenolic compounds, dietary fiber, and prebiotic carbohydrates are the main nutraceutical components that can be obtained from cereal and grain waste. Phytochemical compounds as flavonoids and phenolic acids can be extracted from cereal and grain-waste products. Condensed tannins and phenolic acids like cinnamic, chlorogenic, and ferulic acid are among the most common flavonoids and hydroxycinnamic acids, respectively, found on cereal by-products.

13.6.1 DIETARY FIBER

Dietary fiber defines as indigestible carbohydrates of dietary origin, including resistant starches and lower chain-length saccharides (3–9 units long) (De Menezes et al., 2013; Jones, 2014). Dietary fiber represents a wide spectrum of different compounds with different molecular weight, structure, and physicochemical properties (European Heart Network, 2011).

Added dietary fiber can cause health benefits on the gastrointestinal tract (Little et al., 2007; Bolhuis et al., 2014; Brownlee, 2014). For example:

- Improve bolus production, dietary fiber increases the amount of time food is chewed.
- Improvement of swallowing by modifying the rheological properties of food products by added gel-producing fiber.
- Reduction of gastric emptying rates by increasing the viscosity of food products due to added dietary fiber. Therefore, this can increase satiation more rapidly and reduce caloric intake thus reducing obesity.
- Increased satiation, therefore, lower caloric intake.
- Decreased incidence of cancer in the oropharynx, oesophagus, stomach and large intestine.
- Lower LDL cholesterol.
- Reduction of blood glucose level.

The main major cell wall components of dietary fiber obtained from cereal waste are cellulose and hemicelluloses (Inglett and Carriere, 2001; Nelson, 2001; Tungland et al., 2002; Fahey, 2004).

Dietary fiber obtained from cereal by-products is of great interest in nutrition science because of its potential health effects. Several studies state that soluble dietary fiber can decrease metabolic syndrome in obese patients. Its mode of action is by lowering food glycemix index, interacting with glucose receptors GLUT4 in the human epithelium and binding to sugars of low molecular weight such as glucose, preventing it from being absorbed (Idris and Donnelly, 2009; Krammerer et al., 2014).

The composition of dietary fiber in several cereal grain brans is shown in Table 13.4 (Kahlon and Chow, 2000; Nandini and Silimath, 2001; Ragaee et al., 2001; Esposito et al., 2005; da Silva et al., 2005; Katina et al., 2007; Bilgiçli et al., 2007; Vitaglione et al., 2008).

TABLE 13.4 Amount of Dietary Fiber in the Bran Fraction of Four Common Cereals.

Fiber	Wheat	Rye	Oat	Maize
Total	36.5–52.4	35.8	18.1–25.2	86.7
Insoluble	35–48.4	30.5	14.5–20.2	86.5
Soluble	1.5–4	5.3	3.6–5.0	0.2

Modified from Vitaglione et al. (2008).

13.6.2 ANTIOXIDANTS

The dietary intake of fiber has been related to decreased rate of chronic heart and gastrointestinal diseases. Many studies report that the cereal bran is the key factors related to cereal health benefits. Nevertheless, it has been of recent interest to study the phenolic compounds associated with dietary fiber. This due to their potential liberation in colon by microbiota enzymatic reactions (Erkkilä et al., 2005; Jacobs et al., 2000; Jensen et al., 2004; Koh-Banerjee et al., 2004; Vitaglione et al., 2008).

Most of the antioxidant compounds found in cereals are located in the bran fraction. The major compounds with antioxidant activity that can be obtained from cereal waste are ferulic acid, p-coumaric acid, sinapic acid, and vanillic acid. It is proposed that feruloyl oligosaccharides can be enzymatically extracted from cereal waste. Wheat, maize, barley, and rice are the most commonly used cereals in the extraction of feruloyl saccharides. Table 13.5 summarizes the phenolic acid content of four common cereals commonly consumed in the western diet (Antoine et al., 2003; Kroon et al., 1997; Adom et al., 2005; Andreasen et al., 2001a, 2001b; Gallardo et al., 2006; Vitaglione et al., 2008).

Feruloyl oligosaccharides can exert several gastrointestinal beneficial effects, for example, protect against oxidative stress and may have immune system modulatory effects (Ou et al., 2007; Snelders et al., 2013; Rosa et al., 2014; Yao et al., 2014). Although the composition of the feruloyl oligosaccharides in cereals may be similar, they may vary in side-linking, molecular weight, feruloyl distribution, and length; these differences may have an effect on their antioxidant and prebiotic properties.

TABLE 13.5 Amount of Ferulic Acid (FA), *p*-Coumaric Acid (PCA), Vanillic Acid (VA), Sinapic Acid (SA), and Total Phenolic Content (TPC) in the Bran Fraction of Four Common Cereals.

Phenolic acid	Wheat	Rye	Oat	Maize
FA (mg/kg)	1942–5400	25–2780	–	26,100–33,000
PCA (mg/kg)	100–457	100–190	–	3000–4000
VA (mg/kg)	100–164	10	–	
SA (mg/kg)	300	53–100	–	
TPC (mg gallic acid equivalents/kg)	2800–5643	5840	1950	

Modified from Vitaglione et al. (2008).

In maize, ferulic acid is the major phenolic acid and is mainly bound to cell wall polysaccharides (Fry, 1986). During maize tortilla production or "Nixtamalization," maize is cooked in a calcium hydroxide solution for the production of corn products (tortilla, tortilla chips). This process uses a big amount of water and produces a waste water called nejayote (Liu, 2002; Acosta-Estrada et al., 2014). It is reported that nejayote is a rich source of dietary fiber and phytochemicals. Nejayote solids contain around 45% of dietary fiber; this makes nejayote as an optimal option as dietary fiber additive. Acosta-Estrada et al. (2014) reported that nejayote can be used as bakery additive without affecting its baking properties; it also increased 700 times the antioxidant activity of nejayote-enriched bread. Additionally, Niño-Medina et al. (2009) reported that feruloylated arabinoxylans can be extracted from nejayote, which in turn can be used to obtain gums with cross-linking ferulic acid (Holguin-Acuña et al., 2011).

13.6.3 PREBIOTICS

Prebiotics are a wide range of carbohydrates that are not hydrolyzed by the human digestive enzymes during upper gastrointestinal digestion. Prebiotics are recognized as functional foods because they cause a beneficial effect on human health; prebiotics are reported to promote gut microbiota growth. Gut microbiota is reported to be related to several health benefits, among the most studied is it capacity to decrease/aliviate metabolic disorders (Gibson and Roberfroid, 1995; Charalampopoulos et al., 2002).

From cereal processing waste, some prebiotic components may be obtained with different technological purposes: fermentable substrates for growth of probiotic microorganisms, as dietary fiber promoting several beneficial physiological effects, as prebiotics due to their content of specific nondigestible carbohydrates and as encapsulation materials for probiotic to enhance their stability.

Amongst cereal grains, the content of dietary fiber fluctuates. The majority of dietary fibers generally occur in decreasing amounts from the outer pericarp to the endosperm, except arabinoxylan, which is also a major component of endosperm cell wall materials (Herrera et al., 1998; Nelson, 2001; Charalampopoulos et al., 2002). There are two types of cereal oligosaccharides with prebiotic properties: galactosyl derivatives of sucrose, stachyose and raffinose, and fructosyl derivatives of sucrose, fructooligosaccharides. Arabinoxylan-oligosaccharides and xyloglucan oligosaccharides also play an important role as prebiotics (Broekaert et al., 2011).

Maize is a rich source of arabinoxylans. Maize contains around 39 g/100 g of arabinoxylans. Arabinoxylans are a wide-ranging molecular weight saccharides composed of a linear backbone of β-$(1{\rightarrow}4)$-D-xylopyranosyl units linked to α-L-arabinofuranosyl moieties attached through O-3 and/or O-2,3-positions of the xylose residues (Izydorczyk and Biliaderis, 1995).

High molecular weight arabinoxylans from wheat aleurone have proven to modulate growth of bifidobacteria in rats (Neyrinck et al., 2011).

Cereal by-products bioprocessing through enzymatic reactions or through fermentation can also produce a large range of oligosaccharides with potential prebiotic properties. The α-amylase present in the cereal grain can hydrolyze the gelatinized starch granules, and the extent of the hydrolysis could be regulated through temperature control. The different fractions of oligosaccharides obtained could then be separated and their functionality could be tested. Another alternative for the hydrolysis of the starch would be through fungal fermentation of the use of solid-state fermentation technology. The processing steps prior to the starch hydrolysis could also have an effect in the biotransformation and should be taken into account (Charalampopoulos et al., 2002).

KEYWORDS

- cereals
- dietary fiber
- wheat
- milling process
- antioxidants
- minerals

REFERENCES

Acosta-Estrada, B. A.; Gutiérrez-Uribe, J. A.; Serna-Saldívar, S. O. Bound Phenolics in Foods, A Review. *Food Chem.* **2014**, *152*, 46–55. https://doi.org/10.1016/j.foodchem.2013.11.093.

Adom, K. K.; Sorrells, M. E.; Liu, R. H. Phytochemicals and Antioxidant Activity of Milled Fractions of Different Wheat Varieties. *J. Agric. Food Chem.* **2005**, *53*, 2297–2306.

Aguilera-Carbo, A.; Augur, C.; Prado-Barragan, L. A.; Favela-Torres, E.; Aguilar, C. N. Microbial Production of Ellagic Acid and Biodegradation of Ellagitannins. *Appl. Microbiol. Biotechnol.* **2008**, *78* (2), 189–199.

Ajmal, M.; Butt, M. S.; Sharif, K.; Nasir, M.; Nadeem, M. T. Preparation of Fibre and Mineral Enriched Pan Bread Using Defatted Rice Bran. *Int. J. Food Properties* **2006**, *9*, 623–636.

Andreasen, M. F.; Kroon, P. A.; Williamson, G.; Garcia-Conesa, M. T. Intestinal Release and Uptake of Phenolic Antioxidant Diferulic Acids. *Free Radic. Biol. Med.* **2001a**, *31*, 304–314.

Andreasen, M. F.; Landbo, A. K.; Christensen, L. P.; Hansen, A.; Meyer, A. S. Antioxidant Effects of Phenolic Rye (Secale cereale L.) Extracts, Monomeric Hydroxycinnamates and Ferulic Acid Dehydrodimers on Human Low-Density Lipoproteins. *J. Agric. Food Chem.* **2001b**, *49*, 4090–4096.

Antoine, C.; Peyron, S.; Mabille, F.; Lapierre, C.; Bouchet, B.; Abecassis, J.; et al. Individual Contribution of Grain Outer Layers and their Cell Wall Structure to the Mechanical Properties of Wheat Bran. *J. Agric. Food Chem.* **2003**, *51*, 2026–2033.

Ascacio-Valdés, J. A.; Buenrostro, J. J.; De la Cruz, R.; Sepúlveda, L.; Aguilera, A. F.; Prado, A.; Contreras, J. C.; Rodríguez, R.; Aguilar, C. N. Fungal Biodegradation of Pomegranate Ellagitannins. *J. Basic Microbiol.* **2014**, *54* (1), 28–34.

Batajoo, K. K.; Shaver, R. D. Impact of Nonfiber Carbohydrate on Intake, Digestion, and Milk Production by Dairy Cows. *J. Dairy Sci.* **1994**, *77* (6), 1580–1588. https://doi.org/10.3168/jds.S0022-0302(94)77100-9.

Bilgiçli, N.; İbanoğˇlu, Ş.; Herken, E. N. Effect of Dietary Fibre Addition on the Selected Nutritional Properties of Cookies. *J. Food Eng.* **2007**, *78* (1), 86–89.

Bolhuis, D. P.; Forde, C. G.; Cheng, Y.; Xu, H.; Martin, N.; de Graaf, C. Slow Food: Sustained Impact of Harder Foods on the Reduction in Energy Intake over the Course of the Day. *PLoS One* **2014**, *9* (4), e93370.

Broekaert, W. F.; Courtin, C. M.; Verbeke, K.; Van de Wiele, T.; Verstraete, W.; Delcou, J. A. Prebiotic and Other Health-Related Effects of Cereal-Derived Arabinoxylans, Arabinoxylan-Oligosaccharides, and Xylooligosaccharides. *Crit. Rev. Food Sci. Nutr.* **2011,** *51* (2), 178–194.

Brownlee, I. The Impact of Dietary Fibre Intake on the Physiology and Health of the Stomach and Upper Gastrointestinal Tract. *Bioact. Carbohydr. Dietary Fibre* **2014,** *4* (2), 155–169.

Buenrostro-Figueroa, J.; Ascacio-Valdés, A.; Sepúlveda, L.; De la Cruz, R.; Prado-Barragán, A.; Aguilar-González, M. A.; et al. Potential Use of Different Agroindustrial By-products as Supports for Fungal Ellagitannase Production under Solid-State Fermentation. *Food Bioprod. Process.* **2014,** *92* (4), 376–382.

Butsat, S.; Siriamornpun, S. Phenolic Acids and Antioxidant Activities in Husk of Different Thai Rice Varieties. *Food Sci. Technol. Int.* **2010,** *16* (4), 329–336.

Charalampopoulos, D.; Wang, R.; Pandiella, S. S.; Webb, C. Application of Cereals and Cereal Components in Functional Foods: A Review. *Int. J. Food Microbiol.* **2002,** *79* (1), 131–141.

da Silva, L. P.; Ciocca, M. D. L. S. Total, Insoluble and Soluble Dietary Fiber Values Measured by Enzymatic–Gravimetric Method in Cereal Grains. *J. Food Compos. Anal.* **2005,** *18* (1), 113–120.

De Delahaye, E. P.; Jimenez, P.; Perez, E. Effect of Enrichment with High Content Dietary Fibre Stabilized Rice Bran Flour on Chemical and Functional Properties of Storage Frozen Pizzas. *J. Food Eng.* **2005,** *68,* 1–7.

De Menezes, E. W.; Giuntini, E. B.; Dan, M. C. T.; Sardá, F. A. H.; Lajolo, F. M. Codex Dietary Fibre Definition—Justification for Inclusion of Carbohydrates from 3 to 9 Degrees of Polymerisation. *Food Chem.* **2013,** *140* (10), 1–12.

De Vasconcelos, M. C. B. M.; Bennett, R.; Castro, C. A. B. B.; Cardoso, P.; Saavedra, M. J.; Rosa, E. A. Study of Composition, Stabilization and Processing of Wheat Germ and Maize Industrial By-products. *Ind. Crops Prod.* **2013,** *42,* 292–298.

DeFelice, S. L. FIM Rationale and Proposed Guidelines for the Nutraceutical Research and Education Act—NREA. Foundation for Innovation Medicine, 2002. http://www.fimdefelice.org/archives/arc.researchact.html (accessed Aug. 25, 2015).

Dei-Piu, L.; Tassoni, A.; Serrazanetti, D. I.; Ferri, M.; Babini, E.; Tagliazucchi, D.; Gianotti, A. Exploitation of Starch Industry Liquid By-product to Produce Bioactive Peptides from Rice Hydrolyzed Proteins. *Food Chem.* **2014,** *155,* 199–206.

Delcour-Jan, A.; Hoseney, C. R. Principles of Cereal Science and Technology, third ed.; AACC International: St. Paul, MN, USA, 2010; pp 270.

Di Gioia, D.; Sciubba, L.; Setti, L.; Luziatelli, F.; Ruzzi, M.; Zanichelli, D.; Fava, F. Production of biovanillin from wheat bran. *Enzyme Microb. Technol.* **2007,** *41* (4), 498–505.

Dung, N. N. X.; Manh, L. H.; Udén, P. Tropical Fibre Sources for Pigs—Digestibility, Digesta Retention and Estimation of Fibre Digestibility In Vitro. *Anim. Feed Sci. Technol.* **2002,** *102* (1), 109–124. https://doi.org/10.1016/S0377-8401(02)00253-5.

Dunn, K. L.; Yang, L.; Girard, A.; Bean, S.; Awika, J. M. Interaction of Sorghum Tannins with Wheat Proteins and Effect on In Vitro Starch and Protein Digestibility in a Baked Product Matrix. *J. Agric. Food Chem.* **2015,** *63* (4), 1234–1241.

Duve, K. J.; White, P. J. Extraction and Identification of Antioxidants in Oats. *J. Am. Oil Chem. Soc.* **1991,** *68* (6), 365–370.

ElMekawy-Ahmed, Diels-Ludo, De-Wever H, Pant-Deepak. Valorization of Cereal Based Biorefinery By-products: Reality and Expectations. *Environ. Sci. Technol.* **2013,** *47,* 9014–9027.

Erkkilä, A. T.; Herrington, D. M.; Mozaffarian, D.; Lichtenstein, A. H. Cereal Fiber and Whole-Grain Intake Are Associated with Reduced Progression of Coronary-Artery Atherosclerosis in Postmenopausal Women with Coronary Artery Disease. *Am. Heart J.* **2005,** *150* (1), 94–101.

Esposito, F.; Arlotti, G.; Bonifati, A. M.; Napolitano, A.; Vitale, D.; Fogliano, V. Antioxidant Activity and Dietary Fibre in Durum Wheat Bran By-products. *Food Res. Int.* **2005,** *38* (10), 1167–1173. doi:10.1016/j.foodres.2005.05.002.

European Heart Network. Diet, Physical Activity and Cardiovascular Disease Prevention in Europe; EHN: Brussels, 2011.

Fahey, G. C., Jr.; Flickinger, E. A.; Grieshop, C. M.; Swanson, K. S. The Role of Dietary Fibre in Companion Animal Nutrition. In van der Kamp, J. W., Asp, N. G., Miller Jones, J., Schaafsma, G., Eds.; *Dietary Fibre: Bio-active Carbohydrates for Food and Feed*; Wageningen Academic Publishers: Wageningen, The Netherlands, 2004; pp 295–328.

FAO. 2013 [Internet]. http://www.fao.org/worldfoodsituation/csdb/en/ (accessed Aug. 18, 2015).

FAO. 2015 [Internet]. http://www.fao.org/worldfoodsituation/csdb/en/ (accessed Aug. 18, 2015).

Fava, F.; Zanaroli, G.; Vannini, L.; Guerzoni, E.; Bordoni, A.; Viaggi, D.; Robertson, J.; Waldron, K.; Bald, C.; Esturo, A.; Talens, C.; Tueros, I.; Cebrián, M.; Sebök, A.; Kuti, T.; Broeze, J.; Macias, M.; Brendle, H. G. New Advances in the Integrated Management of Food Processing By-products in Europe: Sustainable Exploitation of Fruit and Cereal Processing By-products with the Production of New Food Products (NAMASTE EU). *New Biotechnol.* **2013,** *30* (6), 647–655.

FDA (Food and Drug Administration). 2014. Determining the Regulatory Status of a Food Ingredient. http://www.fda.gov/Food/IngredientsPackagingLabeling/FoodAdditivesIngredients/ucm228269.htm (accessed July 2015).

Food and Agriculture Organization of the United Nations. 2011 [Internet]. http://www.fao.org/es/faodef/fdef01e.htm (accessed July 2015).

Fry, S. C. Cross-linking of Matrix Polymers in the Growing Cell Walls of Angiosperms. *Annu. Rev. Plant Physiol.* **1986,** *37*, 165–186.

Gallardo, C.; Jiménez, L.; García-Conesa, M. T. Hydroxy-cinnamic Acid Composition and In Vitro Antioxidant Activity of Selected Grain Fractions. *Food Chem.* **2006,** *99*, 455–463.

Gibson, G. R.; Roberfroid, M. B. Dietary Modulation of the Human Colonic Microbiota: Introducing the Concept of Prebiotics. *J. Nutr.* **1995,** *125* (6), 1401–1412.

Gross, G. G. Biosynthesis of Ellagitannins: Old Ideas and New Solutions. *Chemistry and Biology of Ellagitannins: An Underestimated Class of Bioactive Plant Polyphenols*; World Scientific: Hackensack, NJ, 2009; pp 94–118.

Gul, K.; Yousuf, B.; Singh, A. K.; Singh, P.; Wani, A. A. Rice Bran: Nutritional Values and Its Emerging Potential for Development of Functional Food—A Review. Bioact. Carbohydr. *Dietary Fibre* **2015,** *6* (1), 24–30.

Gupta, M.; Abu-Ghannam, N.; Gallaghar, E. Barley for Brewing: Characteristic Changes during Malting, Brewing and Applications of its By-Products. *Comprehens. Rev. Food Sci. Food Saf.* **2010,** *9* (3), 318–328.

Hagerman, A. E. Tannin–Protein Interactions. Phenolic Compounds in Food and their Effects on Health; American Chemical Society: Washington, DC, 1992; Vol 506, pp 236–247.

Hemery, Y.; Rouau, X.; Lullien-Pellerin, V.; Barron, C.; Abecassis, J. Dry Processes to Develop Wheat Fractions and Products with Enhanced Nutritional Quality. *J. Cereal Sci.* **2007,** *46* (3), 327–347.

Herrera, B. I. M.; Gonzalez, G. E. P.; Romero, J. G. 1998. Soluble, insoluble and total dietary fibre in raw and cooked legumes. Arch. Latinoam. Nutr. 48, 179–182.

Holguín-Acuña, A. L.; Carvajal-Millán, E.; Santana-Rodríguez, V.; Rascón-Chu, A.; Márquez-Escalante, J. A.; de León-Renova, N. E. P.; Gastelum-Franco, G. Maize Bran/ Oat Flour Extruded Breakfast Cereal: A Novel Source of Complex Polysaccharides and an Antioxidant. *Food Chem.* **2008,** *111* (3), 654–657.

Holguin-Acuña, A. L.; Ramos-Chavira, N.; Carvajal-Millan, E.; Santana-Rodriguez, V.; Rascón-Chu, A.; Niño-Medina, G. Chapter 14: Nonstarch Polysaccharides in Maize and Oat: Ferulated Arabinoxylans and β-Glucans. In Flour and Breads and their Fortification in Health and Disease Prevention; Preedy, V. R., Watson, R. R., Patel, V. B., Eds.; Academic Press: Cambridge, MA, 2011; pp 153–159.

Homco-Ryan, C. L.; Ryan, K. J.; Wicklund, S. E.; Nicolalde, C. L.; Lin, S.; Mckeith, F. K.; Brewer, M. S. Effects of Modified Corn Gluten Meal on Quality Characteristics of a Model Emulsified Meat Product. *Meat Sci.* **2004,** *67* (2), 335–341.

Hu, G.; Huang, S.; Cao, S.; Ma, Z. Effect of Enrichment with Hemicellulose from Rice Bran on Chemical and Functional Properties of Bread. *Food Chem.* **2009,** *115* (3), 839–842.

Idris, I.; Donnelly, R. Sodium-Flucose Co-transporter-2 Inhibitors: An Emerging New Class of Oral Antidiabetic Drug. *Diabetes, Obes. Metab.* **2009,** *11,* 79–88.

Inglett, G. E.; Carriere, C. J. Cellulosic Fibre Gels Prepared from Cell Walls of Maize Hulls. *Cer. Chem.* **2001,** *78,* 471–475.

Izydorczyk, M. S.; Biliaderis, C. G. Cereal Arabinoxylans: Advances in Structure and Physicochemical Properties. *Carbohydr. Polym.* **1995,** *28* (1), 33–48.

Jacobs, D. R.; Pereira, M. A.; Meyer, K. A.; Kushi, L. H. Fibre from Whole Grains, But Not Refined Grains Is Inversely Associated with All-Cause Mortality in Older Women: The Iowa Women's Health Study. *J. Am. Coll. Nutr.* **2000,** *19,* 326S–330S.

Jankowski, T.; Zielinska, M.; Wysakowska, A. Encapsulation of Lactic Acid Bacteria with Alginate/Starch Capsules. *Biotechnol. Tech.* **1997,** *11* (1), 31–34.

Jensen, M. K.; Koh-Banerjee, P.; Hu, F. B.; Franz, M.; Sampson, L.; Grønbaek, M.; et al. Intakes of Whole Grains, Bran, and Germ and the Risk of Coronary Heart Disease in Men. *Am. J. Clin. Nutr.* **2004,** *80,* 1492–1499.

Jones, J. M. CODEX-Aligned Dietary Fiber Definitions Help to Bridge the 'Fiber Gap'. *Nutr. J.* **2014,** *13,* 1.

Kahlon, T. S.; Chow, F. In Vitro Binding of Bile Acids by Rice Bran, Oat Bran, Wheat Bran, and Corn Bran. *Cer. Chem.* **2000,** *77* (4), 518–521.

Kammerer, D. R.; Kammerer, J.; Valet, R.; Carle, R. Recovery of Polyphenols from the By-products of Plant Food Processing and Application as Valuable Food Ingredients. *Food Res. Int.* **2014,** *65* (Pt. A), 2–12. https://doi.org/10.1016/j.foodres.2014.06.012.

Katina, K.; Laitila, A.; Juvonen, R.; Liukkonen, K. H.; Kariluoto, S.; Piironen, V.; Landberg, R.; Aman, P.; Poutanen, K. Bran Fermentation as a Means to Enhance Technological Properties and Bioactivity of Rye. *Food Microbiol.* **2007,** *24* (2), 175–186.

Kaur, G.; Sharma, S.; Nagi, H. P. S.; Dar, B. N. Functional Properties of Pasta Enriched with Variable Cereal Brans. *J. Food Sci. Technol.* **2012,** *49* (4), 467–474.

Kim, S. P.; Kang, M. Y.; Nam, S. H.; Friedman, M. Dietary Rice Bran Component γ-Oryzanol Inhibits Tumor Growth in Tumor-Bearing Mice. *Mol. Nutr. Food Res.* **2012,** *56* (6), 935–944.

Kissell, L. T.; Prentice, N.; Lindsay, R. C. Protein and Fiber Enrichment of Cookie Flour with Brewers' Spent Grain. *Cer. Chem.* **1979,** *56* (4), 261–266.

Kitrytė, V.; Šaduikis, A.; Venskutonis, P. R. Assessment of Antioxidant Capacity of Brewer's Spent Grain and Its Supercritical Carbon Dioxide Extract as Sources of Valuable Dietary Ingredients. *J. Food Eng.* **2014,** *167* (Pt. A), 18–24.

Koh-Banerjee, P.; Franz, M.; Sampson, L.; Liu, S.; Jacobs, Jr., D. R.; Spiegelman, D.; et al. Changes in Wholegrain, Bran, and Cereal Fibre Consumption in Relation to 8-y Weight Gain among Men. *Am. J. Clin. Nutr.* **2004**, *80*, 1237–1245.

Kong, S.; Kim, D. J.; Oh, S. K.; Choi, L. S.; Jeong, H. S.; Lee, J. Black Rice Bran as an Ingredient in Noodles: Chemical and Functional Evaluation. *J. Food Sci.* **2012**, *77*, 303–307.

Kotlar, C. E.; Ponce, A. G.; Roura, S. I. Improvement of Functional and Antimicrobial Properties of Brewery Byproduct Hydrolysed Enzymatically. *LWT—Food Sci. Technol.* **2013**, *50* (2), 378–385. https://doi.org/10.1016/j.lwt.2012.09.005.

Kroon, P. A.; Faulds, C. B.; Ryden, P.; Robertson, J. A.; Williamson, G. Release of Covalently Bound Ferulic Acid from Fibre in Human Colon. *J. Agric. Food Chem.* **1997**, *45*, 661–667.

Lee, S. C.; Kim, J. H.; Jeong, S. M.; Kim, D. R.; Ha, J. U.; Nam, K. C.; Ahn, D. U. Effect of Far-Infrared Radiation on the Antioxidant Activity of Rice Hulls. *J. Agric. Food Chem.* **2003**, *51* (15), 4400–4403.

Lehtinen, P.; Kaukovirta-Norja, A. Oat Lipids, Enzymes, and Quality. In Oats: Chemistry and Technology, 2nd ed.; Webster, F. H., Wood, P. J.; J. Cereal Sci. 2011.

Li, H.; Qiu, J.; Liu, C.; Ren, C.; Li, Z. Milling Characteristics and Distribution of Phytic Acid, Minerals, and Some Nutrients in Oat (Avena sativa L.). *J. Cereal Sci.* **2014**, *60* (3), 549–554.

Lima, I.; Guaraya, H.; Champagne, E. The Functional Effectiveness of Reprocessed Rice Bran as an Ingredient in Bakery Products. *Die Nahrung* **2002**, *46*, 112–117.

Little, T. J.; Russo, A.; Meyer, J. H.; Horowitz, M.; Smyth, D. R.; Bellon, M. Free Fatty Acids Have More Potent Effects on Gastric Emptying Gut Hormones, and Appetite than Triacylglycerides. *Gastroenterology* **2007**, *133* (4), 1124–1131.

Liu, S. Intake of Refined Carbohydrates and Whole Grain Foods in Relation to Risk of Type 2 Diabetes Mellitus and Coronary Heart Disease. *J. Am. Coll. Nutr.* **2002**, *21* (4), 298–306.

Londono, D. M.; Smulders, M. J.; Visser, R. G.; Gilissen, L. J.; Hamer, R. J. Effect of Kilning and Milling on the Dough-Making Properties of Oat Flour. *LWT—Food Sci. Technol.* **2015**, *63* (2), 960–965.

Matz, S. The Chemistry and Technology of Cereals as Food and Feed, 2nd ed.; 1991; pp 498–499.

McCarthy, A. L.; O'Callaghan, Y. C.; Neugart, S.; Piggott, C. O.; Connolly, A.; Jansen, M. A.; Krumbein, A.; Schreiner, S.; Fitzgerald, R. J.; O'Brien, N. M. The Hydroxycinnamic Acid Content of Barley and Brewers' Spent Grain (BSG) and the Potential to Incorporate Phenolic Extracts of BSG as Antioxidants into Fruit Beverages. *Food Chem.* **2013**, *141* (3), 2567–2574.

Moongngarm, A.; Daomukda, N.; Khumpika, S. Chemical Compositions, Phytochemicals, and Antioxidant Capacity of Rice Bran, Rice Bran Layer, and Rice Germ. *APCBEE Proced.* **2012**, *2*, 73–79.

Moore, G.; Devos, K. M.; Wang, Z.; Gale, M. D. Cereal Genome Evolution: Grasses, Line Up and Form a Circle. *Curr. Biol.* **1995**, *5* (7), 737–739.

Mussatto, S. I.; Dragone, G.; Roberto, I. C. Brewers' Spent Grain: Generation, Characteristics and Potential Applications. *J. Cereal Sci.* **2006**, *43* (1), 1–14.

Nandini, C. D.; Salimath, P. V. Carbohydrate Composition of Wheat, Wheat Bran, Sorghum and Bajra with Good Chapati/Roti (Indian Flat Bread) Making Quality. *Food Chem.* **2001**, *73* (2), 197–203.

Naziri, E.; Nenadis, N.; Mantzouridou, F. T.; Tsimidou, M. Z. Valorization of the Major Agrifood Industrial By-products and Waste from Central Macedonia (Greece) for the Recovery of Compounds for Food Applications. *Food Res. Int.* **2014**, *65*, 350–358.

Ndolo, V. U.; Beta, T. Distribution of Carotenoids in Endosperm, Germ, and Aleurone Fractions of Cereal Grain Kernels. *Food Chem.* **2013,** *139* (1), 663–671.

Nelson, A. L. Properties of High Fiber Ingredients. *Cer. Foods World* **2001,** *46*, 93–97.

Neyrinck, A. M.; Possemiers, S.; Druart, C.; Van de Wiele, T.; De Backer, F.; Cani, P. D.; Larondelle, Y.; Delzenne, N. M. Prebiotic Effects of Wheat Arabinoxylan Related to the Increase in Bifidobacteria, Roseburia and Bacteroides/Prevotella in Diet-Induced Obese Mice. *PLoS ONE* **2011,** *6* (6), e20944.

Niño-Medina, G.; Carvajal-Millán, E.; Lizardi, J.; Rascon-Chu, A.; Marquez-Escalante, J. A.; Gardea, A.; Martínez-López, A. L.; Guerrero, V. Maize Processing Waste Water Arabinoxylans: Gelling Capability and Cross-Linking Content. *Food Chem.* **2009,** *115* (4), 1286–1290.

Okarter, N.; Liu, R. H. Health Benefits of Whole Grain Phytochemicals. *Crit. Rev. Food Sci. Nutr.* **2010,** *50* (3), 193–208.

Ou, S. Y.; Jackson, G. M.; Jiao, X.; Chen, J.; Wu, J. Z.; Huang, X. S. Protection against Oxidative Stress in Diabetic Rats by Wheat Bran Feruloyl Oligosaccharides. *J. Agric. Food Chem.* **2007,** *55* (8), 3191–3195.

Öztürk, S.; Özboy, Ö.; Cavidoğlu, İ.; Köksel, H. Effects of Brewer's Spent Grain on the Quality and Dietary Fibre Content of Cookies. *J. Inst. Brew.* **2002,** *108* (1), 23–27.

Panfili, G.; Fratianni, A.; Di Criscio, T.; Marconi, E. Tocol and β-Glucan Levels in Barley Varieties and in Pearling By-products. *Food Chem.* **2008,** *107* (1), 84–91.

Perretti, G.; Miniati, E.; Montanari, L.; Fantozzi, P. Improving the Value of Rice By-products by SFE. *J. Supercrit. Fluids* **2003,** *26* (1), 63–71.

Ragaee, S. M.; Campbell, G. L.; Scoles, G. J.; McLeod, J. G.; Tyler, R. T. Studies on Rye (*Secale cereale* L.) Lines Exhibiting a Range of Extract Viscosities. 1. Composition, Molecular Weight Distribution of Water Extracts, and Biochemical Characteristics of Purified Water-Extractable Arabinoxylan. *J. Agric. Food Chem.* **2001,** *49* (5), 2437–2445.

Rosa, N. N.; Pekkinen, J.; Zavala, K.; Fouret, G.; Korkmaz, A.; Feillet-Coudray, C.; Atalay, M.; Hanhineva, K.; Mykkänen, H.; Poutanen, K.; Micard, V. Impact of Wheat Aleurone Structure on Metabolic Disorders Caused by a High-Fat Diet in Mice. *J. Agric. Food Chem.* **2014,** *62* (41), 10101–10109.

Rostagno, H. S.; Featherston, W. R.; Rogler, J. C. Studies on the Nutritional Value of Sorghum Grains with Varying Tannin Contents for Chicks: 1. Growth Studies. *Poultry Sci.* **1973,** *52* (2), 765–772.

Schieber, A.; Stintzing, F. C.; Carle, R. By-products of Plant Food Processing as a Source of Functional Compounds—Recent Developments. *Trends Food Sci. Technol.* **2001,** *12* (11), 401–413.

Seetharamaiah, G. S.; Chandrasekhara, N. Studies on Hypocholesterolemic Activity of Rice Bran Oil. *Atherosclerosis* **1989,** *78* (2), 219–223.

Severini, C.; Azzollini, D.; Jouppila, K.; Jussi, L.; Derossi, A.; De Pilli, T. Effect of Enzymatic and Technological Treatments on Solubilisation of Arabinoxylans from Brewer's Spent Grain. *J. Cer. Sci.* **2015,** *65*, 162–166.

Shahidi, F.; Wanasundra, P. K. Phenolic Antioxidants. *Crit. Rev. Food Sci. Nutr.* **1992,** *32*, 67–103.

Sheppard, A. J.; Pennington, J. A. T.; Weihrauch, J. L. Analysis and Distribution of Vitamin E in Vegetable Oils and Foods. *Vitamin E in Health and Disease*; CRC Press: New York, USA, 1993; pp 9–31.

Slavin, J. L.; Gary, F. R.; Marquart, L., Eds. *Whole-Grain Foods in Health and Disease*; American Association of Cereal Chemist: Wisconsin, USA, 2002.

Smith, J.; Hong-Shum, L. Food Additives Data Book. Blackwell Publishing Company, Hoboken, NJ, 2003.

Snelders, J.; Dornez, E.; Delcour, J. A.; Courtin, C. M. Ferulic Acid Content and Appearance Determine the Antioxidant Capacity of Arabinoxylanoligosaccharides. *J. Agric. Food Chem.* **2013**, *61* (42), 10173–10182.

Southgate, D. A. T. Polysaccharide Food Additives that Contribute to Dietary Fibre. Handbook of Dietary Fibre in Human Nutrition, third ed.; CRC Press: Boca Raton, FL, 2001.

Southgate, D. A. T.; Spiller, G. A. Glossary of Dietary Fibre Components. In Handbook of Dietary Fibre in Human Nutrition, 3rd ed.; Hamaker, B.; CRC Press: Boca Raton, FL. 2001.

Sugano, M.; Tsuji, E. Rice Bran Oil and Cholesterol Metabolism. *J. Nutr.* **1997**, *127* (3), 521S–524S.

Sultana, K.; Godward, G.; Reynolds, N.; Arumugaswamy, R.; Peiris, P.; Kailasapathy, K. Encapsulation of Probiotic Bacteria with Alginate–Starch and Evaluation of Survival in Simulated Gastrointestinal Conditions and in Yoghurt. *Int. J. Food Microbiol.* **2000**, *62* (1), 47–55.

Torres, B. R.; Aliakbarian, B.; Torre, P.; Perego, P.; Domínguez, J. M.; Zilli, M.; Converti, A. Vanillin Bioproduction from Alkaline Hydrolyzate of Corn Cob by Escherichia coli JM109/pBB1. *Enzyme Microb. Technol.* **2009**, *44* (3), 154–158.

Townsley, P. M. Preparation of Commercial Products from Brewer's Waste Grain and Trub. *MBAA Tech. Quarterly* **1979**, *16*, 130–134.

Tungland, B. C.; Meyer, D. Non-digestible Oligo and Polysaccharides (Dietary Fibers): Their Physiology and Role in Human Health and Food. *Comprehens. Rev. Food Sci. Food Saf.* **2002**, *1*, 73–92.

Van Nierop, S. N.; Axcell, B. C.; Cantrell, I. C.; Rautenbach, M. Quality Assessment of Lager Brewery Yeast Samples and Strains Using Barley Malt Extracts with Anti-yeast Activity. *Food Microbiol.* **2009**, *26* (2), 192–196.

Veenashri, B. R.; Muralikrishna, G. In Vitro Anti-oxidant Activity of Xylo-oligosaccharides Derived from Cereal and Millet Brans—A Comparative Study. *Food Chem.* **2011**, *126* (3), 1475–1481.

Vitaglione, P.; Napolitano, A.; Fogliano, V. Cereal Dietary Fibre: A Natural Functional Ingredient to Deliver Phenolic Compounds into the Gut. *Trends Food Sci. Technol.* **2008**, *19* (9), 451–463. https://doi.org/10.1016/j.tifs.2008.02.005.

Walton, N. J.; Mayer, M. J.; Narbad, A. Vanillin. *Phytochemistry* **2003**, *63* (5), 505–515.

Wang, W.; Guo, J.; Zhang, J.; Peng, J.; Liu, T.; Xin, Z. Isolation, Identification and Antioxidant Activity of Bound Phenolic Compounds Present in Rice Bran. *Food Chem.* **2015**, *171*, 40–49.

Wang, X.; Brown, I. L.; Evans, A. J.; Conway, P. L. The Protective Effects of High Amylose Maize (Amylomaize) Starch Granules on the Survival of Bifidobacterium spp. in the Mouse Intestinal Tract. *J. Appl. Microbiol.* **1999**, *87* (5), 631–639.

Wilson, T. C.; Hagerman, A. E. Quantitative Determination of Ellagic Acid. *J. Agric. Food Chem.* **1990**, *38* (8), 1678–1683.

Yao, S. W.; Wen, X. X.; Huang, R. Q.; He, R. R.; Ou, S. Y.; Shen, W. Z.; Huang, C.; Peng, X. C. Protection of Feruloylated Oligosaccharides from Corn Bran against Oxidative Stress in PC 12 Cells. *J. Agric. Food Chem.* **2014**, *62* (3), 668–674.

Zheng, G. H.; Rossnagel, B. G.; Tyler, R. T.; Bhatty, R. S. Distribution of β-Glucan in the Grain of Hull-Less Barley. *Cer. Chem.* **2000**, *77* (2), 140–144.

Zheng, L.; Zheng, P.; Sun, Z.; Bai, Y.; Wang, J.; Guo, X. Production of Vanillin from Waste Residue of Rice Bran Oil by *Aspergillus niger* and *Pycnoporus cinnabarinus*. *Bioresour. Technol.* **2007**, *98*, 1115–1119.

INDEX

Printed and bound by CPI Group (UK) Ltd, Croydon, CR0 4YY

23/10/2024

01777705-0005